AF323244

Springer
Berlin
Heidelberg
New York
Barcelona
Budapest
Hong Kong
London
Milan
Paris
Santa Clara
Singapore
Tokyo

Zoltán Molnár

Development of Thalamocortical Connections

Springer

Zoltán Molnár, M.D., D.Phil.
University of Oxford
Oxford, England

ISBN: 3-540-64225-0 Springer-Verlag Berlin Heidelberg New York

Library of Congress Cataloging-in-Publication data

Molnár, Zoltán, 1964- .
Development of of thalamocortical connections / Zoltán Molnár.
p. cm. — (Biotechnology intelligence unit)
Includes bibliographical references and index.
ISBN 1-57059-507-0 (alk. paper)
1. Developmental neurobiology. 2. Cerebral cortex—Growth. 3. Thalamus—Growth. 4. Neural circuitry. I. Title. II. Series.
[DNLM: 1. Thalamus—embryology. 2. Thalamus—growth & development. 3. Cerebral Cortex—embryology. 4. Cerebral Cortex—growth & development. 5. Axons—physiology. WL 312 M727d 1997]
QP363.5.M665 1997
612.6'40181—dc21
DNLM/DLC 97-36608
for Library of Congress CIP

This work is subject to copyright. All rights are reserved, whether the whole or part of the material is concerned, specifically the rights of translation, reprinting, reuse of illustrations, recitation, broadcasting, reproduction on microfilm or in any other way, and storage in data banks. Duplication of this publication or parts thereof is permitted only under the provisions of the German Copyright Law of September 9, 1965, in its current version, and permission for use must always be obtained from Springer-Verlag. Violations are liable for prosecution under the German Copyright Law.

© Springer-Verlag Berlin Heidelberg and R.G. Landes Company Georgetown, TX, U.S.A. 1998
Printed in Germany

The use of general descriptive names, registered names, trademarks, etc. in this publication does not imply, even in the absence of a specific statement, that such names are exempt from the relevant protective laws and regulations and therefore free for general use.

Product liability: The publisher cannot guarantee the accuracy of any information about dosage and application thereof contained in this book. In every individual case the user must check such information by consulting the relevant literature.

Typesetting: R.G. Landes Company, Georgetown, TX, U.S.A.

SPIN 10672451 31/3111 - 5 4 3 2 1 0 - Printed on acid-free paper

PREFACE

Thalamic axons, which later will mediate most sensory information from the environment, are among the first to reach the cerebral cortex, before the majority of cortical neurons have even been born. Anatomical studies on the earliest thalamocortical projections demonstrated their distribution in topographic order and without gross exuberance before entering the cortical plate at embryonic age. Thus, the earliest axonal inputs to the cortex could be capable of conveying an *extrinsic signal* that plays a part in triggering other changes which establish cortical circuitry and lead to specific regional differentiation, perhaps even to most of the repertoire of areal specializations. Indeed, increasing evidence suggests that although a certain basic intrinsic cortical organization can develop without any afferent connectivity, some cytoarchitectural attributes of at least the major sensory cortical regions are somehow imposed by the influence of specific thalamocortical fibers. Therefore it is of major interest to determine the time when the earliest input fibers start to interact with the developing cortex, and to dissect the causal relationships during the assembly of cortical connectivity. Unravelling the mechanism by which developing thalamic axons interact with the forming cortical circuitry in their target regions is important in understanding how cortical organization is determined.

In this book I shall review recent advances in the understanding of the mechanisms by which regional specification, selective thalamic innervation and topographic organization of the rodent neocortex are achieved during development, in the hope of revealing a sequence of simple mechanistic rules whereby the complexity of the adult mammalian neocortex is constructed. I shall discuss in vivo and in vitro approaches used to reveal the possible guiding mechanisms involved in the deployment of the thalamocortical fibers and the earliest interactions between the arriving thalamic fibers and the forming cortical circuitry and their possible role in the cerebral cortical areal specialization. I shall also try to elaborate on the relationships between ontogeny and phylogeny, by examining the possible evolutionary origin of the developmental algorithms employed in mammals. I hope to give a brief introduction to the exciting and growing field of thalamocortical development.

CONTENTS

ACKNOWLEDGMENTS

This work started as part of my D.Phil. thesis (Multiple Mechanisms in the Establishment of Thalamocortical Innervation) in the University Laboratory of Physiology, Oxford. The preparation of my thesis was greatly helped by guidance and advice from my supervisor Colin Blakemore and, although at that time I felt differently, by the criticisms of my examiners Ray Guillery and Raymond Lund. During my subsequent work, I enjoyed the overall stimulating atmosphere at Oxford and in various other laboratories where I collaborated in Kyoto, Brisbane, Lausanne and Davis. I would like to give special thanks to Richard Adams, Gilles Bronchti, Beat Gähwiler, Shuji Higashi, Jon Kaas, Leah Krubitzer, Paul Manger, John Mitrofanis, Elek Molnár, David Price, Norman Saunders, Péter Somogyi, Jeremy Taylor, Ian Thompson, Keisuke Toyama, Egbert Welker and Nobuhiko Yamamoto for their patience while discussing several problems in developmental neurobiology.

I would like to thank the Soros-Hungarian Academy of Sciences Foundation, Wellcome Trust, The Royal Society, Japan Society for the Promotion of Science, CIBA Foundation, McDonnell Pew Centre for Cognitive Neuroscience/Oxford and Medical Research Council for their support. I am grateful to Hertford and Merton Colleges for providing a very pleasant atmosphere and for electing me to a Baring Senior Scholarship and a Junior Research Fellowship.

My colleagues, Alla Katsnelson and Anthony Hannan, helped enormously in the final editing of the manuscript by pointing out the parts in which my expressions showed no resemblance to any Indo-European language. I am grateful to Walter Salinger and Paul Manger for their thoughtful comments on the manuscript and to Nadia Pollini for her Swiss precision and Italian enthusiasm while helping with the references. I also thank Patricia Cordery for her encouragement and for teaching me numerous histological techniques, Lorraine Chapell and Laurence Waters for their excellent photographic assistance, William Hinkes and Richard Adams for help with confocal microscopy, Patricia Owens and Duncan Fleming for taking superb care of the rat colony. I received much encouragement from my 'home university,' the Albert Szent-Györgyi Medical University in Szeged, in particular from Mihály Bodosi (Neurosurgery) and György Benedek (Physiology), to whom I am grateful.

And above all, the ideal working conditions would not have been the same without the constant support and endless distractions of Nadia and Mátyás.

ABBREVIATIONS

ANT anterior
BDA byotinylated dextran amine
CCK cholecystokynin
CP cortical plate
CSD current source density
CTX cerebral cortex
DCP dense cortical plate
DiA 4-[4-(dihexadecylamino) stryryl]-N-methylpyridinium iodide
DiAsp 4-[4-(didecylamino)styryl]-N-methylpiridinium iodide,
DiI 1,1'-dioctadecyl 3,3,3'3'- tetramethylindocarbocyanine perchlorate;
DiO 3,3'-dipentyloxacarbocyanine iodide
DM dorsomedial
DVR dorsal ventricular ridge
E embryonic
FP field potential
GABA gamma amino butiric acid
HIP hippocampus
HRP horseradish peroxidase
IZ intermediate zone
LAT lateral
LGN lateral geniculate nucleus
LP lateroposterior
LTD long term depression
LTP long term potentiation
MAOA monoamine oxidase A
MAM methylazoxymethanol acetate
MED medial
MGN medial geniculate nucleus
MZ marginal zone
NADPH-d nicotinamide-adenosine dinucleotide phosphate diaphorase
NMDA N-methyl-D-aspartate
NO nitric oxide
NOS nitric oxide synthase
NPY neuropeptide-Y
P postnatal
PIC primitive internal capsule

POST	posterior
PRN	perireticular nucleus
PULV	pulvinar
S'P	superplate (in *reeler* mutant mouse)
SP	subplate
SVZ	subventricular zone
THAL	thalamus
TRN	thalamic reticular nucleus
VA	ventro-anterior
VL	ventro-lateral
VPL	ventro-posterolateral
VZ	ventricular zone

“There is but one safe way to avoid mistakes: to do nothing, or at least, to avoid doing something new. This, however, in itself may be the greatest mistake of all. The unknown lends an insecure foothold and venturing out into it, one can hope for no more than that the possible failure will be a honorable one.”

“See what every one sees, but think what no one else thinks.”

“Research is rarely guided by logic. It is guided mostly by hunches, guesses and intuition. All the same, once we get somewhere and present our results, we like to present them as a logical sequence.”

Albert Szent-Györgyi
(1893 - 1986)

CHAPTER 1

General Introduction to the Development of the Cerebral Cortex

The emergence of the mammalian cerebral cortex is arguably one of the most significant events in phylogeny (from a human perspective, at least!), and the development of this complicated structure during individual ontogeny is no less spectacular. A neural sheet, initially just a monolayer of cells, becomes a three-dimensional, laminated, extremely complex structure on which the body musculature, sensory organs and much of the rest of the brain are mapped in intricate ways by the afferent and efferent projections. These connections start to become established while neurons destined to contribute to the cortex are still being generated, and subsequently there is extensive cell migration and cell death. The orchestration of these eventsand how everything arrives in the right place at the right time constitute a fascinating phenomenon.

Interactions of Genetic and Epigenetic Factors During Cortical Development and Specification of Cerebral Cortical Areas

The adult mammalian neocortex consists of numerous areas distinguished from one another largely on the basis of differences in cytoarchitecture, connections and neurophysiological properties. The different areas are specialized to perform different functions, and damage to a particular area usually results in a specific functional loss. The most important problem in the field is to understand the mechanisms by which different cortical areas achieve their identities. Other related questions follow. Are those identities predetermined and immutable? Do the different regions of the neuroepithelium give rise to unique regionally specified lineages of neurons or do they generate a uniform, uncommitted cortex?

Development of Thalamocortical Connections,
by Zoltán Molnár. © 1998 Springer-Verlag and R.G. Landes Company.

The question of how different areas of the cerebral cortex develop their unique characteristics has led to the emergence of two opposing theories that are basically genetic and epigenetic in emphasis.

According to Rakic (1971, 1972, 1988, 1995) and Reznikov et al (1984) the proliferative cells within the ventricular zone constitute a mosaic or fate-map, which generates a *proto-map* of prospective cytoarchitectonic areas (in radial units): the different areas would be predetermined from very early stages, according to the fate of the daughter neurons. Levitt et al (1997) proposes a cell cycle-dependent early commitment of cortical neurons to a specific cortical area. Indeed, there are several examples in the literature for early patterning of the cerebral cortex. Dehay et al (1993) described the rate of cell production in putative striate cortex as higher than in any other cortical regions in primate, leading to a larger number of neurons in the supragranular layer (Rockel et al, 1980). Achlin and van der Kooy (1993) described clonal heterogeneity in the germinal zone of the developing rat telencephalon, suggesting that spatial organization in the germinal zone may determine the production of different cell phenotypes. Barbe and Levitt (1995) transplanted embryonic (E) 12, 13 and 14 presumptive perirhinal or sensorimotor areas of the cerebral wall into newborn hosts in rats and reported that only the young E12/13 donor tissues receive projections that are similar to the host region; the slightly older E14 grafts exhibit both host and original donor characteristics.

Others (e.g. Driesch, 1914; O'Leary and Stanfield, 1989; O'Leary, 1989) suggest that the cortical plate is initially a homogeneous *'protocortex'*, a more or less equipotential structure, committed only

Fig. 1.1. (Opposite) A: Studies which revealed the patterning of cells (cytoarchitecture) demonstrated that all cerebral cortical areas consist of six layers but the relative thickness of the alternating layers of different neuronal populations showed regional variations. 1: motor cortex, 2: frontal, parietal, temporal association cortex, 3: frontal and parietal association cortex, 4: 'polar' cortex in the occipital and frontal lobe, 5: coniocortex of the visual, somatosensory, and auditory projection fields (see von Economo, 1929). B: The cytoarchitectonic map of the brain by Brodmann (1909) indicates the power of anatomical description. His terminology and cytoarchitectonic chart was so precise and correlated with functional areas that it is still used today. Visual cortex: areas 17, 18, 19; Auditory cortex; areas 41 and 42. Wernike's speech area: approximately area 22. Broca's speech area: approximately area 44 (in the dominant hemisphere). Somatosensory cortex: areas 3, 1, 2. Motor cortex: area 4. A: reproduced from Creutzfeldt (1995); B: reproduced from Per Brodal (1992) with kind permission of Oxford University Press.

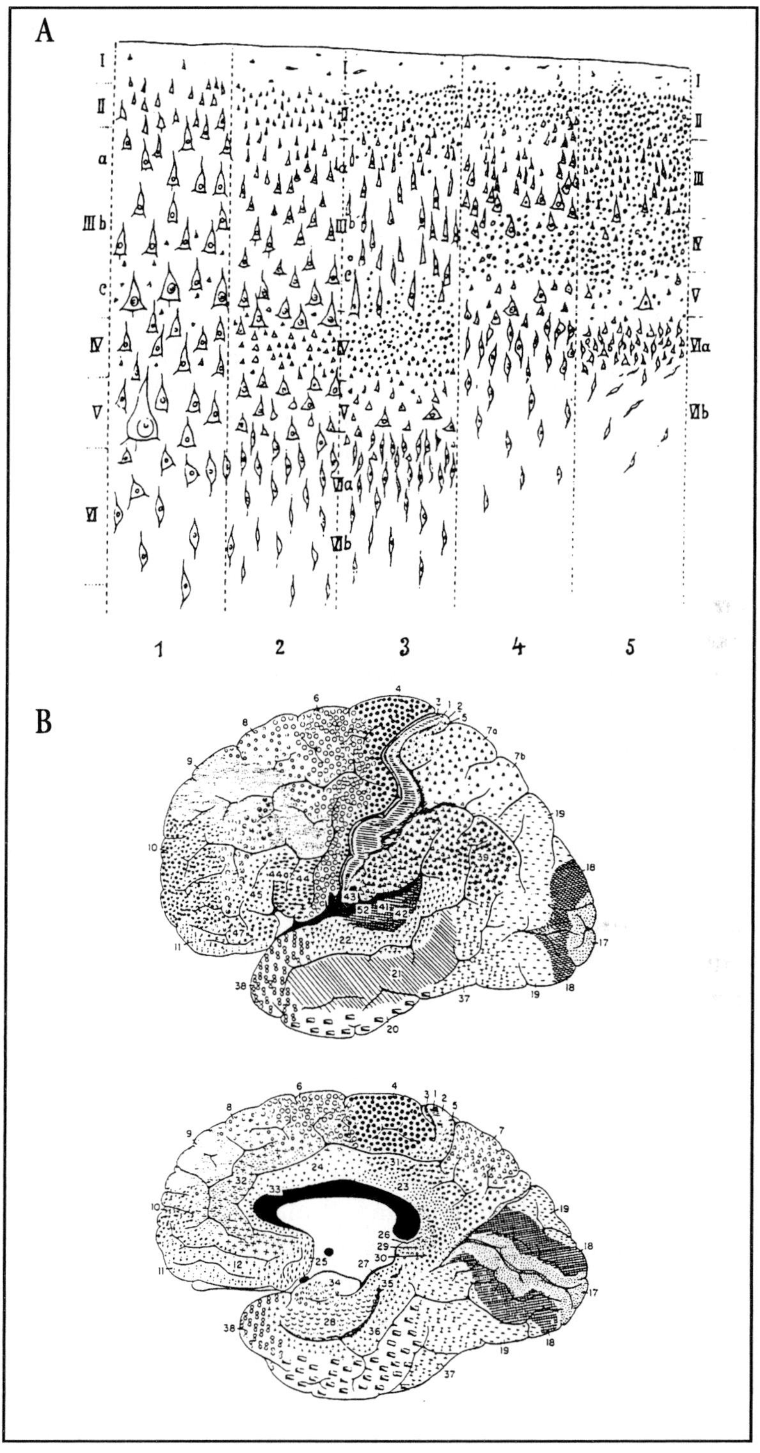
A
I
II
a
III b
c
IV
V
VI
VIa
VIb
1
2
3
4
5
B

to a basic pattern of laminar organization but without any inherent regional differentiation. Numerous in vitro and in vivo manipulation experiments demonstrate that forebrain precursor cells are multipotential and can adopt various phenotypes depending on their location (Fishell, 1995; Brustle et al, 1995; Campbell et al, 1995). Area specificity would depend on some *extrinsic signal*, which would trigger regional differentiation and determine the characteristic local discrimination and target-specific outgrowth from the cortex.

Both theories have gained support from recent progress in the field, suggesting a complex interplay between genetic and epigenetic factors during development. Although the cortical lamina for which each immature neuron is destined is apparently determined very early, during mitosis (McConnell, 1988, 1989, 1995), no molecular or genetic manifestation of regional parcellation has been described prior to the arrival of the first afferent input that reaches the cortex during and after neural migration. By marking clonally related neocortical cells with retrovirus-mediated transfer of a histochemical marker gene, Walsh and Cepko (1992) demonstrated widespread dispersion of neuronal clones across different functional regions of the cerebral cortex. This and other findings, together with the more recent developmental studies on the dispersion of proliferative layers of the cerebral cortex (Fishell et al, 1993; Walsh and Cepko, 1993; Reid et al, 1995), also argue against the 'proto-map hypothesis' of cortical generation and against the idea that the cortex is generated as strict radial units. Tan and Breen (1993) investigated cortical cell dispersion using transgenic mice in which half of the brain cells express the transgene after random inactivation of the X chromosome. The inactivation is completed by day 12.5 of gestation (E12.5). The initial strict radial cell dispersion (where all cells in a radial column were still either labelled or unlabelled) was later degraded by an even distribution of tangentially spreading cells, but the radial bands were still apparent. This and other transgenic experiments suggest that both radial mosaicism and tangential cell migration are involved in cerebral cortical development (Tan et al, 1995; Soriano et al, 1995) and indeed some cortical neurons maintain their spatial relationship as they migrate to the cortical plate (Kornack and Rakic, 1995; Rakic, 1995). O'Rourke et al (1992, 1995) demonstrated with time-lapse videomicroscopy that the majority of cell movement in the cerebral cortex is radial. There is evidence for early areal commitment in the cerebral cortex from transplantation experiments in an enhancer-trap transgenic mouse by Cohen-Tannoudji et al (1994). In this transgenic mouse the receptor gene *lac Z* was expressed primarily in layer 4 neurons of the somatosensory cortex. Putative somatosensory cortex grafts from E14-15 transgenic mouse embryo transplanted to various locations in neonatal hosts maintained the donor transgene identity, suggesting some form of early specification which could not be changed in the new location.

Numerous examples support the notion that different areas of the embryonic cortex are not specified. One important epigenetic mechanism is believed to be mediated through the thalamocortical afferentation. The importance of afferents in generating certain cortical properties is increasingly apparent, at least for the sensory areas in the posterior half of the cerebral hemisphere. Just a few examples will be mentioned here to demonstrate the importance of the input in promoting cortical differentiation.

In rodents the whiskers provide a very important sensory input. Each whisker has its own well defined representation (a neural "barrel") in the somatosensory cortex and these barrels do not develop fully until several days after birth, after the arrival of thalamic afferents (Van der Loos and Woolsey, 1973; Erzurumlu and Jhaveri, 1990; Jhaveri et al, 1991). Changes in the sensory input during the first postnatal week (produced by cauterizing some of the whisker follicles or blocking the inflow of sensory afferentation by cutting the trigeminal nerve of the mouse) prevents formation of those barrels corresponding to the missing sensory input (Van der Loos and Woolsey, 1973; Woolsey and Wann, 1976; Welker and Van der Loos, 1986). Schlaggar and O'Leary (1991) transplanted late embryonic visual cortex into the position of somatosensory cortex and demonstrated barrel formation within the transplanted cortex suggesting that the site of transplantation rather than the site of origin is important in the final differentiation.

Rakic (1988) produced another striking example of the organizing influence of the sensory input. Removal of the eyes very early in embryonic life in primates results in the degeneration of a large proportion of the cells of the lateral geniculate nucleus (LGN). The shrunken LGN innervates the occipital cortex, but the size of the primary visual cortex (where LGN fibers terminate), recognized by its distinctive cytoarchitecture, is substantially reduced, and the entire sulcal pattern of the occipital lobe is changed. We should note that in all these experiments the effect of interference with afferent input (through whisker removal or enucleation) has a dramatic effect on the developing brain only during a certain period of vulnerability.

The patterns of output characteristic of different cortical regions in adults is also not an inherent property of the cortical tissue. In the rat, layer 5 cells from all cortical regions initially project to the spinal cord, but, with subsequent development, the occipital cortex loses these projections through axon withdrawal (O'Leary and Stanfield, 1986), and only the collateral innervation of the pontine nuclei and the superior colliculus remains. But, when a piece of cortex from the occipital region of a late embryonic rat is transplanted to another region of the hemisphere of a neonatal host, it adopts the output characteristics of the host site, retaining its projection to the spinal cord (O'Leary and Stanfield, 1989).

These results emphasize the influence of the afferentation on the formation of local cytoarchitecture and the determination of output projections. The afferent input to each cortical area from the thalamus might provide the extrinsic signal that determines regional differentiation of sensory areas, though it must be emphasized that these experiments do not allow us to say whether it is merely the *presence* of afferent axons in the cortex that triggers local differentiation, or whether the pattern of *activity* in those axons plays a crucial role.

Importance of Thalamocortical Innervation in the Development of the Cerebral Cortex

The thalamus is a major relay structure in the diencephalon. It transmits nearly all sensory input destined for the cortex. Each sense organ (except the olfactory) and major subcortical motor center provides input to one or more *specific* nuclei of the thalamus, and these nuclei have well defined reciprocal interconnections with particular cortical areas (Fig. 1.2).

There is a strict relationship between the various specific thalamic nuclei and the cortical fields with which they are interconnected. The connections in both directions between the major thalamic nuclei and the cerebral cortex are specific not only for the area of cortex that receives or sends them, but also for the laminae in which they terminate or originate. Layer specificity is remarkably similar for all cortical territories and is conserved between species. In all areas in which the early stages of thalamic innervation have

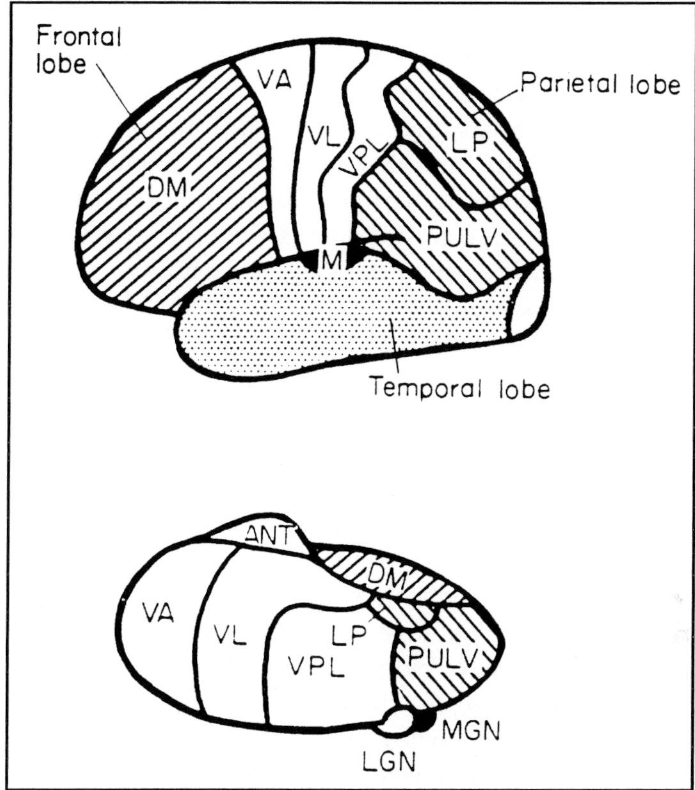

Fig. 1.2. Schematic lateral view of cerebral cortex and thalamus, showing corresponding regions of each. DM, dorsomedial; LP, lateroposterior; PULV, pulvinar; VA, VL, VPL, ventro-anterior, -lateral, and -posterolateral; MGN, LGN, medial and lateral geniculate nuclei; ANT, anterior. VA also projects diffusely to the frontal lobe. Reproduced from Carpenter (1984) with the permission of Edward Arnold Ltd.

been studied, the specific thalamo-cortical input terminates mainly in layer 4 (with some termination in layers 1 and 6). A high proportion of neurons of layer 6 of each area send corticofugal projections back to the corresponding thalamic nucleus.

The most obvious way in which events in the outside world could influence the development of the cortex is through the pattern of neural activity arriving in each cortical region through the thalamic afferents. Over the past 30 years or so, there has been great interest in the extent to which such influences actually occur and the ways in which they might be mediated (e.g., Hubel and Wiesel, 1970; Blakemore, 1978; Movshon and van Sluyters, 1981; Rakic and

Singer, 1988; Rauschecker and Maler, 1987; Singer, 1987; Merzenich et al, 1990; Shatz, 1990; Katz and Shatz, 1996). There is now a consensus that activity-dependent plasticity is a fundamental property of sensory cortical areas, that it takes place during quite well defined 'sensitive periods', and that some plastic phenomena are due to modulation of the strength of individual synapses through forms of long term potentiation and/or depression (see Toyama et al, 1991). However, before such activity-dependent plasticity can occur, thalamic axons must obviously be in place and have formed synaptic connections in an appropriately organized area of cortex.

There is also the possibility that thalamic afferents exert influences on cortical regional differentiation even at very early stages, shortly after the arrival of the first cells that contribute to the cortex proper. Recent work has established that thalamic axons are already distributed in remarkable topographic order by this stage, directly below the cortical plate, and are able to transmit to, and elicit activity patterns within, the cortex (see chapters 4 and 9). Moreover, thalamic axons invade the putative sensory regions of cortex before the regional differentiation of cytoarchitectonic and output characteristics. In the broad sense, the thalamic input is part of the environment of the developing cortex and indeed the arrival of thalamic axons is the first such environmental event that is regionally specific. Thus, the thalamic afferentation could constitute the extrinsic signal required, in the uncommitted 'protocortex' hypothesis, to trigger many of the early morphogenetic processes in the cortex. This gives special importance to the study of the early development of the thalamocortical system and especially the mechanisms by which these axons are guided to the appropriate cortical area. The entire process of regional differentiation of the cortex could depend on the way in which thalamic axons become distributed beneath the developing cortical plate and the way in which they invade it and terminate in it.

Recent work demonstrates that proliferative cells of the cortical ventricular zone possess receptors for GABA and glutamate, and activation of these receptors causes their depolarization, a rise in intracellular free calcium and a decrease in DNA synthesis (LoTurco et al, 1995). We are beginning to understand some of the local signals that influence early cortical neurogenesis, cell migration and differentiation, but it is conceivable that early thalamocortical and corticocortical connections might influence them.

Overview

The aim of this book is to describe and discuss several factors involved in the establishment of specific interconnections between the thalamus and cerebral cortex.

For non-specialist readers chapter 2 provides a general introduction to axonal pathfinding and cell-cell recognition during development and chapter 3 gives an overview of the principles of cerebral cortical and thalamic development in mammals. Readers with basic background in developmental neurobiology may choose to skip this part.

Chapter 4 describes the overall pattern of thalamocortical development in mammals and the establishment of reciprocal connections between different thalamic nuclei and cortical areas. The development of lipid soluble fluorescent dyes for axonal tracing (Honig and Hume, 1986, 1989; Godement et al, 1987) now makes it possible to study the ordering of fibers in the earliest projections without performing intrauterine surgery or using immunohistochemical markers. In this chapter I shall focus on the basic question of orderliness of the earliest projections. Are they organized topographically from the very beginning or are they initially exuberant and lacking precise organization? At what time is the adult-like areal distribution of thalamic fibers achieved?

In numerous systems the temporal sequence of neuron production and differentiation in the interconnecting structures is correlated with the temporal sequence of formation of connections between them. The translation of a temporal gradient into topography of connectivity has been proposed for a number of structures (see Bayer and Altman, 1987). Bayer and Altman (1987) pointed out that many of the major anatomical interconnections of the telencephalon can be related to the order of neurogenesis and to neurogenetic gradients. What is the case for thalamocortical connections? Is there a spatial gradient in the establishment of thalamic connections with different cortical areas? The timing of establishment of initial thalamocortical interconnections as well as other cortical connections (callosal, corticospinal and mature cortico-thalamic projections) will be discussed and their ordering and relationship to each other will also be considered in chapter 4. The plausibility of the proposal that these temporal gradients are indeed important in the formation of orderly interconnections rests on showing that the maturational gradients are matched within the various interconnecting structures, that fiber outgrowth follows a spatio-temporal gradient, and that fibers are ordered during their outgrowth and arrival. I shall consider the possibility that, at the site of innervation, the arriving thalamic fibers occupy their place on a 'first-come-first-served' basis, establishing as initial map based on chronotopy.

Chapter 5 gives a review of in vitro development of thalamocortical connections and of the recent advances made using co-culture techniques. I shall focus especially on the regional selectivity and laminar specificity of thalamic innervation. I shall consider the mechanisms that might be responsible for the development of layer-specific innervation patterns and also examine the possibility that during development, different cortical areas exert a specific attraction on thalamic axons from the distinct thalamic nuclei with which they will establish connections. I shall address these questions by reviewing results of in vitro organotypic culture experiments in which different parts of the thalamus were cultured together with various cortical areas, in order to create many paradigms under controlled conditions.

Some of the earliest generated cells of the mammalian forebrain are in the perirhinal cortex, cortical preplate, thalamic reticular nucleus and the cells of the ganglionic eminence (see Bayer and Altman, 1991, for references). Initially, these cells form a continuous layer from the diencephalon through the primitive internal capsule to the perirhinal cortex and cortical preplate. These cells have numerous similarities, and are the first to develop connections in the forebrain in carnivores and rodents (McConnell et al, 1989; Mitrofanis and Guillery, 1993; Métin and Godement, 1994, 1996). The aim of chapter 6 is to review the recent understanding of the initial phases of the establishment of early connectivity in the forebrain and to elucidate the way in which early telencephalic and diencephalic axons grow and are guided. The chapter discusses new results and ideas of the possible role of the early generated diencephalic and telencephalic cells in escorting early thalamic connections. I shall then discuss recent findings which strongly support a role for these first postmitotic cells in the development of thalamocortical fiber ordering.

Chapter 7 examines the hypothesis that subplate cells play a central role in the deployment of thalamocortical connections (Shatz et al, 1990; Blakemore and Molnár, 1990; Molnár and Blakemore,

1995a,b) by observations on an 'experiment of Nature'. The autosomal recessive mutation in the *reeler* mouse has its primary effect on the migration of immature neurons (Caviness and Sidman, 1973), disrupting the process that normally leads to the layers of the cortex being formed in an 'inside-out' sequence (Berry and Eayrs 1963; Rakic 1974; see chapter 3). The first-generated postmitotic neurons migrate to the cortical surface, as in normal mammals, and form the primordial plexiform zone (the preplate). But, in the *reeler* mutant, the cells of the cortical plate itself fail to invade and split the population of first-generated polymorph cells, which stay together at the top of the cortex as a single layer (termed the 'superplate'). The cortical plate develops below the superplate, with its layering inverted. In spite of this gross disturbance of cortical genesis, *reeler* develops relatively normal thalamocortical connectivity. Previously, this was taken as an example of axons finding malpositioned targets on the basis of remote chemoaffinity, despite their having to traverse abnormal pathways (Caviness et al, 1988). I shall argue that the *reeler* mutant actually provides strong evidence for the mechanisms of thalamocortical development proposed in normal animals (Molnár and Blakemore, 1995a,b). I shall examine the development of the reciprocal thalamocortical connections in *reeler* step-by-step from early embryonic ages and shall show that precisely the same developmental program occurs as in normal animals, and the same pathways are followed, the only difference being the relative positions of cortical plate cells and the cells that should have formed the subplate.

In chapters 8-10 the various mechanisms of the establishment of early topographically ordered thalamocortical connections are divided into two phases, one which is autonomous and not dependent on the sensory periphery, another which is activity dependent. In the activity dependent phase the early thalamocortical afferents are beginning to govern their own topography within the subplate and cortical plate by transmitting activity patterns and thereby influencing the early cortical circuit formation. Recent results obtained with electrophysiological and optical recording techniques will be reviewed to illustrate the various forms of interactions between the arriving thalamic fibers and the forming cortical circuitry. The relative contribution of the autonomous and periphery dependent processes will be examined in various mutants and experimental paradigms.

Chapter 11 elaborates on the relationship between ontogeny and phylogeny of the thalamocortical interactions, and on the possible evolutionary origin of the developmental algorithms employed during the establishment of the thalamocrtical connections. I shall compare ontogeny and phylogeny to try to get an insight into the evolutionary origin of the subplate system and the way in which it has come to play its different roles in cortical development. Early cell death is now known to be a very widespread phenomenon in the developing brain and it is clearly more exaggerated in some areas and cell types than for others. I shall argue that specific populations of transient cells in the thalamoreticular and perireticular nuclei (Mitrofanis, 1992a) and subplate might have evolutionary origin. Their examination is important in the understanding of the general logic of developmental mechanisms.

By examining these questions I hope to give an overview of the *multiple mechanisms* involved in the areal and laminar specification of thalamocortical connections, focusing on the periods during which extrinsic mechanisms can build on and interact with intrinsic ones.

Chapter 2

Overview of Axonal Pathfinding and Cell-Cell Recognition During Development

The cerebral cortex might not be the simplest or most easily accessible model system in which to ask developmental questions, but it is certainly a very important one. The differentiation of the cerebral cortex presumably underlies the highest human cognitive functions and therefore the solution to the puzzle of how each region of the cortex becomes specialized may provide insight into the mechanisms of perception, thought and language and the way in which they have evolved. Specification of the structure of a functional brain, able to interact meaningfully with the outside world, involves a great deal of information. Millions of neurons have to interconnect in highly specific patterns. The genome surely cannot hold all the information required as a blueprint for the precise wiring of every single cell in the brain. Rather than a genetic programme determining every connection and every cellular commitment, broad emergent principles seem to characterize the development of the nervous system. In the brain, and especially the cerebral cortex, remarkable amplification and supplementation of the information in the genome are brought about by self-organizational phenomena. They result mainly from the regulation, in space and time, of the expression of genes, influenced by other events and signals unfolding in the immediate environment of the structure in question. To reduce the genetic load, nature does not construct invariant patterns of organization; to some extent it creates redundancy and later eliminates the unwanted. The precise nature and timing of developmental events is not always entirely genetically determined, they often depend on an interaction of genetic predispositions and epigenetic influences, such

Development of Thalamocortical Connections,
by Zoltán Molnár. © 1998 Springer-Verlag and R.G. Landes Company.

as external chemical signals and mechanical forces (many of those dependent on other events in, and influences on, the growing embryo). Only a part of the sequence of these signals is specifically coded genetically, giving a certain stability and flexibility to the system at the same time. To get a better understanding of the enduring question of "Nature or Nurture"—to understand the contribution of precise genetic commands and those that are modulated by environmental effects (in the broadest sense)—we have to challenge the system.

Our goal is to understand the mechanisms that underlie cell-cell interactions during the development of neural connections, with special attention to the steps that might be influenced by the extracellular environment of those cells, including diffusible substances, neighboring surface properties and electrical and chemical signals from neighboring neurons, which in some cases are related to or initiated by the outside word. This book considers the mechanisms involved in early cortical development, focusing on the significance of thalamocortical connectivity in the development and differentiation of the cortex.

Mechanisms Involved in Pathfinding During Development

The ontogenesis of the nervous system relies on numerous cell-cell interactions. The appropriate cells have to recognize each other during their development. The morphogenetic role of the preference shown by any type of cell for associating with other cell types is an old concept (Wilson, 1907; Humphreys, 1967; Burger, 1974).

There have been many direct experimental approaches, in vivo and in vitro, to the study of axonal path finding, target recognition and synapse formation in the nervous system (see Jacobson, 1991; Lumsden and Cohen, 1991; Goodman and Shatz, 1993).

As early as the forties, Weiss (1941, 1947) proposed that the intercellular recognition of nerve cells during the assembly of nerve circuits is the consequence of their cytochemical properties. Steinberg (1963) observed that chicken embryonic cells from various tissues, reaggregating in vitro, tend to regain the intercellular relations typical of the original tissue, through a hierarchical order of affinity. This concept formed the basis for Sperry's (1950, 1963) influential chemoaffinity theory of neuronal specificity, which suggested that selective innervation in the developing nervous system is due to selective chemical signals, attracting growing axons towards their appropriate targets. Because of the improbability of every target cell carrying its own unique chemical tag, it was proposed that gradients of such affinity substances across target tissues interact with whole populations of incoming fibers, imposing topographic order on them, rather than defining precisely where each should terminate. These questions have been most intensively studied in the retinotectal system. The dominating general view is that growing retinal axons are indeed sensitive to a gradient distribution of guidance cues across the tectum (for a critical review see Sterling, 1991).

Chemotaxis, Contact Guidance, Guide Post Cells and Scaffolds

Distant cells can influence each other's growth by releasing diffusible factors. The idea of *chemotaxis* was originally proposed by Cajal (1909, 1911). Target cells are thought to produce some chemical substance that diffuses to the growing cells and stimulates their outgrowth. Trophic and tropic effects are distinguished. *Trophic* factors enhance the *growth* and the *survival* of cells and axons (NGF, BDNF, CNTF, NT3 and FGF are trophic factors). Unfortunately the nomenclature does not include separate expressions for survival-promoting and growth-promoting effects and the usage of the term trophic (literally = 'nourishing') is therefore ambiguous and can be misleading. In order to distinguish the different meanings of the term 'trophic', two separate expressions have been suggested: 'survival-promoting' or '*sótéric*' and 'growth-promoting' or '*auxétic*' (Molnár, 1994). Both short forms have Greek origin; soteric derives from σωζω 'to be saved or survive' and auxetic comes from αυξω 'to cause, to increase or grow'. The somato-, dendro-, or axo-prefixes, could be used with them to describe the specific sites of the soteric or auxetic effects, depending whether they are exerted on the cell body or different cell processes.

According to the classical view, *tropic* factors cause directed axonal outgrowth specifically towards the source by producing a directional response of growth cones towards the vector of the concentration gradient (no specific tropic factor, chemoattractant without trophism, has yet been identified.)

In the literature the difference between these two classical concepts (trophism and tropism) is usually sharp, and it is widely accepted that the two can be distinguished experimentally by the observation of neurite growth unconstrained by surface properties in three-dimensional collagen gels, which are believed to capture diffusible factor gradients (Lumsden, 1991). In such cultures, tropic and trophic interactions are assumed to produce different outgrowth patterns (Fig. 2.1).

However a very steep concentration gradient of a purely trophic factor might produce asymmetric, directed growth. In the three dimensional collagen matrix, the effects on outgrowth can appear either tropic or trophic depending on the *size* of the explant, outgrowth from large explants appearing very asymmetric presumably because of the difference in concentration of the growth promoting substance on the two sides of the explant. By culturing dorsal root ganglion with a gel containing different concentrations of conditioned media or NGF, Pini demonstrated that the phenomenon is indeed concentration-dependent (Pini, 1993, personal communication). NGF is probably the best studied growth factor (Levi-Montalcini, 1952, 1982). One aspect of NGF behavior that is still difficult to clarify; however, it is the mode by which NGF directs axonal outgrowth. Various studies demonstrated that NGF acts as both a tropic and trophic factor (discussed in Hankin and Silver, 1986; Hankin and Lund, 1991). The sharp distinction between tropic and trophic factors has to be reconsidered, at least temporarily, until a genuine tropic factor (different from any of the known trophic molecules) is definitively identified.

Even the advocates of chemotropism recognize the influence of local cues: "The distant chemotaxic attractions act in the mechanical guides and constraints imposed by tissue spaces and planes" (Lumsden, 1991). The earliest axons are probably directed towards their target by chemoattractants, but the later-growing projections may simply follow the surfaces of pre-existing ones, using guidance cues of the immediate neighborhood. This form of guidance has been termed *contact guidance*. Cajal himself was very much aware of the potential guiding role of the physical environment for growing axons. He made fundamental observations about the surrounding milieu in regenerating axonal pathways (Cajal, 1928). Indeed some of the pre-existing fibers can be essential for the further development. If: (1) a fiber tract extends the first axons along a particular trajectory through the extracellular space; (2) later-growing axons follow the same pathway; and (3) the pre-existing pathway is essential for the later-growing axons to find their targets, then the first axons are called pioneers (for more details see chapter 6). Strictly, the term pioneer

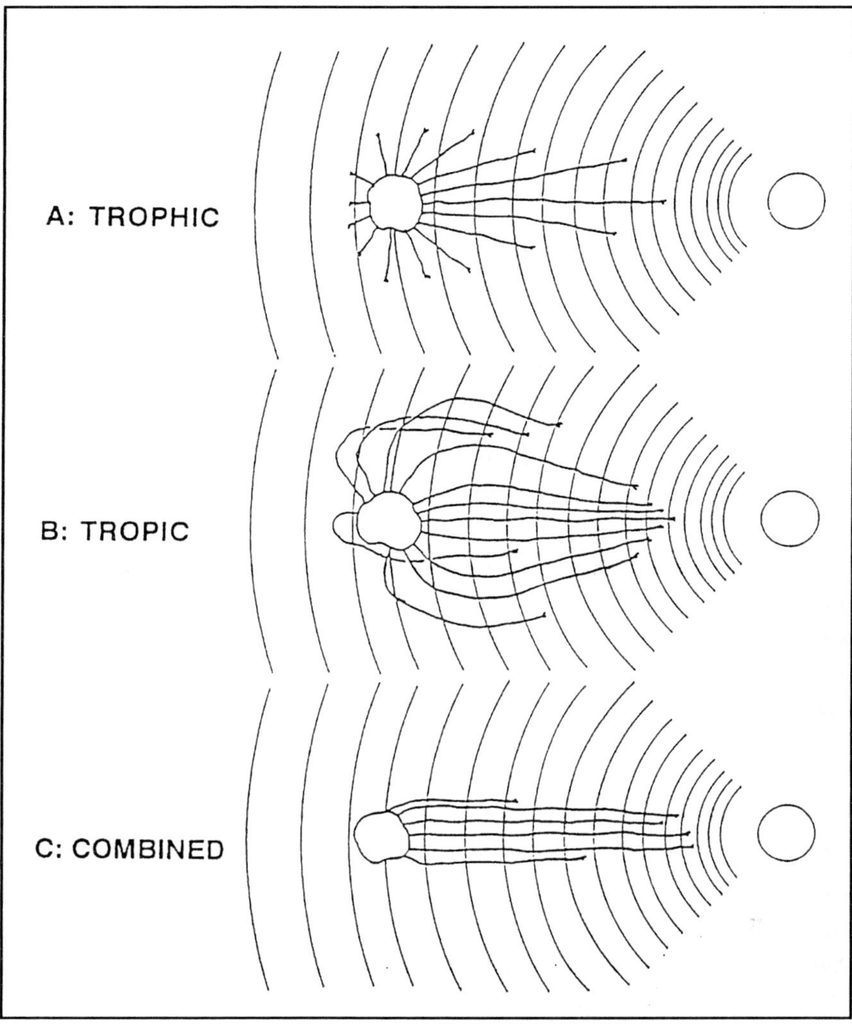

Fig. 2.1. According to the pattern of the fiber outgrowth co-cultures in 3-dimensional collagen gels, Lumsden (1991) proposed that trophic, tropic and combined (trophic and tropic) effects can be distinguished. A: A trophic agent emanates fom one of the co-cultured explants, diffuses within the collagen matrix and establishes a concentration gradient (represented by the concentric lines). Because of the concentration drop across the tested explant, the axons of cells closer to the target explant (receiving more concentrated trophic influence) extend longer fibers than those on the opposite side. The fiber outgrowth is simply radially directed. B: In the presence of a hypothetical purely tropic (or tactic) agent, there is no distinction in the length and intensity of fiber outgrowth throughout the explant and the tactic factor has no effect on cell survival. Only the *direction* of fiber outgrowth is affected. C: Lumsden (1991) distinguishes a third combined (trophic and tropic) effect, but in my opinion a trophic (growth-promoting) factor alone (depending on the distance, steepness of the concentration gradient, etc) could produce such an outgrowth pattern, without a specific tropic (purely directional) influence. Reproduced from Lumsden (1991) with kind permission of Raven Press Ltd, New York.

should be used only for initial fibers that have a subsequent guiding role. Although the term pioneer is quite widely used, the guidance function of pre-existing fibers has only been sufficiently demonstrated in insects where the study of mutants allowed the underlying genetic mechanisms to be worked out (Bate, 1976; Bentley and Keshishian, 1982; Goodman et al, 1984).

Occasionally fibers make characteristic sharp changes of directions as they reach specific groups of target cells or 'stepping stones' (Bate, 1976). The literature uses the expression guide-post cells for particular cell-types exerting a guiding role at a particular circumscribed point, during the course of the establishment of a pathway. These 'stepping stones' or 'guidepost' neurons (Bentley and Keshishian, 1982) are in strategic positions to orchestrate the pathfinding of developing projections through sequential target recognition.

The commonly used term 'scaffold' has a less well defined meaning (see McConnell et al, 1989). The term implies a three-dimensional structure of axons that could mediate a variety of different types of interaction with subsequently growing axons (e.g., growing over the scaffold in the same direction as the pioneer fibers or in the opposite direction). The term scaffold is used in various way by different authors.

Morphogenetic Gradients

The generation of cells in a given structure is usually orderly and follows a certain spatio-temporal gradient (Bayer and Altman, 1991). Fiber outgrowth in many neuronal structures follows the neurogenetic and morphogenetic gradient (e.g., spinal cord and retina) and fibers can occupy positions in a projecting bundle according to their time of generation within the whole population (His, 1904; Herrick, 1941, 1942; Gaze and Grant, 1978; Easter et al, 1981; Reh et al, 1983; Walsh and Guillery, 1985). This type of fiber ordering pattern in which position is a reflection of time of birth has been termed *chronotopic*. The extent to which chronotopic fiber ordering plays a part in the initial sorting of the connectivity and in establishing topographic order in different target structures is still a controversial issue, but it probably contributes to various extents in different systems to the multiple mechanisms by which the final mapping patterns are established (Horder and Martin, 1978; Hankin and Silver, 1986; Blakemore and Molnár, 1990; Shatz, 1992a; Molnár and Blakemore, 1995a,b).

Chronotopy

In many systems the temporal sequence of neuron production and differentiation in interconnecting structures is correlated with the order of establishment of connections between the structures. The translation of a temporal gradient into topographically ordered connectivity was first suggested for the retina after its neurogenetic and morphogenetic gradients were revealed (Horder and Martin, 1978; Martin and Perry, 1983). Bayer and Altman (1987) pointed out that many of the major anatomical interconnections of the telencephalon can be related to the order of neurogenesis and to neurogenetic gradients of either the source neurons (entorhinal cortex, olfactory bulb mitral cells), receiving neurons (lateral septal nucleus, cortical nuclei of amygdala, anterior olfactory nucleus, primary olfactory cortex) or both (magnocellular basal nuclei, striatum) (Bayer, 1980, 1985, 1986; Bayer and Altman, 1987; Bayer and Altman, 1991). The attractive possibility that fiber ordering is based on the relative time of their establishment, and that temporal events are translated into spatial distributions, is termed *chronotopic* fiber ordering. It could provide (or at least contribute to) a mechanism for the establishment of mapped projections without the need for strict specific chemical matching between the interconnecting structures.

The possibility that chronotopy plays a part in the establishment of ordered thalamocortical projections assumes greater significance, in view of the co-culture experiments reviewed in chapter 5, which suggest that in vitro there are no powerful, regionally specific attractive signals between cortex and thalamus. However, the plausibility of the chronotopic hypothesis rests on showing that:

(1) Fiber outgrowth follows a spatiotemporal gradient.
(2) Fibers remain ordered, according to their times of generation, during their outgrowth and arrival.

Only if these conditions exist would it be conceivable that at the site of innervation the arriving fibers could simply occupy their places on a 'first-come-first-served' basis, establishing a map based on chronotopy.

The Development of Innervation Patterns and 'Maps' Depends on Cascade of Mechanisms

Topographic order is a very common feature of the brain. Information becomes distributed and represented *spatially* across structures of the central nervous system. This tendency of the brain to be organized in 'maps', as they are commonly known, is particularly clear in the major sensory and motor areas of the cerebral cortex. In visual, auditory and somatosensory cortical areas, the representation of the receptor surface is arranged across the cortex in an *isomorphic* fashion. This is not to say that cortical maps are topologically simple, linear projections of receptor sheets: rather they are distorted and sometimes fragmented, in various interesting ways. In particular, a general feature of central sensory maps is that densely innervated regions of the receptor surface occupy a larger fraction of the map than regions that are more sparsely innervated. This anisotropy in central representation means that the local scale of the mapping is determined by the density of nerve fibers arriving from that particular region of the periphery. This in turn suggests that the scale of the cortical representation is related to the density and spatial arrangement of afferent fibers bringing information into the cortex from the periphery. In motor areas of the cortex too, regions devoted to the richly innervated musculature of the hands and face (in man) occupy larger regions than those concerned with the muscles of the trunk. This anisotropy obviously reflects the different volumes of neuronal machinery required for the control of richly, as opposed to sparsely, innervated muscle. The development of these maps is gradual and probably based on a sequence of mechanisms.

The obvious conclusion we can draw by examining the development of various systems is that none of the above mentioned mechanisms (affinity gradients, chemotaxis, contact guidance, spatio-temporal waves in neurogenesis or ordered fiber outgrowth, selective cell survival and death, activity dependent development and maintenence of neural connectivity) can *alone* adequately explain all aspects of the formation of orderly interconnections and map formation. Tables 1.1 and 1.2 summarize the mechanisms involved in axonal pathfinding during development and the most common experimental techniques used in developmental studies

This is not to say that one of the distinct factors can not be dominant at a particular stage (the initial establishment of a fiber path may be governed by different mechanisms from those determining the final synaptic targeting, and different again from mechanisms controlling regenerating pathways). But in general it is likely that multiple mechanisms act simultaneously or in an overlapping cascade of mechanisms. To understand the final result we can examine the development of a pathway step by step.

Table 2.1. Mechanisms involved in axonal pathfinding during development

Chemoaffinity, chemotaxis	surface bound affinity substrates
	trophic factors
	tropic factors
Contact guidance	guide-post cells
	pioneers
	scaffolds
Activity-dependent regulation	cell survival and death
	formation and maintenance of neuronal connectivity

Table 2.2. Common experimental techniques in developmental neurobiology

Anatomical	tracing in living and fixed tissue (carbocyanine dyes)
	3-D laser scanning confocal microscopy
	organotypic co-culturing
	immunocytochemistry
	in situ hybridization
Physiological	optical recording (voltage- and Ca^{2+}-sensitive dyes)
	long term observation of neurons with time-lapse video microscopy
	electrophysiology
Molecular biological	mutants
	transgenic animals
	gene knockout

Since thalamic axons in all mammalian species reach the cortex at a very early stage in cortical development, there is the possibility that those axons not only determine the positions and boundaries of major cortical regions but even impose on sensory and motor areas their initial internal topography. The mapping within a cortical region may be fundamentally dependent on the array of thalamic axons that reach it. The aim of the work described in this book is to understand more of the initial phases of the establishment of thalamocortical connectivity and to elucidate the way in which thalamic axons grow and are guided towards the cortex. If the distribution of thalamic axons does indeed influence the parcellation of the cortex and the formation of its maps the guidance of thalamic axons is the key to the understanding of early cortical differentiation.

CHAPTER 3

General Introduction to Development of the Cerebral Cortex and Thalamus

Neurogenesis and the Formation of the Cortical Plate

The basic principles of cortical neurogenesis are very similar in all mammals. The cells are born in the ventricular and subventricular zone. From this 'mitotic factory' the first-generated postmitotic neurons migrate to the cortical surface and form the primordial plexiform zone (the preplate). The subsequently formed neurons migrate and take up their position in the middle of the preplate, splitting it into the marginal zone at the top and the subplate at the bottom (nomenclature: Boulder Committee 1970; Marin-Padilla, 1971; Luskin and Shatz, 1985a,b; Valverde et al, 1995) (Fig. 3.1). The cortical plate proper, sandwiched between the two components of the preplate (Luskin and Shatz, 1985a,b), steadily thickens as subsequent waves of arriving neurons migrate through the existing cells of the plate and take up their position under the marginal zone in an inside-first outside-last fashion (Angevine and Sidman, 1961; Berry and Eayrs, 1963; Berry and Rogers, 1965; Rakic, 1974; Lund and Mustari, 1977; Luskin and Shatz, 1985a,b)(Fig.3.2).

In the embryonic cortex of mammals, the subplate forms a substantial layer (Kostovic and Rakic, 1980; Luskin and Shatz, 1985a). In the adult, however, layer 1 and the white matter under the cortex contain few neurons (Cajal, 1909, 1911). The relative paucity of neurons in the adult in regions corresponding to the original preplate is surely one of the reasons why the subplate, and subplate cells, were largely ignored until relatively recently. Luskin and Shatz (1985a,b) in the cat confirmed Marin-Padilla's original suggestion

Development of Thalamocortical Connections,
by Zoltán Molnár. © 1998 Springer-Verlag and R.G. Landes Company.

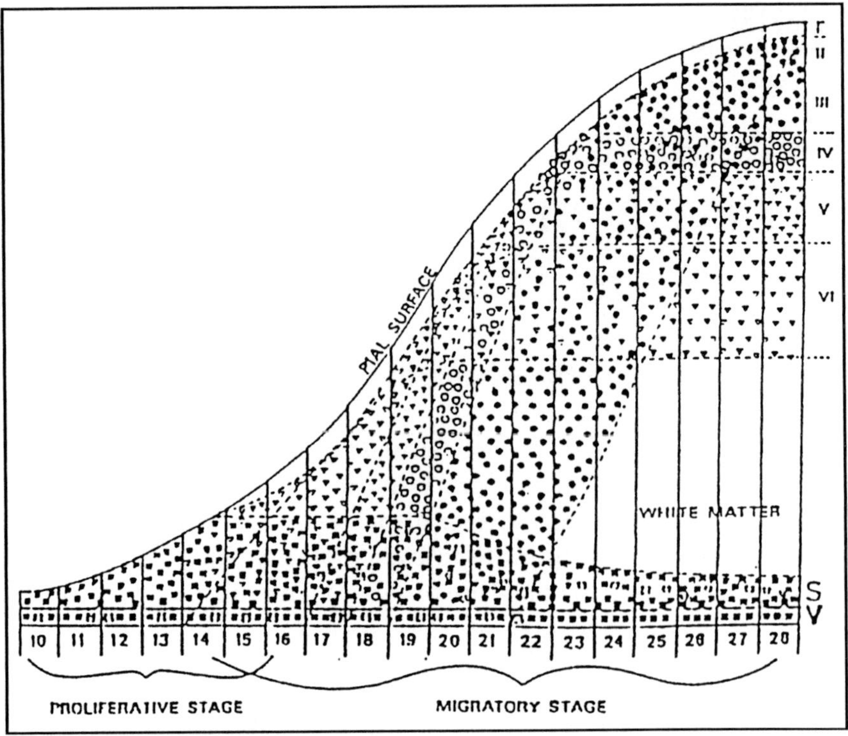

Fig. 3.1. The pattern and timing of cortical neurogenesis. Early thymidine autoradiographic birthdating studies revealed the *inside-out pattern of cortical assembly* (Sidman et al, 1959; Uzman, 1960; Angevine and Sidman, 1961). The earlier generated neurons migrate to the most superficial position, but they are subsequently displaced to deeper levels by those that arrive later. Berry and Rogers (1965) summarized the inside-out sequence of the development of cortical layers in this diagram, but at that time the existence and special nature of the preplate had not been recognized. Each column represents a narrow sector of the rat cortex of a given age (the ages are marked under the bottom of each column). Cell proliferation, which extends from E12 to E21 in the rat, (Berry and Rogers, 1965; Berry et al, 1964, Hicks and D'Amato, 1968) is over at birth, but cell migration continues until the end of the first postnatal week (the gestation period is 21/22 days in the rat). Reproduced with the permission of Cambridge University Press from Berry and Rogers (1965).

(Marin-Padilla, 1971) that cells of the subplate and the marginal zone are co-generated and form the primordial plexiform zone. This was the first study to demonstrate that the subplate and marginal zone cells have a common origin and they both decline in number after birth compared with cortical plate cells (Fig. 3.3).

In the rat, thymidine birthdating at E12 labels only cells of the subplate and layer 1 (Lund and Mustari, 1977; Valverde et al, 1989, 1995a,b; Bayer and Altman, 1990). Recent work shows that these ear-

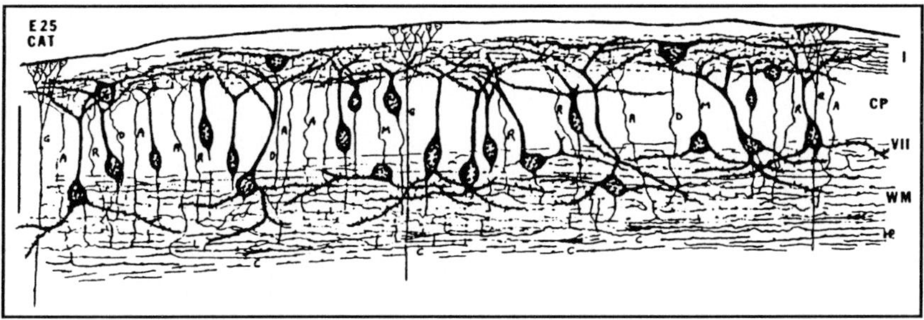

Fig. 3.2. Subsequently formed cortical plate neurons migrate and take up their position in the middle of the preplate, splitting it into the marginal zone at the top and the subplate at the bottom. The simple inside-out concept of cortical development had to be modified. Marin-Padilla, based on his Golgi studies (Marin-Padilla, 1971, 1972, 1978, 1988), first suggested that the cortical plate develops *within* the preplate. Luskin and Shatz's (1985a,b) autoradiographic studies confirmed that the *cortical plate* does indeed *split the first-generated cell layer* of the primordial plexiform zone (or preplate) into two, as Marin-Padilla originally suggested. This camera lucida drawing of a Golgi preparation (25-day-old cat embryo cortex, taken from Marin-Padilla, 1988) demonstrates the splitting of the preplate into marginal zone (I) and subplate (VII) by the developing cortical plate (CP). Reproduced with the permission of Plenum Press, New York from Marin-Padilla (1988).

liest born cells in rodents (originally residing in the primordial plexiform zone) also die to a greater extent than cells of the cortical plate (Wood et al, 1992; Derer and Derer, 1992; Gilles and Price, 1993, Price et al, 1997), although relatively more of them seem to survive than in cats and primates (Kostovic and Rakic, 1990). Unfortunately, in primates there is no birthdating study similar to the one performed by Luskin and Shatz (1985a,b in the cat; Fig. 3.3). This makes it difficult to compare the amount of cell death in cortical plate and subplate in different species. In rodents there are conflicting results between rat, hamster and mouse (see Valverde et al, 1995b). Valverde et al (1995b) found in rat that the cell death in layer 6b, a layer regarded as a homolog to the subplate of primates and carnivores, is neither particularly prominent nor significantly different from other cortical layers. This finding however was very recently repeated and questioned in mouse (Price et al, 1997).

In the rat, the first cells of the cortical plate itself are born on fetal day 14, the last at E21. Thus cortical cell generation finishes before birth and the subventricular zone disappears by postnatal day 3 (P3), but cell migration into the upper layers continues until the end of the first postnatal week (Berry and Rogers, 1965; Lund and Mustari,

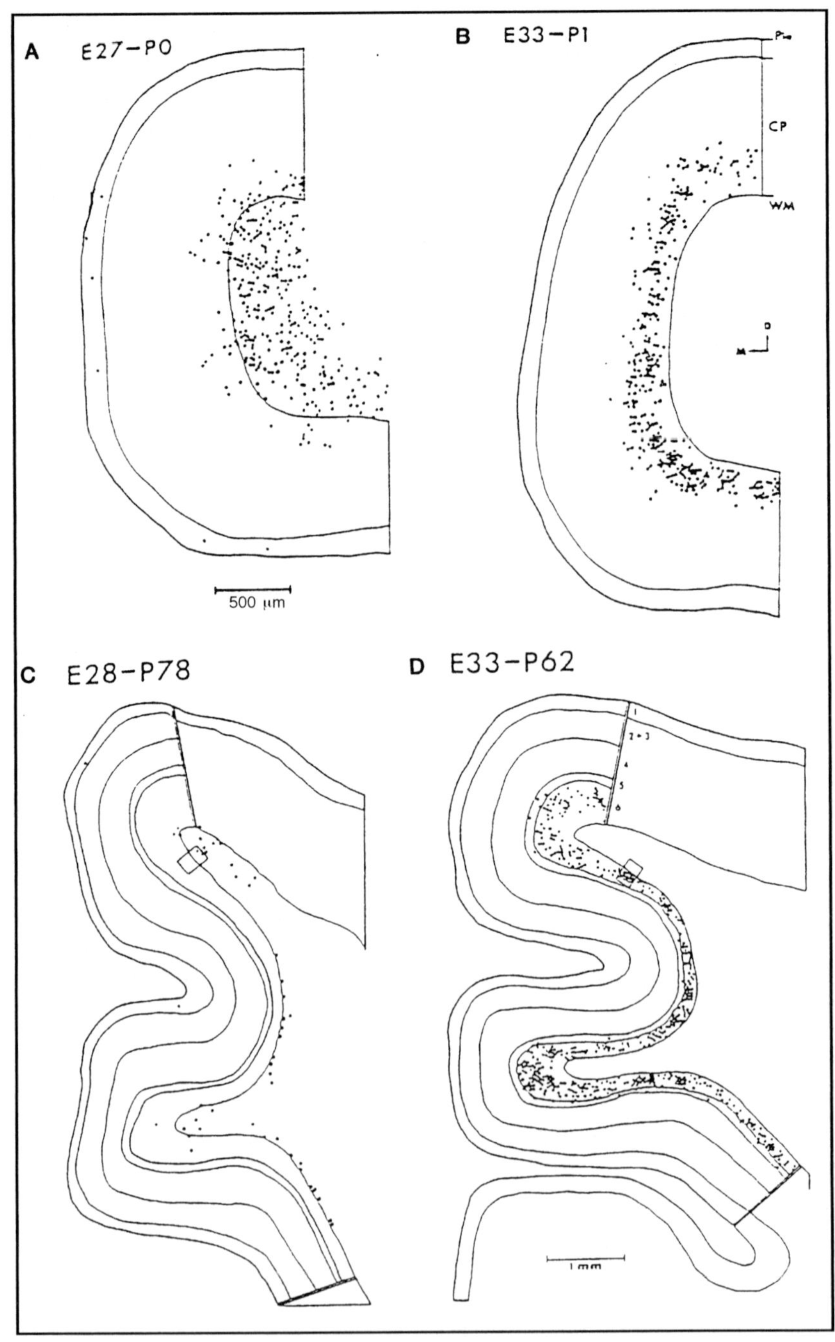
A E27–P0
B E33–P1
CP
WM
500 µm
C E28–P78
D E33–P62
1
2+3
4
5
6
1 mm

1977; Miller, 1988). Clonal relationship and areal dispersion of cortical neurons is beginning to be understood by using retrovirus mediated gene transfer experiments (Walsh and Cepko, 1993; Mione et al, 1994; Götz et al, 1995) and transgenic techniques (Tan et al, 1995; Soriano et al, 1995). I would like to refer to some comprehensive reviews on this topic (Rakic, 1995; McConnell, 1995; Levitt et al, 1997).

In the tangential direction across the developing cerebral cortex, there is another distinct neurogenetic gradient. The anteroventral areas antecede the caudodorsal in maturity (Berry and Rogers, 1965; Lund and Mustari, 1977). The most detailed studies on these gradients were performed in marsupials because of the prolonged neurogenesis of the cortex (Sanderson and Weller, 1990; Mark and Marotte, 1992) (Fig. 3.4). Autoradiographic birthdating studies in these species with protracted cortical neurogenesis suggested that the cortex can be divided according to the time of its generation into areas that develop synchronously (Sanderson and Weller, 1990). Corticogenesis begins first in the ventromedial region adjacent to the paleocortex and archicortex, and there is a simple orderly wave of maturation, from rostral and ventral to caudal and medial. Lines defining the locus of points at an identical stage of development lie

Fig. 3.3. (Opposite) Subplate and marginal zone cells have a common origin and they both decline in number after birth compared with cortical plate cells (Luskin and Shatz, 1985b). A and B: Camera lucida drawings of representative autoradiographs to show the distribution of heavily labelled cells (black dots) in the visual cortex of the cat at birth resulting from [^{3}H]thymidine injection on E27 (A) or E33 (B), as seen in coronal sections coinciding with the midpoint of the lateral geniculate nucleus. Roughly comparable numbers of labelled cells are found after an injection on E27 (332) and E33 (335), although they are situated within the cortical plate (CP)/(layer 6) after the E33 injection (B), and predominantly below it (subplate), with a few in the marginal zone, after the E27 injection. Labelled cells are also present in layer 1 after the E27 injection, but not after the injection on E33. Dorsal (D) is up and medial (M) is to the left. WM, white matter. Calibration bar at A refers to both drawings. C and D: Camera lucida drawings of comparable representative autoradiographs of adult cat primary visual cortex (delimited by radial double lines) to compare the distribution and density of heavily labelled cells (black dots) resulting from [^{3}H]thymidine injection on E28 (C) or E33 (right). Note the much greater density of heavily labelled cells remaining in layer 6 below the cortex after the E33 injection, despite the fact that the E28 section was purposely chosen because an unusually *large* number of labelled cells was present compared with other sections. Both sections were selected from around the coronal plane passing through the midportion of the lateral geniculate nucleus. Numbers indicate layers 1 to 6. Boxes indicate the locations of photomicrographs in Figure 2 of the Luskin and Shatz (1985b) paper. The figure was adapted from Luskin and Shatz (1995b) with kind permission of The Journal of Neuroscience, Washington DC.

roughly parallel to the rhinal fissure and are termed *isogenetic strips* (Sanderson and Weller, 1990)(Fig.3.4).

Introduction to the Development of Thalamic Nuclei

It was Edinger (1885) and His (1893, 1904) who established the first terminology for the gross differentiation of the diencephalon, based on their work on human embryos (Jones, 1985). They described the way in which the hypothalamic sulcus (an early sulcus on the medial wall of the diencephalon) divides the diencephalon into dorsal and ventral parts. The ventral part later develops into hypothalamus, the dorsal forms the epithalamus, thalamus and metathalamus (metathalamus = geniculate bodies; Edinger, 1885).

Based on comparative studies in amphibian embryos, Herrick (1918) made similar, but slightly different descriptions. To emphasize the different developmental history of the thalamus and metathalamus from the epithalamus and hypothalamus he divided the thalamus into dorsal and ventral thalamus. Herrick's (1918) fundamental subdivision of the developing diencephalon (Fig. 3.5) was confirmed for mammals (in the rabbit) by Droogleever-Fortuyn (1912) and it eventually became widely accepted and is still used (Jones, 1985).

Even before the application of autoradiographic birthdating techniques two phases of thalamic growth were distinguished: (1) an early phase of cellular differentiation during which the epithalamus, dorsal thalamus and ventral thalamus differentiate from one another, followed by (2) a later phase during which the individual nuclei differentiate within the three larger subdivisions:

(1) In the rat, the thalamus and the hypothalamus can be distinguished after the appearance of a ventricular groove (anterior diencephalic wall) at the 12th embryonic day of gestation (Cogeshall, 1964). On E13, a wedge-shaped enlargement appears in the diencephalic wall with further furrows defining the developing dorsal and ventral thalamus. These furrows become less apparent as the thalamus further increases in size.

(2) In the rat, at embryonic day 16/17 nuclear differentiation is commencing in the epithalamus and ventral thalamus but not in the dorsal thalamus (Cogeshall, 1964; Jones, 1985). Within the thalamus nuclear differentiation starts posteriorly and laterally and proceeds anteriorly and medially (Rose, 1942a,b; Cogeshall, 1964; Jones, 1985). In view of the growing evidence that there

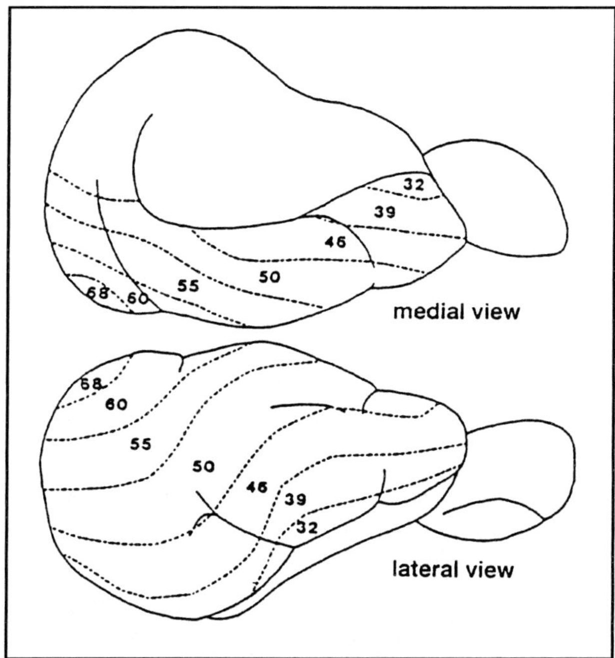

Fig. 3.4. Maturation gradients in cortical neurogenesis. The early autoradiographic studies on cortical cell generation revealed the relative delay in development of the medial-occipital cortex compared with the lateral-frontal (Berry and Rogers, 1965; Lund and Mustari, 1977). In the rat there is a maturational difference of 1.5-2 days, but in at least some marsupials it is more than 30 days. This relatively slow development of the marsupial cortex was exploited by Sanderson and Weller (1990) when they described the *areal pattern of the histogenesis* of the marsupial brain. The diagram (taken from Sanderson and Weller, 1990) demonstrates the regional pattern of neurogenesis in the opossum cortex (which is probably more or less similar in sequence to that of the rodent cortex). The strips represent the ages (numbers in postnatal days) at which neurogenesis was complete for all neurons in the given segment (called isogenetic stripes). The strips of synchronously generated cells run roughly parallel to the rhinal fissure. Reprinted from Sanderson and Weller (1990) with kind permission of Elsevier Science, Amsterdam.

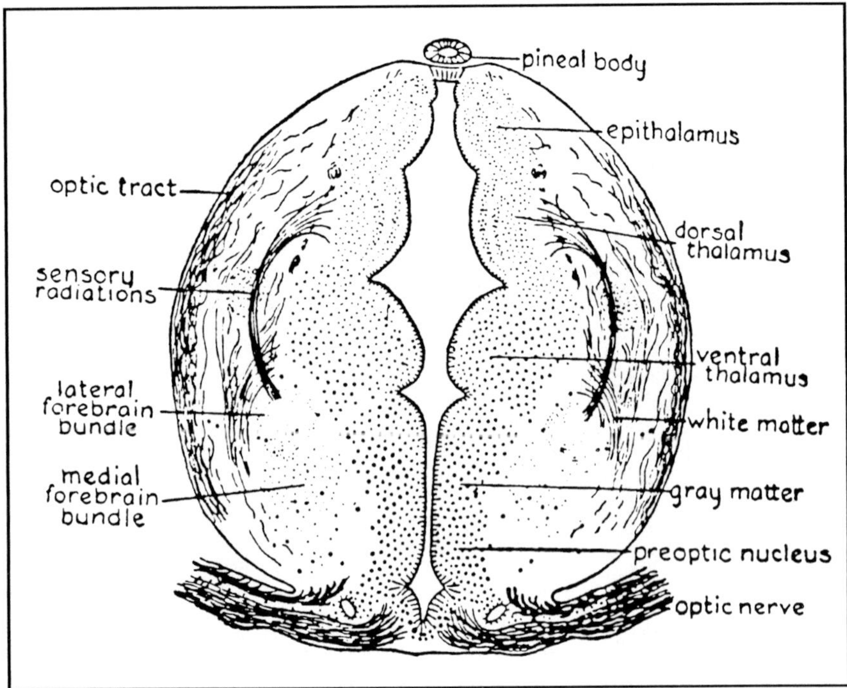

Fig. 3.5. The subdivisions of the developing diencephalon by Herrick (1926). Coronal section of a salamander diencephalon demonstrating the epithalamus, dorsal thalamus, ventral thalamus, and hypothalamus, separated by the sulci on the medial diencephalic wall. Reproduced from Jones (1985) with kind permission of Plenum Press, New York.

are transient cell populations in the developing thalamus itself (Mitrofanis, 1992a), it is important to re-examine the development of the thalamus in relation to the birthdates of its cells in embryonic and early postnatal material rather than drawing conclusions in the mature brain. These studies should be readdressed in detailed autoradiographic developmental series to define the nature, position and possible function of transient cells in the developing thalamus.

Autoradiographic studies show that cellular proliferation precedes nuclear differentiation by several days. Virtually all thalamic neurons are born in the rat between E13 and E19 (McAllister and Das, 1977; Altman and Bayer, 1979) coinciding with the period of generation of cells of the preplate of the occipital cortex. The postmitotic neurons migrating outwards from the ventricular and subventricular zones come to rest in regions that are determined by their time of birth, and not by nuclear boundaries. The lateral geniculate nucleus (LGN) is generated over 2 days between E12 and E14 (Brückner et al, 1976; Lund and Mustari, 1977). Jones (1985) describes three neurogenetic gradients in the developing thalamus:

(1) *posterior to anterior* in that the cells of the posterior nuclei are born before those of anterior nuclei (referring to the relative position of nuclei in the adult thalamus).
(2) A second operates from *lateral to medial*, with lateral nuclei born before the medial ones (e.g., neurons in the reticular nucleus are born before those of the mediodorsal nucleus).
(3) The third thalamic neurogenetic gradient operates from *ventral to dorsal* (Jones, 1985).

It is not yet established whether these gradients exist within individual nuclei as well as across the thalamus as a whole (Hickey and Cox, 1979; Hickey, 1980; Shatz, 1981, 1983; Hickey and Hitchcock, 1984; Altman and Bayer, 1979). Also it is important to know how these gradients are distributed across the *embryonic* thalamus, since there are considerable changes in the orientation of the thalamus as a whole during development in some species (Rakic, 1977; Jones, 1985) (Fig. 3.6).

Adult Thalamocortical Projections

Briefly, I would like to draw attention to some basic aspects of thalamocortical projections. Only dorsal thalamus sends projection to the cerebral cortex, epithalamus and ventral thalamus do not (Jones, 1985). In addition to the cerebral cortical projections, dorsal thalamus sends axons to striatum, olfactory tuberculum, parts of the amigdala, piriform cortex and hippocampal formation (Jones, 1983, 1985; Macchi, 1983). Every dorsal thalamic nucleus projects to the cerebral cortex and all cortical areas receive projections from the dorsal thalamus (Caviness and Frost, 1980; Jones, 1983, 1985; Caviness, 1988). The literature distinguishes two types of projections, specific and nonspecific (Jones, 1985). The specific projections terminate within the borders of one or a few cortical fields, the nonspecific or diffuse projections project over a wide area of the cortex. The same dorsal thalamic nucleus can establish both types of projections (Jones, 1985). The majority of thalamocortical axons terminate on layer 4 and to a lesser extent on 6, 3, 1 of the cerebral cortex (Jones, 1985).

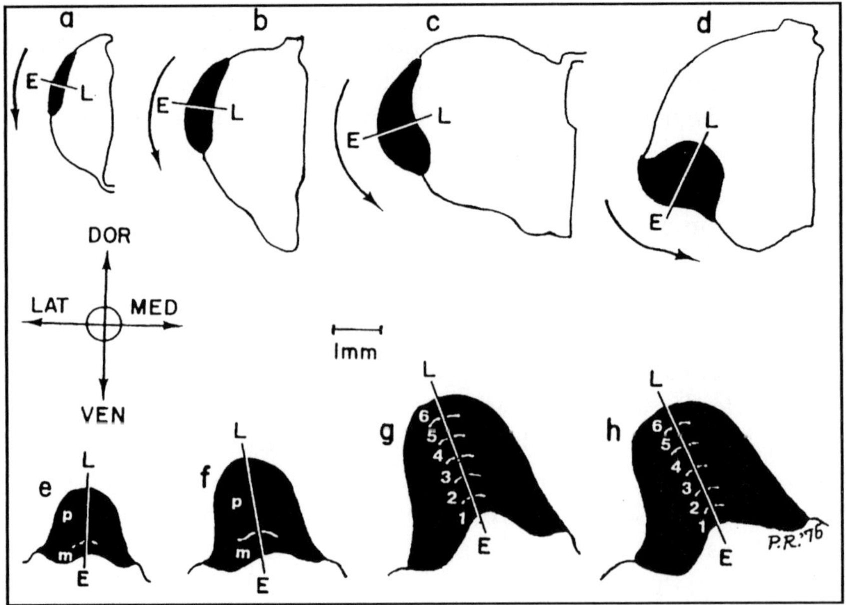

Fig. 3.6. The development of the lateral geniculate nucleus of the rhesus monkey (Rakic, 1977). After cell generation the neurons migrate from the ventricular and subventricular zones (medial wall) to the lateral wall of the thalamus and take up their position in an outside first inside last fashion (E, early; L, late) within the nucleus (black). Cells of the magnocellular (m) layers (1 and 2) are born and differentiate into layers before the parvocellular (p) layers (3 to 6). During development there are substantial rearrangements in the entire thalamus and the lateral geniculate nucleus rotates from a lateral to a ventral position (arrows). Ages: a, embryonic day (E) 48; b, E58; c, E77; d, E84; e, E91; f, E97; g, E112; and h, P60. Reproduced from Rakic (1997) with kind permission of John Wiley & Sons Inc, New York.

Most of input to thalamus comes from the cortex (Mitrofanis and Guillery, 1993). Guillery (1995) proposed two types of thalamic nuclei. "First order nuclei", which receive primary afferent fibers from the periphery and receive their corticothalamic afferents from layer 6 pyramidal cells, and "higher order nuclei" which receive their input from pyramidal cells in cortical layer 5. Only layer 6 projections send branches to the thalamic reticular nucleus and are believed to have modulatory function, whereas layer 5 projections are proposed to transmit information about the output of one cortical area to another, and thus are thought to be involved in corticocortical communication (Sherman and Guillery, 1996; see chapter 11).

Thalamocortical connections are generally ipsilateral although contralateral connections exist in several species (for references see Dermon and Barbas, 1994).

In the following chapters I shall examine the factors involved in the development of the area and layer specific thalamocortical connectivity.

CHAPTER 4

The Establishment of the Early Topographically Ordered Thalamocortical Connections in Mammals

Adult Topography of the Thalamocortical Projections

Thalamocortical projections have distinct and characteristic areal and laminar specificity (Jones, 1985). The representation of the three-dimensional thalamic volume on the continuous planar sheet of the cortex has been studied extensively in the rodent (Caviness and Frost, 1980; Crandall and Caviness, 1984; Jones, 1985). These studies demonstrated that all areas of the rodent neocortex receive a thalamic projection (Caviness and Frost, 1980). In the *adult*, neighborhood relationships within the thalamus do not always correspond precisely to the distribution in the cortex. In most cases adjacent neocortical fields receive their projections from adjacent thalamic nuclei, and proximity relationships correspond in the cortex and the thalamus; but occasionally, non-adjacent thalamic nuclei project to neighboring distinct cortical regions, with no overlap, separated sharply at the borders of their cortical regions (Caviness, 1988). The relative positions and neighborhood relationships of the fibers originating from different thalamic regions are, to a considerable extent, preserved on their course to the cortex within the characteristic fan-shaped radiation of thalamic fibers (Caviness and Frost, 1980; Caviness, 1988; Agmon and Connors, 1991), but this does not appear so convincing at the level of individual fibers, which exibit mixing and twisting along their path (Bernardo and Woolsey, 1987). However, these studies do not necessarily provide information about the

---

Development of Thalamocortical Connections,
by Zoltán Molnár. © 1998 Springer-Verlag and R.G. Landes Company.

mechanism of establishment of connections. First, all of these studies used only a single tracer, but the synchronous application of two or more distinguishable tracers into neighboring thalamic or cortical regions is needed to reveal topographic relations on a more global scale. More importantly, all these observations were made in the adult (Caviness and Frost, 1980; Bernardo and Woolsey, 1987) or in the postnatal animal (Crandall and Caviness, 1984). Therefore they do not address the crucial question of topographic relationships within the embryonic thalamus during the differentiation of the individual nuclei nor the mapping of the ascending embryonic thalamic fibers at the earliest stages of outgrowth. It might be misleading to infer from adult to a developing system. One has to consider the dynamics and has to examine the real developmental sequence.

The Timing and Early Pattern of Thalamic-Axon Outgrowth Carbocyanine Dyes in Embryonic Tracing

Until recently, the only information about the development of thalamocortical connectivity came from studies using degeneration techniques (e.g., Lund and Mustari, 1977) or transneuronal transport of label from the eye (Rakic, 1976). In the past couple of years, however, the use of fluorescent carbocyanine dyes (Honig and Hume, 1986) has revolutionized the tracing of embryonic pathways, because they can be employed in fixed tissue (Godement et al, 1987). Although some aspects of the issue of the development of thalamocortical projections were addressed more than a decade ago (e.g., Lund and Mustari, 1977; Wise and Jones, 1978) the greater versatility and sensitivity of the new techniques promised a greater depth of understanding.

The ingrowth of thalamic afferents has now been studied with carbocyanine dyes in fixed brains by several groups (Catalano et al, 1991; Reinoso and O'Leary, 1990; Erzurumlu and Jhaveri, 1990, 1992; Molnár and Blakemore 1990a,b, 1992; Kageyama and Robertson; 1993; Miller et al 1993). In most respects this recent work has confirmed the essential findings of the earlier studies, but has added further precise information. I shall review the basic pattern and order of thlamocortical innervation in embryonic ages revealed by using carbocyanine dye tracing in embryonic and early postnatal animals.

In the rat, small carbocyanine dye crystal implantation into the dorso-lateral part of the thalamus reveals thalamic fibers reaching the internal capsule in an organized fashion at E14 (Fig. 4.1).

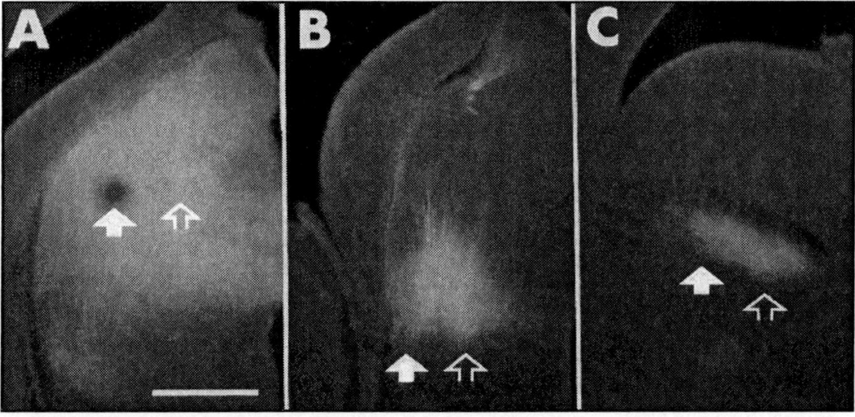

Fig. 4.1. A: DiI and a DiA crystal were inserted 200 μm apart (DiI lateral to the DiA) into the dorsal thalamus of an E14.5 brain. After 3 weeks incubation at room temperature, 100 μm-thick coronal sections were cut and counterstained with bisbenzamide. Fluorescent micrographs were taken by exposing the same film three times using rhodamine, fluorescein and ultraviolet filters. The labelled thalamic fiber bundles remain separated as they descend in the ventral thalamus. B: and as they pass through the internal capsule. Fibers originating from more lateral thalamic regions (filled arrow) lie below the fibers originating from a more medial thalamic region (empty arrow) as they pass under the striatum. C: and they head more caudally within the hemisphere. Bar: A: 100 μm; B-D: 500 μm

The thalamic fibers run through the primitive internal capsule and under the developing corpus striatum but do not reach the intermediate zone of the ventralmost cortical areas until E15.5 days of gestation. In addition to staining thalamocortical fibers, DiI implantation in the thalamus labelled numerous cells within the ventral thalamus (within the thalamic reticular nucleus) and a few within the internal capsule, in the so-called perireticular nucleus (Mitrofanis, 1992a, 1994). I shall discuss the possible role and evolutionary origin of these cells in chapters 6 and 11 respectively.

Patterns of Thalamic Fiber Ordering Is Different as Fibers Travel Through the Diencephalon, Primitive Internal Capsule and in the Cortical Intermediate Zone

In rat, fibers from the dorso-lateral thalamus reach the intermediate zone beneath the ventral cortex around E15.5, and the occipital cortex itself at E16 (Fig. 4.2). As they leave the diencephalon,

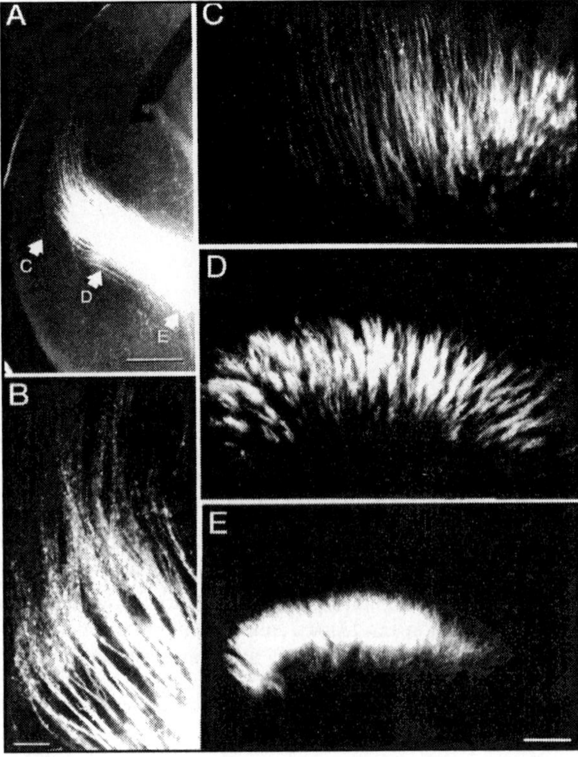

Fig. 4.2. A: Crystal placement into the dorsal thalamus reveals the first fibers reaching the intermediate zone beneath the ventralmost part of the intermediate zone by E15.5 (half a day before they reach the occipital pole itself). Leaving the diencephalon, the fibers form fascicles which open up in a fan shaped fashion. Viewed here in a coronal section A, the ordering of the fibers is such that the inferior-lateral fascicles are destined for the more ventral part of the cortex, while the superior medial fascicles turn upwards and head towards more dorsal cortical areas. The trajectories of the fascicles remain parallel as they turn up into the intermediate zone: they do not cross each other extensively. B: Same as A, but taken with higher power from the region where the fascicles break up into individual fibers as they reach the intermediate zone. To examine the fiber ordering further the contralateral hemisphere (with identical crystal placement to that in A, was sectioned perpendicular to the fiber path. C, D and E are sections at levels corresponding to the labelled arrows in A. E: The fibers leave the diencephalon through the primitive internal capsule. The bundle is at its narrowest at this point. No fasciculation is apparent in this region. D: As the fibers run under the corpus striatum, they form 30-50 μm-thick fascicles, slightly separated from each other. The labelled fiber array loosens up, and expands compared to E. C: The fiber bundles defasciculate as they reach the intermediate zone and turn under the cortical plate. The right side of the section still contains some fascicles, while the left demonstrates the defasciculated fibers running parallel to each other in the intermediate zone. Scale bars: A: 300 μm; B: 50 μm; C-E: 100 μm.

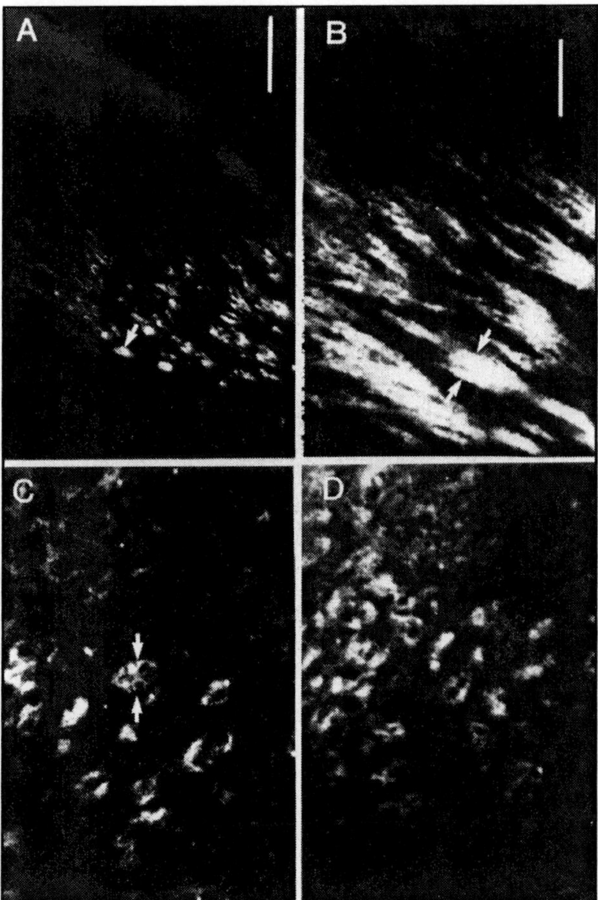

Fig. 4.3. The individual fascicles of a preparation very similar to the previous figure are seen here in cross-section with confocal microscopy. A: Optical cross-section through the fascicles labelled from the dorsal thalamus at E15.5. White arrow indicates a single fascicle. B, C and D: Optical sections along (B) and perpendicular (C,D) to the trajectories of labelled thalamic fascicles. It is apparent that within each fascicle the axons are not homogeneously distributed and are not compact. The profiles contained unlabelled 'holes.' Scale bar: A: 100 μm; B-D: 50 μm. From Molnár (1994).

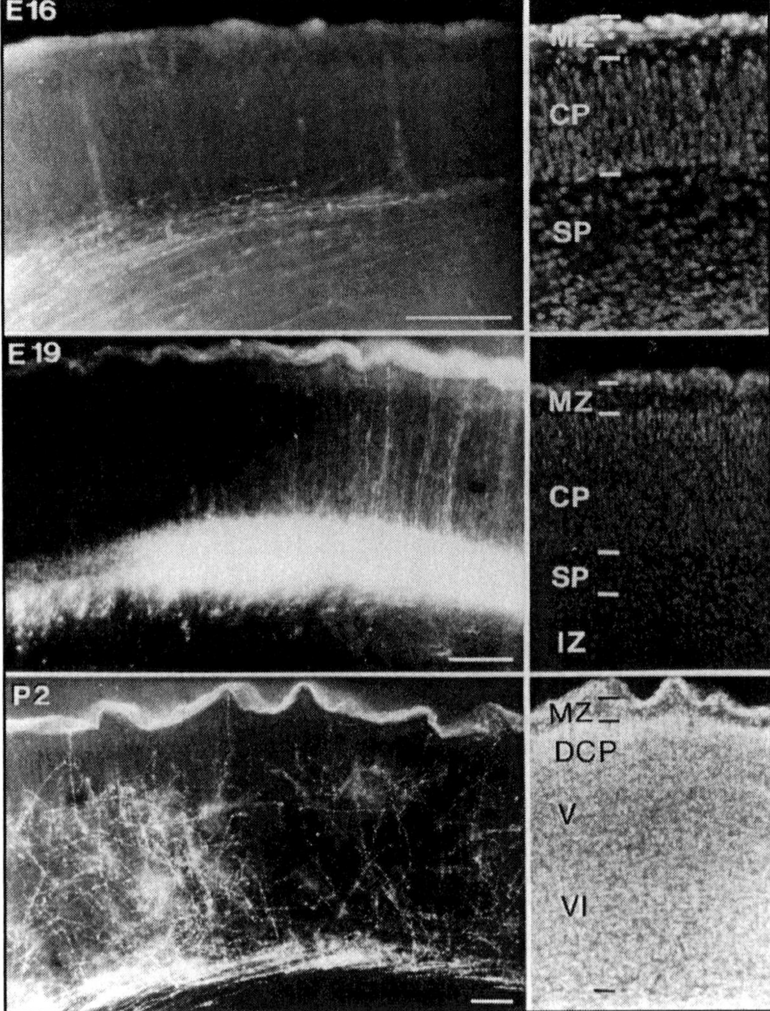

Fig. 4.4. Three stages of thalamic fiber ingrowth, at E16, E19 and P2. The thalamic fibers were stained with a DiI crystal placement into the dorsal thalamus of fixed brains. After 3-6 weeks incubation at 37°C, 100 μm coronal sections were cut and counterstained with bisbenzimide. They were viewed under rhodamine illumination (left panels) to reveal DiI labelling and under UV illumination (*right* panels) to reveal the bisbenzimide staining, showing the layering of the cortex. The first thalamic fibers arrive in the intermediate zone of the occipital pole at E16 (top row) and accumulate within the subplate (SP) (from E16-19) but do not enter the cortical plate (CP) before E19 (B). In more anterior and ventrolateral areas this happens a day earlier. The thin radial processes seen extending up to the marginal zone (MZ) at E19 do not have the appearance of axons and are presumably the processes of radial glia since they are identical in appearance to the processes of such cells labelled by DiI staining of the cortex itself. After E19 (occipital cortex) or E18 (ventrolateral cortex) the lower layers

the thalamocortical fibers form distinct fascicles, which open up in a fan shaped fashion. Viewed in coronal sections the ordering of the fibers is such that the inferior lateral fascicles are destined for the ventral part of the cortex, while the superior medial fascicles turn upwards and head towards more dorsal cortical areas. The trajectories of the fascicles remain parallel as they turn up into the intermediate zone; they do not cross each other extensively, although individual axons and small axon bundles do appear to switch from one fascicle to another along their course (Fig. 4.2). Examination of cross sections of these fascicles (labelled from the dorsal thalamus at E15.5) with a confocal microscope shows that they are not entirely compact: their optical cross sections reveal holes and the bundles have rather irregular edges (Fig. 4.3).

At the border of the corpus striatum and the intermediate zone the thalamic fibers defasciculate and break up into a fairly uniform array of individual fibers. Within the intermediate zone the fibers run parallel to each other in an organized fashion. Similarly to the fascicles, the ordering of the individual fibers is such that the inferior lateral fibers are destined for the more ventral cortical segment, while the more medial fibers, situated deeper in the white matter, head towards more dorsal cortical segments (Fig. 4.2). This behavior of the thalamic axons is very general in all mammalian species including humans (see Fig. 6.6 in chapter 6) and might reflect different extracellular environments along the pathway.

Although thalamic fibers from the region of the LGN reach the occipital cortical regions around E16, they do not substantially invade the cortical plate for some time. Most of the labelled fibers remain restricted to the intermediate and subplate zones until E19. During this period, thalamic fibers accumulate and develop local branches, but these are also mainly restricted to the subplate region (Fig. 4.4).

of the cortical plate are suddenly invaded by the thalamic fibers, although axons still stop short of the dense cortical plate (DCP). Very few fibers reach the marginal zone (MZ), but most of them terminate and arborize about 100-200 μm below. The bottom row demonstrates the distribution of thalamic fibers at P2. Note the very regular, parallel axon arrays in the white matter, visible in the examples at E16 and P2, compared with the somewhat disorganized growth of the axons into the cortical plate at P2. Bars are all 100 μm. From Molnár (1994).

The Waiting Period

Although there is pressure on thalamic axons to establish their projections to the cortex as early as possible (because the route between is increasing in structural complexity all the time), there is nowhere for them to go when they arrive (Rakic, 1977; Lund and Mustari, 1977; Shatz et al, 1988). In the rat, on E16, when LGN fibers reach the occipital cortex, only the lowest part of the true cortical plate is in place and the cortical plate is not yet permissive to axonal ingrowth (see Fig. 4.4 and chapter 5). Moreover, most of the main target cells of the thalamic axons, the cells of layer 4, do not take up their position in the plate until after birth (Lund and Mustari, 1977). Wise and Jones (1976) first used the term 'waiting period' to describe the behavior of commissural axons in the rat, which accumulate for a protracted period in the intermediate zone before they invade the cortical plate. A similar phenomenon has been reported for thalamocortical axons in the monkey (Rakic, 1976, 1977), cat (Shatz and Luskin, 1986) and rat (Lund and Mustari, 1977). Thalamic axons arrive below the correct regions of cortex before the cortical plate has fully formed and they accumulate and 'wait' over the subplate layer (for several weeks in cats and monkeys) before invading the plate proper. Catalano et al, (1991, 1996) have recently questioned the existence of a waiting period in the rat. Using DiI application in the ventral thalamus, they demonstrated that some thalamocortical afferents invade the deep part of the cortical plate of the somatosensory cortex on E18/19. They suggest that thalamic axons simply move up through the cortical plate progressively, as the layers differentiate, always extending up to the lower edge of the dense cortical plate of newly arrived, immature neurons. However, the data of others (Erzurumlu and Jhaveri, 1990; Reinoso and O'Leary, 1990; Molnár and Blakemore 1990a, 1995a,b) indicate that, although thalamic fibers indeed invade the rodent cortex before birth, there is nevertheless a waiting period of a few days, not as long as was originally suggested by Lund and Mustari (1977). In the occipital cortex, thalamic fibers arrive on E16, but do not invade the cortex significantly until around the 19th day of gestation. In marsupials, however, the pattern of thalamocortical development is indeed different: the front of thalamic axons move up through the cortical plate progressively as the layers differentiate, always extending up to the lower edge of the dense cortical plate of newly arrived, immature neurons (Molnár et al, 1995b, 1997c; Knott et al, 1997).

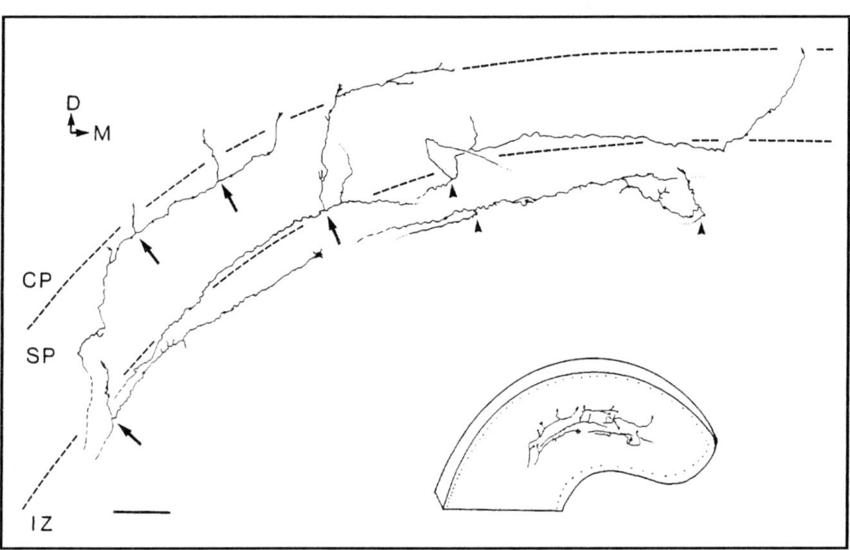

Fig. 4.5. Reconstructions of HRP-filled axons in hamster visual cortex shortly after birth (P0-P3). The thalamic fibers grow in the intermediate zone and subplate forming numerous side branches which extend into the subplate (SP) and penetrate the lower parts of cortical plate (CP). Arrows indicate widely spaced side branches. Some branches change direction, some form U-turns (indicated by arrowheads). Inset: Low power view of axons in visual cortex. Scale: 50 μm. Reproduced from Naegele et al (1988) with kind permission of John Wiley & Sons Inc., New York.

Naegele et al (1988) in the hamster and, more recently, Ghosh and Shatz (1992a) in the cat have shown that during a limited period some thalamic fibers send transient side branches towards inappropriate regions of cortex on their way to their correct target area. Initially the growing thalamic fibers each consist of a single parent axon tipped with a growth cone. Perhaps some of the first fibers entering the cortex at E18 are side branches of principal axons.

In chapters 8 and 9, I shall discuss the possibility that the development and extension of side branches might be caused by activity patterns set up by the thalamic axons themselves within the subplate region. Shatz and her colleagues (Herrmann et al, 1991; Friauf et al, 1990; Friauf and Shatz, 1991) have provided considerable evidence that at least some thalamic axons in the cat actually terminate and form functional synapses on subplate neurons. These connections might relay activity into the subplate sheet, which is in a strategic position to influence further development of thalamocortical and intracortical circuitry (Higashi et al, 1996;

Molnár et al, 1996b). I shall discuss the possible functional role of the waiting period in chapters 8 and 9.

Topography of the Early Thalamocortical Projections

In rat, both anterograde and retrograde tracing experiments demonstrated that thalamic axons advance through the internal capsule and by E16 arrive under the entire cortex with a reasonable topographic order (Molnár et al, 1993, 1994; Molnár and Blakemore, 1995a,b; Catalano et al, 1996). The discrete fiber bundles, containing both corticofugal and thalamofugal axons, do not cross each other and maintain their neighborhood relationships along the entire path between thalamus and cortex. Unfortunately it is rather difficult to draw firm conclusions about the organization of individual fibers within each bundle, even by examining fine 3-D reconstructions of confocal microscopic sections. We shall need to develop new techniques (e.g. making even smaller multiple injection sites) to resolve individual fibers along the whole pathway. Nevertheless, the results of these multiple cortical crystal experiments are very clear. Even after E16 (until E18/19), the bundle (including subplate and thalamic fibers) labelled by a small dye placement in the cortex remains discrete and single throughout its trajectory through the intermediate zone and primitive internal capsule, with no tendency for the fibers to separate into two distinct fascicles at any point in the telencephalic portion of the pathway (Fig. 4.6). Axons (including corticofugal fibers) within the white matter display an organized, parallel distribution with apparently little rearrangement. Within the corpus striatum the axons reorganize into numerous fascicles. There is some exchange of individual axons between adjacent fascicles but no gross intermingling or crossing. Entering the internal capsule the fascicles labelled by a single crystal in the cortex coalesce into a tight bundle of closely associated, strictly parallel axons. The gross topography is maintained along the entire fiber pathway for both corticofugal and corticopetal axons (see Figs. 4.6 and 4.7).

The most compelling evidence for the topographic arrangement comes from experiments involving multiple cortical dye placements (Fig. 4.7), in which the arrays of fibers labelled by each crystal can be followed towards the primitive internal capsule and back to the thalamus. Each distinct, single axon bundle is clearly separate from the others, even as they converge on and pass through the primitive internal capsule. The ascending thalamic axons appear, within the

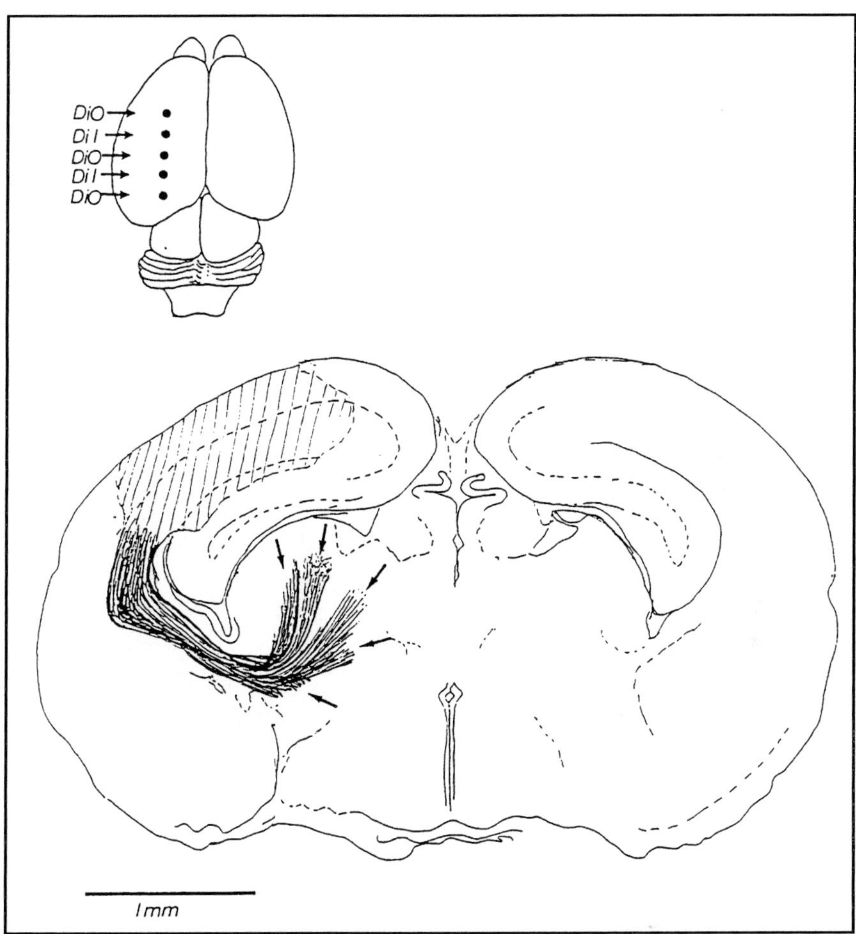

Fig. 4.6. Bundles of axons linking cortex and thalamus at E20 (containing both corticofugal and thalamocortical axons) were revealed by application of five discrete crystals of carbocyanine dye at points in a parasagittal row along the left hemisphere (indicated by arrows on inset diagram of surface view). After 4 weeks for anterograde and retrograde diffusion, 250 μm-thick coronal sections were examined with different filters in the fluorescence microscope to reveal the various dyes. Five distinct bundles were clearly visible passing through the primitive internal capsule without obvious mixing or crossing, and running to different thalamic nuclei: the tip of each bundle is marked with an arrow. This coronal section was at the level of termination of the bundles from the two caudal crystal placements, and retrogradely labelled thalamic cells are visible at the end of those bundles. The tracts labelled from the more anterior crystals ended at more ventral and medial levels in the thalamus, labelling thalamic cells in different regions. Bar, 1 mm. Reproduced from Blakemore and Molnár with kind permission of Cold Spring Harbor Laboratory Press, NY (1990).

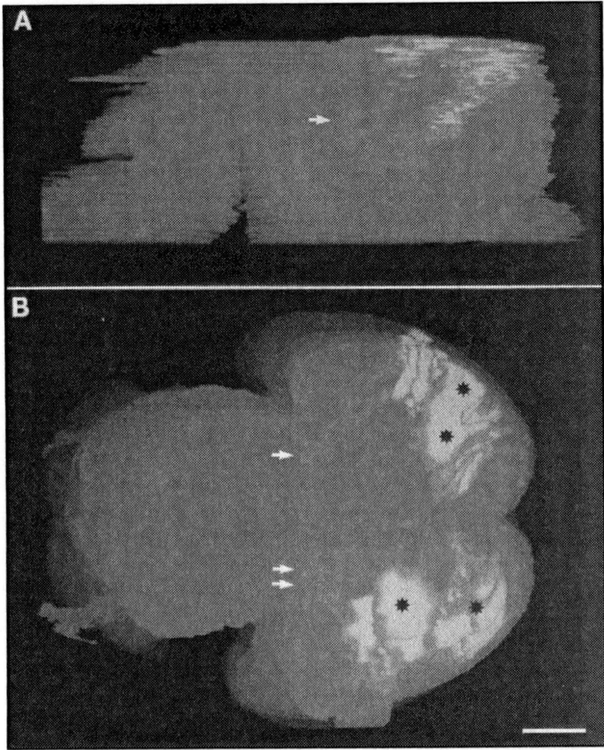

Fig. 4.7. Bundles of axons linking frontal cortex and thalamus at E16 (containing both corticofugal and thalamocortical axons) were revealed by a pair of DiI crystals placed parasagittally from each other in the right hemisphere and coronally in the left hemisphere. After 6 weeks incubation to allow anterograde and retrograde diffusion, 75 μm-thick horizontal sections were cut and counterstained with acridine orange. From the series (of 43 sections) each section was imaged in a fluorescent microscope both with rhodamine and fluorescein filters and 3-D datasets were constructed from the images. A: Viewing the superimposed sections from the side (rostral to the right), the contours of the sections outline the characteristic shape of an E16 brain (with its cortical surface, olfactory bulb, pons etc). B: Viewing the reconstructions from above, two distinct bundles were noticeable on both sides passing through the primitive internal capsule without mixing, running to different thalamic regions. In the right hemisphere the two groups of backlabelled thalamic cells are situated side by side, forming two parallel slabs, whereas in the left hemisphere only one slab was observed, each half of which was labelled with a different dye: each slab is marked with a white arrow. The reconstructions demonstrate the angle (approximately 45°) of the oblique path of the thalamo-cortical connections linking frontal cortex with thalamus. Scale bar: 1 mm. From Molnár et al (1993a).

accuracy of this method, to follow exactly the same topographically ordered paths through the telencephalon as their counterparts descending from the subplate.

Important inferences can be drawn from the patterns of retrograde labelling of thalamic cell bodies (Fig. 4.8). In horizontal sections the thalamic cells labelled by a single crystal in the cortex line up along an antero-posterior longitudinal slab. With different crystals placed along a parasagittal line as early as E16, tracing revealed that more anterior cortical regions are connected with more medial thalamic slabs, whereas more posterior cortical regions were connected to more lateral slabs. The two backlabelled cell groups revealed by any two implantation sites along a coronal line seem to form a continuous corresponding slab within the thalamus. These experiments reveal that adjacent cortical fields receive their projections from adjacent groups of thalamic cells, although the arrangement of corresponding regions on the surface of the cortex and within the volume of the thalamus is rather different. The interlinking fibers (within the accuracy of the technique) preserve their neighborhood relationships along the whole length of the pathway. As soon as thalamic fibers arrive below the cortical plate the general ordering of thalamocortical connections is determined.

Invasion of the Cortex and Establishment of Laminar Termination Patterns

The migration of cortical plate neurons is at its peak at the beginning of this 'waiting period' for thalamic axons. The migrating cells therefore have to pass through the mass of accumulating fibers waiting in the subplate layer, as well as between the subplate cells themselves. Before about E19-20 very few thalamic axons appear to have grown into the occipital cortical plate itself. Invasion suddenly begins, on a massive scale, at about late E20, just a day or so before birth. The course taken by these invading thalamic axons is distinctive and highly ordered: they virtually make a 90° turn from their trajectory in the white matter and the vast majority of the fibers initially grow radially, straight up into the cortex directly above the region of the subplate in which they have been waiting (see Agmon et al, 1993). Only a small portion take more erratic routes, running somewhat obliquely upwards.

By birth a substantial number of thalamic fibers has invaded the cortical plate. Within the cortical plate the fibers follow a much

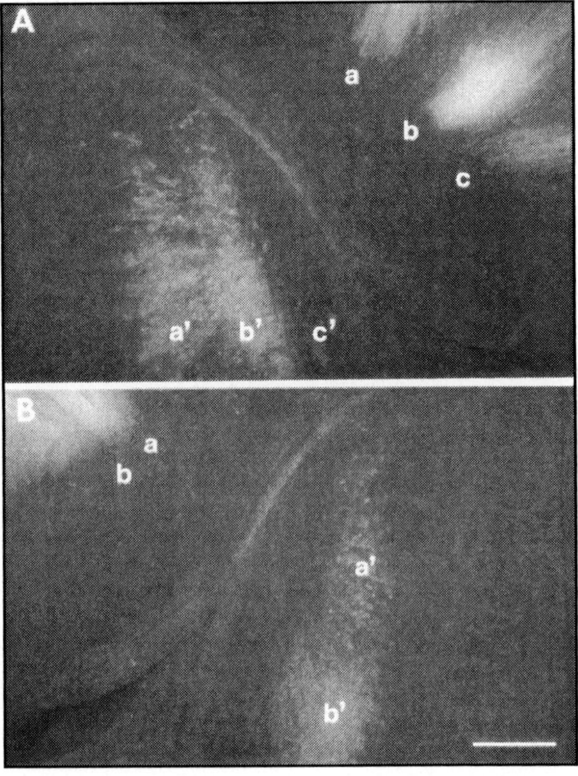

Fig. 4.8. Fluorescent photomicrograph taken from a 100 μm thick horizontal section of an E16 brain by exposing the same film three times using rhodamine, fluorescein and ultraviolet filters (rostral, up), after crystal placements in the cortex. Since thalamic fibers have already arrived at the cortex at E16, the crystal placements revealed both thalamocortical and corticofugal axons linking cortex and diencephalon. 5 weeks incubation was used at room temperature to allow full anterograde and retrograde diffusion. Three labelled bundles (containing both corticofugal and thalamocortical axons) were revealed by application of three crystals of carbocyanine dyes (DiA, DiI and DiA) in a parasagittal row along the right hemisphere (A), and two bundles from two crystals placed coronally into the left hemisphere (DiA ventral to DiI) (B). A and B were taken with the same magnification from the different sides of the same horizontal section.
A: Three distinct bundles are clearly visible within the ventral telencephalon (a,b and c to the right) approaching the primitive internal capsule and running towards different thalamic nuclei. The labelled thalamic cells line up along antero-posterior lines forming slabs (a', b' and c' on the left) such that the more anterior cortical crystal placement (a) revealed a more medial thalamic region (a). More posterior cortical crystal placements (b and c) were connected to more lateral slabs (b and c). The antero-posterior sequence of the carbocyanine dye implantations in the cortex (a,b,c) is reflected in a medio-lateral sequence in the thalamic labelling (a',b' and c'). B: In the left hemisphere two crystals were placed along a coronal line (DiA ventral to DiI). The two backlabelled cell groups line up along a continuous anteroposterior slab in the thalamus, where the ventral crystal placement (DiA) labelled more anterior thalamic cells (a'). Ventro-dorsal sequence of the carbocyanine dye implantation along a coronal line on the cortical convexity seems to correspond to an antero-posterior sequence of thalamic labelling along a continuous slab. Scale bar: 300 μm. Part A of the figure is reproduced from Molnár and Blakemore (1995b) with kind permission of John Wiley and Sons Ltd, Chichester, England and the CIBA Foundation, London.

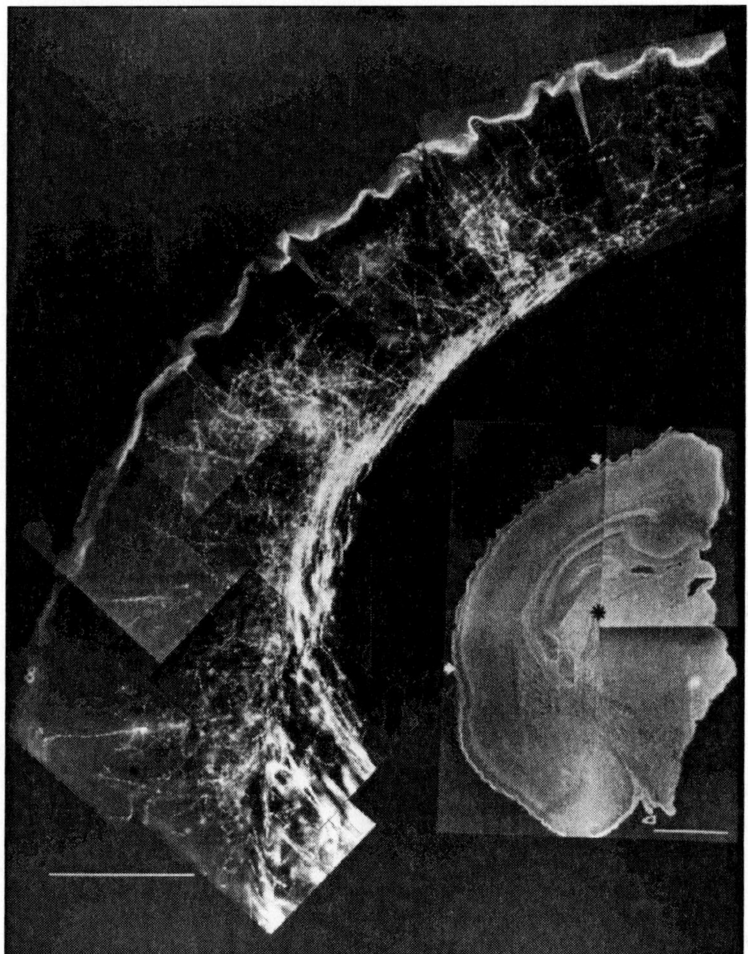

Fig. 4.9. Montage of photomicrographs taken (under epifluorescent illumination) from a coronal section of a P2 hemisphere containing the caudalmost part of the somatosensory cortex. A DiI crystal was placed in the thalamus, then after 6 weeks of incubation at 37°C, 100 μm coronal sections were cut. The insert (*lower right corner*) represents the entire coronal section of the hemisphere, and was photographed at low power, under UV illumination. The bisbenzimide counterstain reveals the outline of the section and the contours of various distinctive structures (layer 1, hippocampus, etc). The limits of the cortical territory shown at high power under a rhodamine filter (*upper left montage*) are indicated on the insert by white arrows at the pial surface. The black star indicates the site of the crystal placement.
The labelled fibers form a narrow band in the white matter immediately below the cortex. At the white matter/cortical plate border the fibers turn up into the cortex and enter the cortical plate at approximately 90° to their trajectory in the white matter. They then follow less regular patterns of

less regular pattern of ascent than was apparent in the white matter. A few fibers reach the marginal zone, but the vast majority stop below the uppermost sector of the cortex, the dense cortical plate (Fig. 4.4). Individual axons examined in 200 μm-thick sections and drawn with a camera lucida: at P2 showed arborizations typically extending over an area with a diameter of about 300 μm.

This extensive lateral spread of the terminal arbors of thalamic axons contrasts with the remarkable order with which they are arranged at the junction between white and grey matter. In favorable preparations whole arrays of individual axons can be seen entering the cortex, and their strict parallel arrangement is most impressive. The fibers heading for ventral areas run superficial to fibers destined for more dorsal cortical segments. Since the earlier fibers enter the more ventral cortical areas first, and since these areas develop earlier, this type of fiber ordering reflects the timing of its establishment. It is as if the topographic array of arriving fibers, established in the intermediate zone and held during the waiting period in the subplate layer, is 'read out' as the fibers invade the cortex, to transform the array into a precise two-dimensional topographic pattern of axons entering the cortical sheet (see chapter 8).

Some individual thalamic axons can be seen to give branches in layer 6a and within the lower third of the dense cortical plate, but most of the terminal arbors are in the putative layer 4. By about P2, most axons have bifurcated at least once at around the top of layer 5, have arborized quite widely directly below the dense cortical plate in what is presumably destined to become layer 4, and appear to have lost their growth cones. A few fibers grow all the way up through the dense cortical plate to the marginal zone, where they branch and appear to terminate within a fairly narrow laminar range. Between P2 and about P8 (when the cortex has achieved its mature lamina-

ascent. A few fibers reach the marginal zone, but the vast majority avoid the uppermost 200-250 μm of the cortex. At the relatively more mature ventral part, the superficial layers, above the level of branching axons, are relatively thicker than in the more dorsal sectors, indicating the ventrodorsal temporal gradient of maturation of the cortex. The fiber-dense areas in the dorsolateral aspect of the section may represent developing whisker barrels. Only two backlabelled cells (lower left corner) were observed in the entire section, which suggests that the crystal placement site was not penetrated by corticofugal projections.
Bar in the insert is 1 mm, in the montage is 0.5 mm. From Molnár (1994).
(Figure on previous page; figure caption continued from previous page.)

tion) the arborizations of thalamic afferents in putative layer 4 become progressively displaced downwards as migrating neurons continue to arrive and the supragranular layers form.

These experiments demonstrate clearly that thalamic fibers arrive in the subplate layer and (at least for the occipital cortex) wait under the cortical plate in a topographically organized fashion. This is not to say that this topography cannot and will not change during the periods of thalamocortical fiber accumulation below the cortex and invasion of the cortical plate. This question will be addressed in chapters 8 and 9.

CHAPTER 5

A Cascade of Signals from the Cortex Is Revealed In Vitro

Aim of Studying the In Vitro Development of Thalamocortical Connections

The various forms of cell-cell interaction during the formation of connections can be studied in several different ways. Description of the anatomy of normal development gives much useful background information but is often not sufficient to reveal developmental mechanisms. Transplantation and lesion studies in vivo can provide substantial information, but their interpretation can be complicated by variability in the survival of grafts and the extent of and reaction to lesions. The technique of organotypic co-culture makes it possible to create conditions under which neurons can develop, differentiate and interconnect, at the same time allowing very specific manipulations.

Our overall goal is to gain insight into the nature of the mechanisms responsible for the development of specific interconnections between thalamus and cerebral cortex. Obviously this is a complex process and many different mechanisms are likely to be involved. There is a general interest in seeing what features of the timetable of normal development could be replicated in vitro, in the hope that one could gain insights into the mechanisms of the development of area- and lamina-specific connections in the living animal. In this chapter I am particularly concerned with examining whether the 'waiting period' and the layer-specific termination pattern within the cortex (see chapter 4) might be duplicated in culture and whether different thalamic nuclei and cortical areas when co-cultured demonstrate regional selectivity in vitro. The use of an in vitro model

Development of Thalamocortical Connections,
by Zoltán Molnár. © 1998 Springer-Verlag and R.G. Landes Company.

system simplifies the task of studying axonal guidance, neuron-neuron recognition and the possible role of chemoattractants and trophic interactions during the formation of the thalamocortical connections.

My plan in this chapter is to to examine the results of numerous paradigms of co-culture, combining different thalamic regions with different cortical areas, taking the explants at different ages, arranging the cultured structures in many different orientations with respect to each other or giving choice to the thalamic or cortical piece by culturing three structures at once.

Historical Remarks: Development of Organotypic Tissue Culturing

Wilhelm Roux succeeded in maintaining the medullary plate of a chick embryo in vitro as early as 1881. However, it later became possible not only to keep the tissue alive, but also to promote cell growth and differentiation. Harrison reported the first successful, reproducible technique for growing nerve cells under in vitro conditions in 1907. The techniques for the maintenance of organotypic cultures and co-cultures of explants of immature nervous tissue are now well developed (see Bornstein, 1958, 1973; Crain, 1974, 1976; Soblowski et al, 1968; Gähwiler, 1988).

Organotypic cultures (the term was originally introduced by Maximov in 1925) preserve the basic structural organization of their tissue of origin. In such cultures of nervous tissue, cells not only differentiate but also form neural circuits typical of the structure in question (for review, see Gähwiler, 1988 and Gähwiler et al, 1997). Numerous techniques have been developed to maintain the necessary nutrition and gas exchange within the tissue piece and to allow development of the whole organotypic tissue block at the same time: Maximov depression slide chambers (for so-called 'lying-drop' preparations); roller-tube techniques; polyamide gauze carriers in plastic culture dishes (for a practical manual, see Freshney, 1986). Each technique has its own advantages and shortcomings. In my work I used the *interphase-type, stationary culture system* (Fig. 5.1), for the following reasons:

1) The tissue does not flatten to monolayer thickness even after several weeks in vitro. This is important because neural growth and the formation of local circuitry are likely to be more natural in three-dimensional substrates than in flattened monolayers.

2) The cultures are stationary, so, as in vivo, the fluid environment of the tissue is not mixing continuously. This is especially important when considering whether diffusible trophic or tropic factors are present and whether they produce directed axonal outgrowth (Lumsden, 1991), though it must be said that diffusional characteristics are likely to be rather different within fluid culture medium compared to the extracellular space of the developing brain.
3) The conditioned medium can be collected, and the medium can easily be changed.
4) One has direct access to observe the specimen during culturing, allowing the possibility of time-lapse photographic techniques, electrophysiological recording, optical recording, etc. (however, the cultures have to be resectioned for conventional light microscopy and can be transferred into an electrophysiological recording set-up only after cutting the membrane around them if not cultured on a second additional membrane piece.)

Development of Serum-Free, Chemically Defined Nutrient Media for Culturing of Brain Tissue

Up to the 1950s there was no alternative to the use of media containing blood serum, amniotic fluid, or embryo extracts to provide the rich nutritional environment needed for neural culture. In vivo, the central nervous system (with the exception of few areas) is normally protected from direct contact with serum by the blood-brain barrier and therefore serum almost never comes into close contact with nerve cells. Nevertheless, serum has been the most widely used culture medium supplement, although there are considerable disadvantages to its use. There is frequently variation in growth-promoting properties between different serum batches. Serum contains a large number of active factors, some inhibitory or toxic to neurons. It contains neurotoxic and neurite outgrowth inhibiting factors whose effects become more apparent when serum is applied in high concentration. Paradoxically, in lower concentration serum stimulates excessive neurite outgrowth. Cytostatic agents must be applied to long term cultures in serum-enriched medium to prevent uncontrolled proliferation of glial cells (for review, see Romijn, 1988). Culturing with serum always involves striking a balance between its toxic and its growth-promoting properties. When culturing

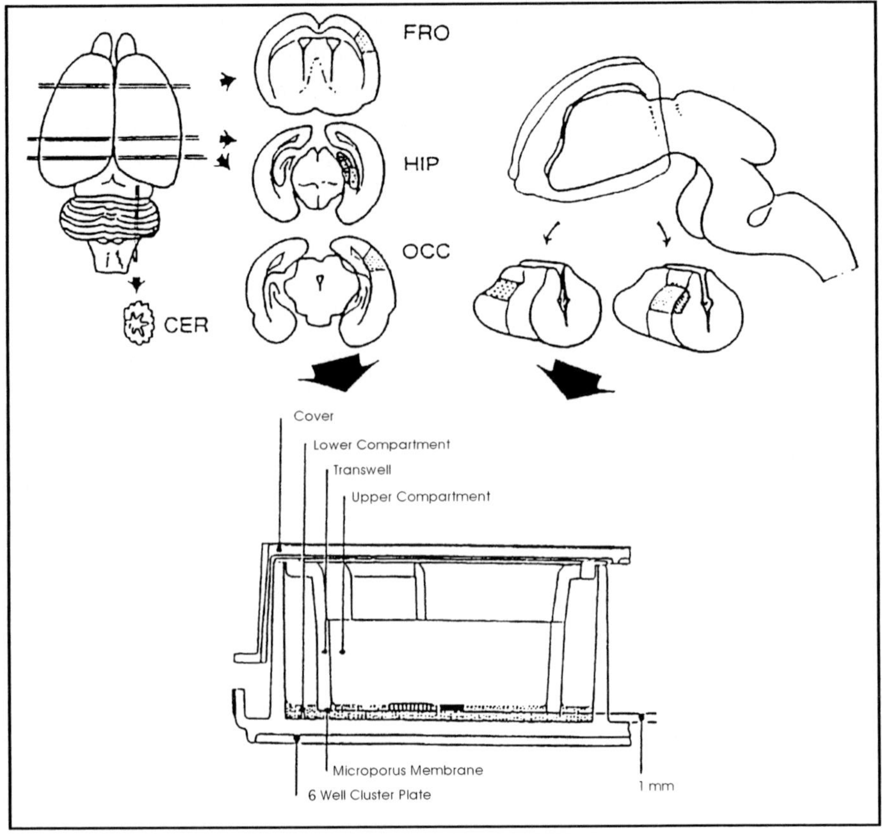

Fig. 5.1. Schematic diagram of the procedure for the collection of tissue for co-culture. At the top are drawings of the neonatal rat brain (*left*) and the fetal brain (*right*). Small thalamic blocks were taken from E14.5-17 rat fetuses, usually from the dorsolateral aspect of the diencephalon, as shown at the top right. The posterior blocks (shown very schematically in light stippling) contained almost exclusively LGN, whereas the more anterior blocks (coarse stippling) definitely excluded LGN and consisted of fragments of thalamic nuclei that would normally have connected to more frontal areas of cortex. In some cases, more ventral and lateral thalamus was explanted. The thalamic explant was placed on the collagen-coated microporous membrane of a Costar Transwell-COL culture chamber (*lower diagram*) close to one or more slices of tissue from fetal (E14.5-21) or postnatal (P0-11) rats. The upper left illustration of the neonatal rat brain shows the positions from which slices were cut. Slices of occipital (OCC) cortex (putative area 17) frontal (FRO) cortex (putative sensorimotor cortex) or hippocampus (HIP) were excised, as shown by the stippled regions, from coronal sections cut at the anteroposterior levels indicated. In some experiments sagitally cut slices of cerebellum (CER) were used. The lower diagram shows a 6-well tissue culture dish. The diameter of each well is 3.5 mm. Reproduced from Blakemore and Molnár, (1990) with kind permission of Cold Spring Harbor Laboratory Press, NY.

with medium containing serum enriched with growth factors one has to be especially cautious about the interpretation of 'spontaneous' axonal outgrowth (Hankin and Lund, 1991).

To avoid these shortcomings in the use of serum-enriched media, chemically defined, serum-free growth media have been developed. After the pioneering work of Evans et al (1956), Eagle (1959), and Ham (1965), Ludena (1973) performed the first successful culturing of neural tissue under serum-free conditions. Many different chemically defined media were subsequently developed to provide optimal conditions for growth of particular tissue or cell types. The most widely used media for culturing neural tissue were developed by Bottenstein and Sato (1979), Snyder and Kim (1979), Honegger et al (1979) and Romijn et al (1988). For rat neocortical cultures the N-2 and the R-16 media proved to be the best (Bottenstein and Sato, 1979; Brunner et al, 1982; Romijn et al, 1988; Yamamoto et al, 1989, 1997). For the co-culture experiments described here, I used N-2 medium.

Application of the Technique (Advantages and Limitations)

Only relatively recently has it become possible to maintain embryonic and neonatal cerebral cortex in culture (Romijn et al, 1988 and Caeser et al 1989; Yamamoto et al, 1989; Klauer, 1991). In successful cultures, neurons mature remarkably normally, developing characteristic morphological, electrophysiological and pharmacological properties (e.g., Caeser et al, 1989). The electrophysiological properties of cultured cortical slices do, however, show interesting differences when compared to acute preparations. For example, after several days in vitro, cortical cultures develop rhythmic, multiunit discharges with high levels of activity not seen in the initial period of culturing (Wolfson et al, 1989; Gutnick et al, 1989; Baker et al, 1989). The time-course over which this activity appears might reflect the process of synaptogenesis in vitro but the selective loss of GABA-ergic neurons (because of their particular sensitivity to hypoxic damage) has also been suggested as the cause of the high spontaneous activity (Caeser et al, 1989).

In some co-culture systems, selective interconnection, following general rules of specificity similar to those operating in the living animal, has been demonstrated (e.g. Smalheiser et al, 1982; Frotcher and Gähwiler, 1988; Gähwiler, 1988). The first successful co-cultures of thalamus and neocortex were reported by Yamamoto et al (1989).

There are several reasons for the increasing popularity of organotypic co-culture techniques. Certain questions that cannot be addressed in acute slice preparations (because of the inevitable de-afferentation during the cutting of the slices or the need for a long period of study) become accessible in co-cultures. Long-term, controlled pharmacological manipulations and gene transfer during synapse formation are feasible only in co-cultures (Gähwiler et al, 1997; Wilkemeyer and Angelides, 1995). It is relatively easy to examine interactions between remote areas of the brain or interactions between areas at different developmental stages. In acute slice preparations recordings are made from neurons that have just undergone mechanical damage and extreme alterations in their immediate environment. Cultures provide mechanical stability, which facilitates intracellular recording as well as direct microscopic observation, and the application of activity-imaging techniques.

Organotypic co-culture techniques now provide a potentially powerful method to address basic questions about the nature of neuronal specificity and the cellular and molecular mechanisms underlying the formation of afferent and efferent cortical connections (Gähwiler et al, 1997).

Long Range Effects Between Thalamus and Cortex In Vitro

Chemotropism as a mechanism of axon guidance was originally suggested by Cajal (1893, 1909-1911). The theory was reformulated as the specific chemoaffinity hypothesis by Sperry (1963). Recently, interest in chemotropism has re-emerged with quite convincing examples of this mechanism operating in various parts of the peripheral and central nervous system (see Lumsden, 1991). In the light of recent work with co-culture we are in a position to ask questions about the nature of the cell-cell interactions underlying the formation of area- and layer-specific thalamocortical connections.

Yamamoto et al (1989, 1992) reported that LGN and occipital cortex form layer-specific interconnections after 2 weeks in vitro, with geniculate axons terminating in layer 4 and cells of layer 6 sending axons into the LGN. In their study, P1 to P3 cortical slices were combined with small E16 thalamic blocks. We examined thalamic ingrowth mostly after 2 to 5 days in vitro and the age at which the cortical slice was taken covered a much larger range (Figs. 5.2 and 5.3; Molnár and Blakemore, 1991; 1995a,b; Blakemore and Molnár, 1990). In cortex of different ages, thalamic fibers exhibited different behaviors that could be related quite convincingly to the situation in vivo. Culturing with Transwell culture chambers (see Fig. 5.1), few axons grew out of thalamic or cortical explants, on either collagen or laminin, in the absence of an appropriate neighbor (Blakemore and Molnár, 1990). This allowed us to assess the presence of remote influences of one explant on another and we found that cortex does exert a clear growth-stimulating influence on thalamus at appropriate ages (Molnár and Blakemore, 1991, 1995a,b).

The interaction seems best described as trophic rather than tropic (see Lumsden, 1991). Although classically, trophism implies an interaction that is essential not only for the growth but for the survival of the cells in question (see chapter 2). A survival-promoting influence has not been shown by any of the studies in co-culture (though it would be an interesting topic for future investigation with more sensitive viability assays; see Adams R et al, 1993). Cunningham et al (1987) have provided evidence for a classical (death-preventing) trophic influence of cortex on thalamus in the rat in vivo. They showed that the degeneration of the LGN that normally follows removal of the occipital lobe in newborn rats could be partially prevented. This was done by replacing the ablated cortical tissue with a gel impregnated with medium conditioned by cultured cortex. The soluble fraction responsible was heat-labile and they believed it to be proteinaceous.

In vitro results suggest that a cortical slice in culture produces some substance that diffuses through the medium, and if the thalamic explant is close enough for the concentration of the substance to be sufficiently high, it stimulates axons to leave the thalamic explant (Molnár and Blakemore, 1991; Rennie et al, 1994). Outgrowth from the explants can be rather different from the growth of thalamic axons *within* the block (see Placzek et al, 1990; Tessier-Lavigne and Placzek, 1991). Whether the effect is *trophic* (simply enhancing the growth and/or survival of those axons closest to the source of the factor) or *tropic* (causing growth specifically towards the source by producing a directional response of growth cones up the concentration gradient) is problematic to determine in an interphase-type culture system, but obviously of considerable importance and could be addressed in other culture systems (Lumsden, 1991). In principle,

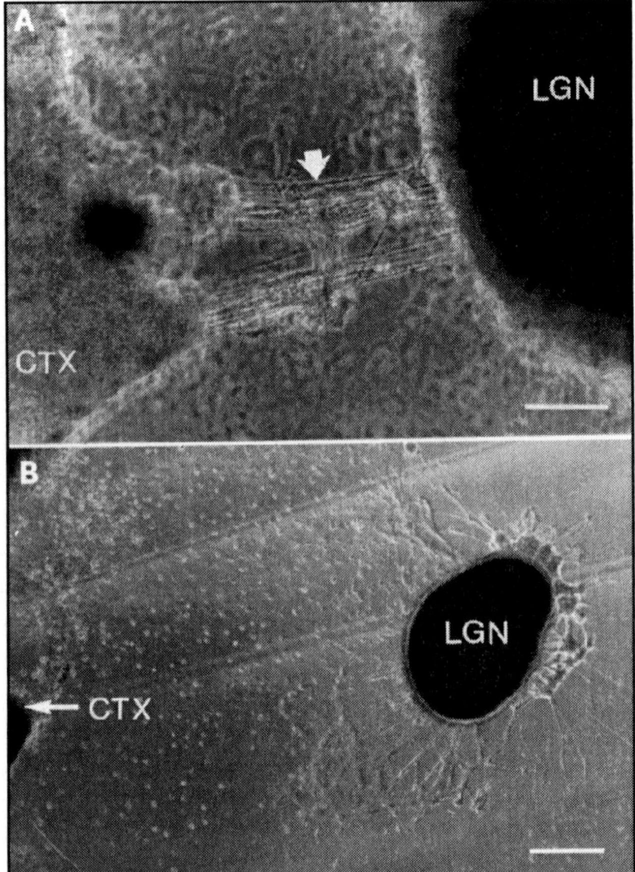

Fig. 5.2. Under phase-contrast microscopy the fiber outgrowth outside from tissue blocks could be examined and followed in living slices. A: A 'tissue bridge' (arrow), containing cells and fibers, formed between the co-cultured explants after 6 days in vitro on a laminin-coated Petriperm tissue culture dish. The E16 LGN block is to the right (LGN), the ventricular surface of a slice of P0 occipital cortex (CTX) is to the left. There is a considerable spread of the cells on the membrane below the plane of the tissue bridge (not in focus). B: Cells migrated into the space between the explant of a P0 cortex (CTX) and a E16 LGN (LGN; 6 days in vitro) on a laminin-coated Petriperm membrane. There is more cell spread from the LGN in the direction of the cortical slice, whereas fibers leave the LGN block in all directions, though more densely from the side facing the cortical slice. Scale bar in A: 100 µm; in B: 500 µm. From Molnár (1994).

the difference between the two classical concepts is quite sharp, and they can be distinguished experimentally, for instance by the observation of neurite growth in three-dimensional collagen gels, unconstrained by surface properties (Lumsden, 1991). Such experiments are required on thalamocortical co-cultures to elucidate how the growth-promoting factor actually works (see Fig. 2.1 of chapter 2). However, very steep concentration gradients of a purely trophic factor, in vivo or in vitro, can produce directed growth, and therefore mediate a form of chemotaxis.

The functional significance of soluble growth-promoting substances is especially difficult to interpret for explants in culture, because the tendency to establish gradients will obviously depend on the extent to which the medium is mixing or being changed. With the roller-tube technique, the medium is constantly in motion. Indeed, Bolz et al (1992), have reported that they see no directed outgrowth from thalamus to cortex in co-culture, using the roller-tube method. Bolz and colleagues even suggested that there is no diffusing influence at all, because they observed no difference in axon outgrowth for thalamic explants with and without a neighboring piece of cortex. However, serum-enriched medium, of the type used by Bolz et al (1992) stimulates considerable (and variable) axon outgrowth even from an isolated thalamic explant, which makes it difficult to interpret the effect of any added cortical slice. Since, in the conditions that we and Rennie et al (1994) have used, little or no outgrowth occurs from a thalamic explant without a nearby cortical explant, these experiments seem to provide firm evidence for a remote influence. Rennie et al (1994) observed and quantified thalamic axon outgrowth under phase-contrast microscopy in living cultures and concluded that it is dependent on the distance of the cortical explant, up to a distance of 2 mm on collagen. Therefore, with a sufficient gap between the two explants, outgrowth from the thalamic explant is denser and axon elongation faster on the side facing the cortical slice. Price and his colleagues described that thalamic neurite outgrowth in vitro is stimulated by cortex-derived growth factors and their level depends on the age and activity of the cortex (Price et al, 1995; Lotto and Price, 1995).

The Age-Dependent, Ingrowth-Permissive Property of Cortical Slices

Bolz et al (1992) described the failure of thalamic axons to invade very young cortex and have reported that older cortex has a membrane-bound factor that encourages axon ingrowth (Götz et al, 1992; Bolz and Götz, 1992; Hübener et al, 1995; Tuttle et al, 1995; Bolz et al, 1997). Bolz and colleagues deposited a membrane suspension extracted from cortical explants onto the surface of a culture chamber on which they grew thalamic explants. If the cortex came from E16 embryos, there was little or no axon outgrowth, but if it came from P7 cortex there was florid outgrowth on the membrane-treated surface.

There is evidence that the molecule or molecules responsible for the growth-permissive influence are not the same as the diffusible factor. When E16 LGN was cultured with explants of occipital cortex taken from embryos younger than about E19, axons grew out of the LGN block but few invaded the cortical fragment (Fig. 5.5). Instead they mostly ramified around the edges and to some extent over the surface of the cortical explant without penetrating it (see below). This implies that a diffusible factor is produced by the cortex at least as early as E16, but that some additional property is required of the cortex to encourage axon invasion, and this is not expressed until about E19.

In co-culture with young embryonic cortex, thalamic fibers grow around the edges of the cortical slice but rarely enter it. In late fetal and early postnatal cortex, thalamic fibers grow through the whole thickness of the cortical slice but still fail to terminate within the cortex and tend to grow up to the pial surface and fasciculate over it. But in slices taken from animals older than P3 they stop and arborize near the middle of the cortical plate after 4 days in vitro, and they do so even when the fibers enter through the supragranular layers (see Figs. 5.3 and 5.4; for a schematic summary diagram; see Fig. 5.5).

This sequence correlates convincingly with that which occurs in vivo. Axons do indeed start to grow out of the thalamus at around E14 (chapter 4) and it is conceivable that this is at least partly due to a remote (but not regionally specific) influence from the subplate. At E14, subplate cells are in place and they (or some other element in the cerebral wall) might generate a diffusible growth-promoting factor capable of producing some degree of stimulus for growth in culture. Equally, the synchronized outgrowth of subplate axons towards the thalamus in vivo might be influenced by a (non-region-specific) growth-promoting factor from the thalamus (Molnár and Blakemore, 1991; Lotto and Price, 1996) or from the internal capsule (Richards

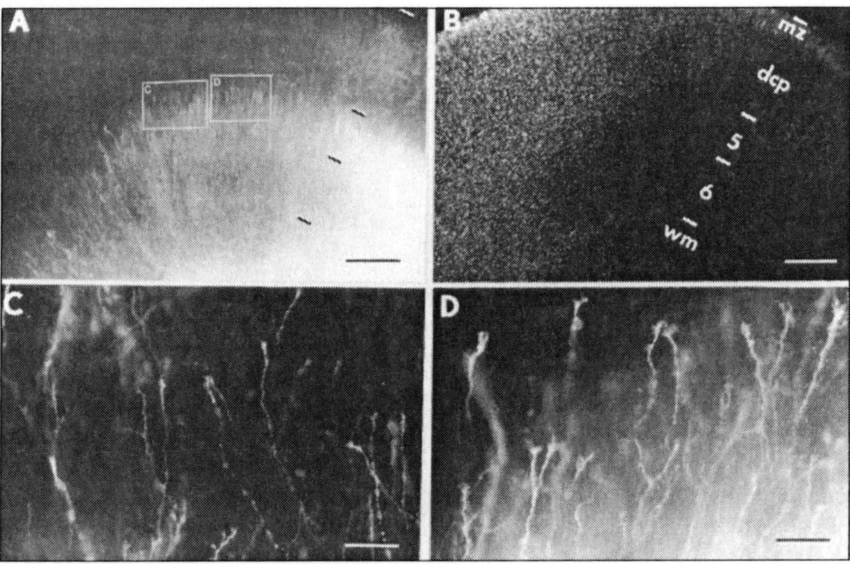

Fig. 5.3. Darkfield micrographs under epifluorescent illumination, showing DiI-labelled axons from an E16 LGN explant invading a cortical slice (taken at P3) after 1.5 days in culture. A: The thalamic fibers have invaded the P3 occipital cortex densely. Axons are running radially upwards through the cortical plate, and many have growth cones at their tip (see C and D). The majority of the fibers have reached layer 4 within 1.5 days in vitro. B: The same as A but viewed under ultraviolet illumination to reveal the cortical layering and the pial surface after staining with bisbenzimide (mz: marginal zone, dcp: dense cortical plate, wm: white matter). C,D: High-power view from the regions indicated with small squares at A showing growth cones at the tips of many axons. Scale bars: A and B: 250 µm; B and C: 50 µm. Reproduced from Molnár and Blakemore (1995b) with kind permission of John Wiley & Sons Ltd, Chichester, England. and the CIBA Foundation, London.

et al, 1997). In my own experiments I found that early corticofugal projections did not find the embryonic thalamus permissive for ingrowth, although a few fibers entered the thalamic blocks (Molnár, 1994). This outgrowth however was not area specific, as frontal and occipital cortical blocks sent axons to both anterior and posterior thalamus (Molnár and Blakemore, 1991).

There is, of course, the possibility that position signals normally expressed by cortex in vivo are lost or inadequately expressed in vitro. But if these in vitro results do apply in vivo, some additional mechanism must account for the topographic nature of the early thalamocortical projections. Thalamic axons arrive under the correct region of cortex at about E16—a stage at which the cortex itself is not yet permissive for ingrowth of axons in culture. Indeed, in vivo, the thalamic axons accumulate below the cortical plate and

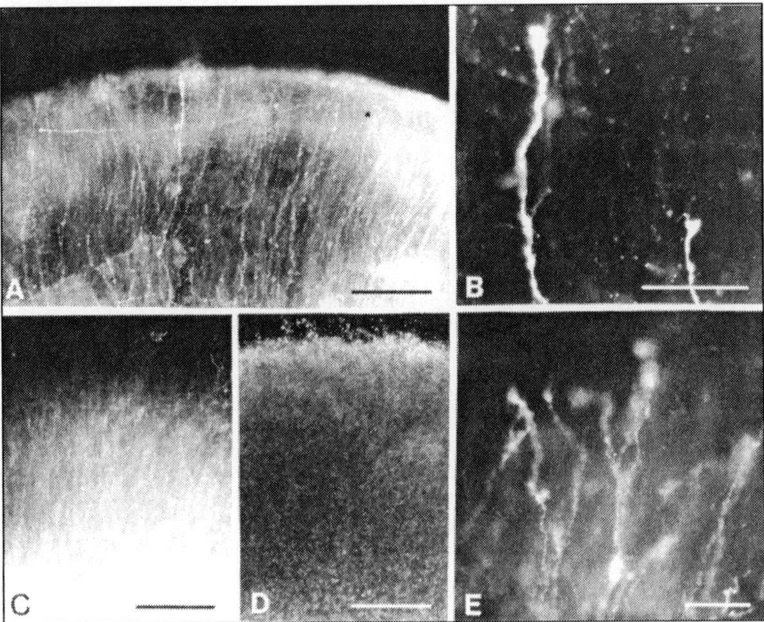

Fig. 5.4. Darkfield micrographs under epifluorescent illumination, showing DiI-labelled axons from an E16 LGN explant invading cortical slices after 4 days in culture. A: In P0 occipital cortex, axons are seen running radially or obliquely up through the cortical plate to the marginal zone, where some have turned laterally to run over or beneath the pial surface. One axon (far right) has turned 90° and coursed horizontally more than 1.5 mm: at its tip is a growth cone. Scale bar: 0.25 mm. B: High-power view of growth cones at the tips of axons below the cortical surface in the same (A) co-culture. Scale bar: 50 μm. C: With P6 occipital cortex, axons from an E16 LGN explant have invaded densely and most are branching and losing their growth cones about 300 μm below the pial surface. Scale bar: 0.25 mm. D: The same as C but viewed under ultraviolet illumination after staining with bisbenzimide, to reveal the pial surface and cells within the slice. Scale bar 0.25 mm. Comparison between Nissl-stained sections of such cortical cultures and fixed occipital cortex from P6 animals indicated that the depth of termination of thalamic axons corresponded to layer 4. E: High-power view from C, showing individual axons branching and apparently terminating. Scale bar: 50 μm. Reproduced from Molnár and Blakemore with kind permission of Macmillan Magazines Limited, London (1991).

may form temporary synapses on cells of the subplate (Shatz et al, 1988; Herrmann et al, 1991; Friauf et al, 1990; Higashi et al, 1996). At about E19, as judged by its properties in culture, the cortex itself starts to become permissive to thalamic axon invasion and, in vivo, small numbers of axons do grow into the cortical plate proper at this stage, but the majority remain in the subplate for 2-3 days after the

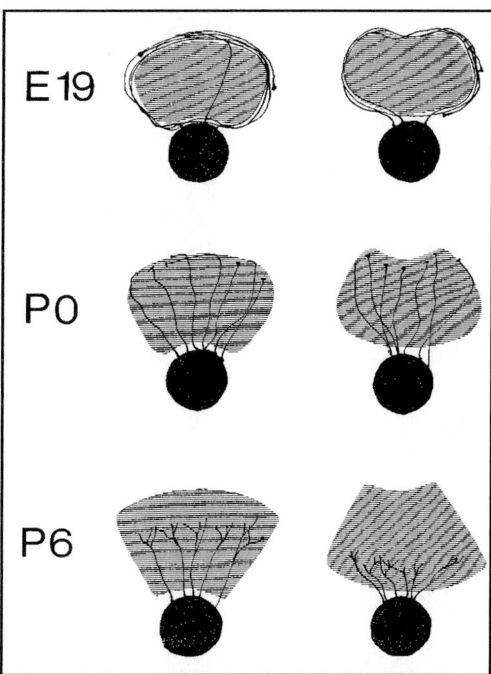

Fig. 5.5. Summary diagram of the thalamic fiber ingrowth patterns in thalamocortical co-cultures. Occipital cortical slices were taken at E19, P0 and P6 and cultured for 5 days with E16 LGN blocks. In the left column the schematic drawings demonstrate the fiber ingrowth from the ventricular surface, and in the right column from the pial surface. Depending on the age of the cortical slice different fiber ingrowth patterns were observed.
E19: Cortical slices at E19 (or younger) do not accept innervation by thalamic fibers. Most grow around and not within the cortical slice. The distribution is similar in the slices regardless of whether the thalamic block was placed on the ventricular side and on the pial surface. P0: In newborn cortical slices the thalamic fibers run radially or obliquely through the cortical plate to the marginal zone, where some turn laterally to run over or beneath the pial surface. The ingrowth pattern of the thalamic fibers entering from the pial surface was similar to that of axons entering from the ventricular surface: the fibers grew through, all the way to the opposite side. P6: In cortical slices taken from a six day old pup, most thalamic fibers branched and terminated some 250-300 μm below the pial surface and very few invaded the upper layers. This sub-pial region does not seem to be actively inhibitory for axon ingrowth because it was invaded by the thalamic fibers when the thalamic block was placed against the pial surface. The fibers grew in and branched and arborized within the same termination region, 250-300 μm below the pial surface. These experiments imply that the cortex starts to become ingrowth-permissive at about E19 and that a layer-specific stop signal is expressed before the end of the first postnatal week. Reproduced from Molnár and Blakemore (1995b) with kind permission of John Wiley & Sons Ltd, Chichester, England and the CIBA Foundation, London.

cortical plate becomes permissive to invasion (E19-birth). This waiting period is one characteristic of the natural pattern of innervation that I could not duplicate in culture. Perhaps thalamic axons have to arrive at the subplate cells by fasciculated growth along their particular axons in order to be able to form synapses on them and hence to be temporarily captured in that layer (see chapters 6 and 7 for further discussion).

The experiments in which thalamic explants were co-cultured with ribbons of subplate or cortical plate (see Fig. 5.6) suggest that

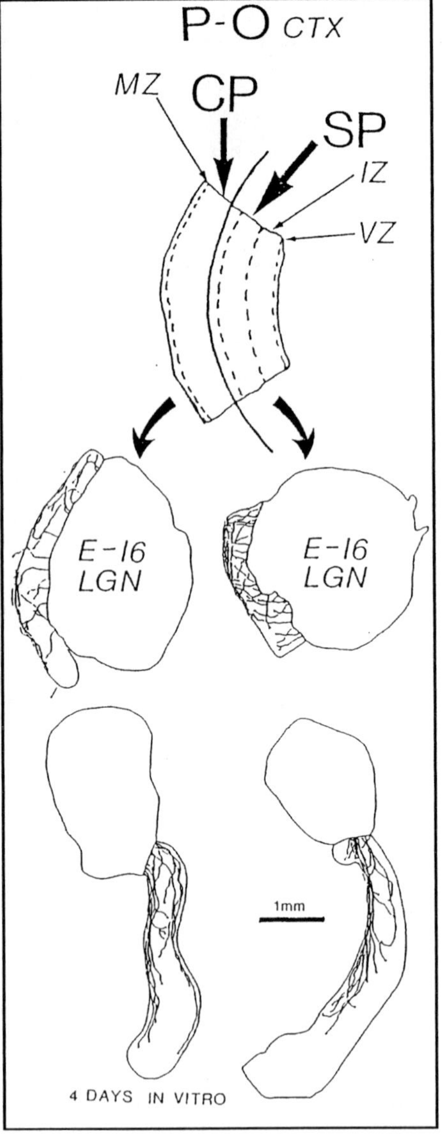

Fig. 5.6. The upper schematic diagram demonstrates the procedure for making a pia-parallel cut through the cortical plate (CP) along a line definitely above the top of the subplate (SP). This cut produced a 'cortical plate ribbon' (consisting of upper cortical plate with no subplate; only the marginal zone contaminates the purity of the cortical plate) and a 'subplate ribbon' (also containing a thin segment of the lowest part of cortical plate, plus intermediate and ventricular zones). These ribbons then were co-cultured with E16 LGN explants in a variety of configurations. The LGN block was placed against the lower surface of the 'cortical plate ribbon' (upper-left) or the 'subplate ribbon' (upper-right). In others it was placed against the radially cut end of the ribbon (lower diagrams). Then after 4 days in culture the thalamic fiber outgrowth was examined and camera lucida drawings were made. I saw no obvious difference in the amount or velocity of outgrowth of the thalamic axons or the pattern of their invasion into the two different Po cortical fragments. From Molnár (1994).

in postnatal cortex both subplate and upper cortical layers are capable of remotely stimulating thalamic outgrowth and permitting ingrowth, and that prior exposure to the environment of the subplate is not necessary in order for axons to enter the plate proper. The subplate and the interactions between subplate and the forming cortical plate might be essential for the upregulation of the ingrowth-permissive properties of the cortical plate (Bolz et al, 1993). Watanabe et al (1995) proposed that neuronal stimuli through early thalamocortical fibers might cause reduced expression of neurocan mRNA in neurocan, producing cortical cells in the presumptive barrel centers in newborn rats. Neurocan expression might be regulated by thalamocortical fibers before and during their ingrowth to the cortical plate. Perhaps without early cortical activity patterns the development of the permissive property of the cortical plate would be delayed. One possible explanation for the subplate lesion experiments of Ghosh et al (1990) is that in the absence of subplate cells these activity patterns do not spread into the cortical plate and the thalamocortical projections fail to enter the still non-permissive cortical plate above the lesion. However, the fact that thalamic axons do grow out when combined with cortical slices as young as E16 (see Fig. 5.6), when mostly subplate cells, ventricular and subventricular cells are present, strongly suggests that the subplate alone can generate a remote growth-stimulating factor (even though thalamic axons are reluctant actually to *enter* and grow through cortical explants at such young ages).

Emerling and Lander (1994, 1996) examined the attachment and neurite outgrowth of embryonic thalamic neurons on embryonic (E15-16) cortical slices and found that there were substantial layer specific differences. They proposed that the subplate and intermediate zone possess thalamic neurite outgrowth stimulatory activity and the embryonic cortical plate inhibitory activity. The layer specific differences in neurite adhesion and neurite outgrowth could be eliminated by the enzymatic removal of chondroitin sulfate.

Layer-Specific Innervation and the Stop Signal

The parallel between the natural pattern in vivo and events in culture becomes convincing again after birth. In vivo, the majority of thalamic axons start to detach from the subplate, to turn sharply and grow into the cortical plate shortly before birth (at least for the occipital cortex: see chapter 4)—a stage at which the cortex is indeed highly attractive to thalamic axons in vitro. In vivo, they grow up through the deep layers and reach layer 4 by about P2-3. When most of them reach that level layer 4 is starting to express its stop signal in vitro. Thalamic axons that enter somewhat precociously in vivo tend to grow up to the cortical surface just as they do with cortex younger than P3 in vitro; odd axons are seen in vivo extending into the marginal zone, as if they have managed to grow through the cortex before the stop signal appears. This tendency for early thalamic axons to avoid being 'captured' by layer 4 may be the origin of the disproportionate early innervation of layer 1 of the kitten cortex (Kato et al, 1984; Henderson and Blakemore, 1986). Possible mechanisms will be discussed further in chapters 6 and 7.

When thalamic explants are cultured close to the pial surface (rather than the ventricular surface) of a cortical slice, they grow down through the superficial layers and either terminate in layer 4 (after about P3; see Fig. 5.7) or continue through into the intermediate zone (before P3). Therefore the upper cortical layers are not inhibitory to axon growth and the signal expressed around P3-4 is not one that simply discourages axons entering the upper layers. It seems likely that neurons of layer 4, which have reached their correct position and matured considerably by that age, express a specific *stop signal*, which encourages thalamic axons to arborize and terminate. The molecular nature and localization of this signal deserve further investigation.

Even though the first experiment strongly suggests that the superficial layers do not entirely resist thalamic ingrowth, it is still conceivable that growth inhibition (perhaps within layer 4 itself) plays a part in encouraging termination in layer 4. There might be some kind of interaction between local growth-terminating and growth-inhibiting signals.

Appearance of Layer 4 Coincides with the Stop Signal

One requirement for this kind of tight comparison between culture and real life is that the upper laminae of the cortex should continue to develop, at least to some extent, in culture. There are few systematic studies on this matter. Bolz et al (1990) have shown, by means of birthdate labelling with bromodeoxyuridine, that cells born on E16 and E18 have distributions through the cortex that look fairly similar in normal animals more than 12 days old and in slices taken shortly after birth and cultured (by the roller tube technique) for 2-3 weeks. We were especially interested in the position of layer 4 cells, but unfortunately there is a huge overlap in the generation of the cells contributing to different layers in rodents (Miller, 1988). Although the peak production of cells destined for layer 4 occurs on E16, cells born between E14.5 and E18 contribute to this layer (Lund and Mustari, 1977; Miller, 1988). Therefore it cannot be completely and exclusively labelled by thymidine or BrDU birthdating, and

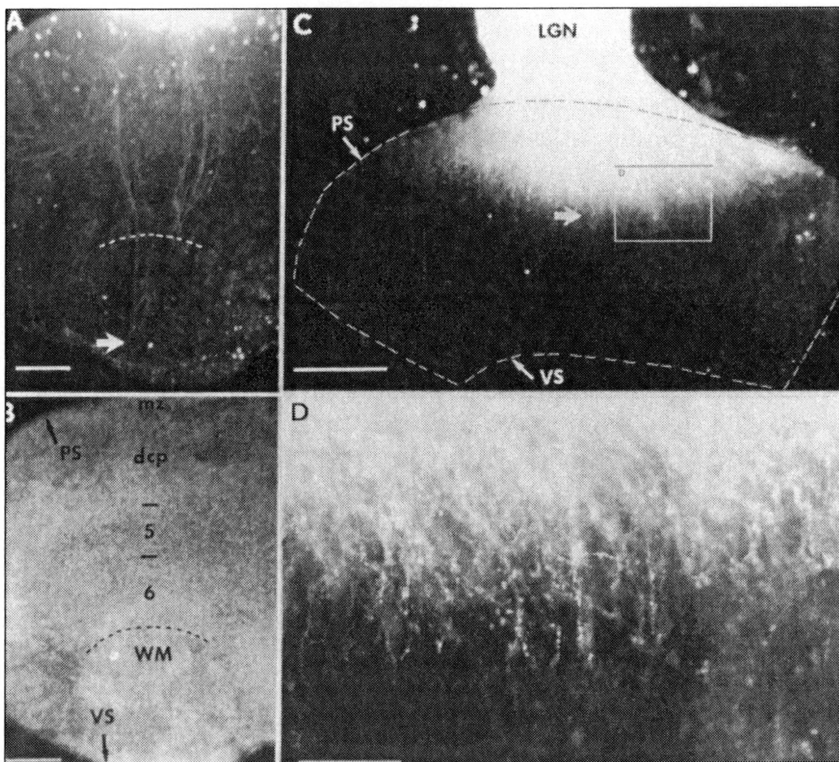

Fig. 5.7. An E16 LGN explant was placed next to the cleaned pial surface of a P0 (A and B) and a P6 (C,D) occipital cortical slice and cultured for 5 days. A: In the P0 cortex, the thalamic fibers pass through the entire depth of the cortex from the pial surface and reach the ventricular zone (white arrow) from the pial surface. Dotted line marks the cortical plate (white matter border). B: The same field as A, but viewed under UV illumination to reveal the bizbenzimide counterstaining. Scale bar: 100 µm. C: Dark-field micrograph under epifluorescent illumination, showing the front of the DiI-labelled axons from the LGN explant invading a P6 cortical slice from the pial surface (PS) after 5 days in vitro. The fibers do not reach the ventricular surface (VS, lower white arrow), but stop about 300 µm below the pial surface (the zone of fiber termination is indicated with the upper white arrow). Bar: 500 µm. D: High-power view from the region labelled with a white box in C, showing the fibers at their termination zone. Only a few of the DiI-labelled fibers are in focus since the culture is more than 150 µm thick. Bar: 100 µm. From Molnár (1994).

therefore, the prenatal lesion experiments with X-ray (Ito, 1995) or with methylazoxymethanol acetate (MAM) should also be interpreted with caution (Molnár, 1994; Woo and Finlay, 1996).

At E20, only the cells of the uppermost layers are still being born in the rat. By injecting BrDU on this day and culturing the cortical slices from P3 for several weeks, Bolz et al (1990) showed that there is indeed migration in culture, and that these cells take up their expected position below the marginal zone. However, slices taken at or before birth and cultured for several weeks show a retarded distribution in similar experiments, indicating that migration and the maturation of laminae is disturbed in such young tissue in vitro (Bolz et al, 1990). De Jong et al (1988) examined the cytoarchitecture of cultured rat neocortex explants taken at P6 and cultured (by the interphase-type tissue culture technique) up to 13 days. They reported major changes in the layering of the cortical slices during the first 2 days in culture. The thickness of the pial-ventricular axis increased due to an increase in the proportion of layers 2, 3 and 4. In spite of this evidence of continued migration and/or enlargement of cells and neuropil in the upper layers, a distinct layer 4 could not be consistently identified in that study, perhaps because of the degeneration of its thalamic afferents in the cortex-alone culture. In addition, neuronal migration has been observed directly in similar cultures (O'Rourke et al, 1991), but its extent is difficult to judge.

All these experiments were performed with cortical slices which were removed from normally developing brains at various ages. Some of these cortices had already been exposed to thalamic fibers and their development could therefore have been affected by those thalamic fibers. It would be interesting to repeat some of these experiments with 'virgin' cortical slices of various ages, which have never been exposed to thalamic fibers. It would be very informative to coculture thalamic explants with the subplate ablated cortical slices of Ghosh et al (1990), which do not accept thalamic fibers after an early injection of excitotoxin in vivo. One also has to keep in mind that during the preparation of cultures, explants suffer extreme alterations in their environment. It is therefore important to consider to what extent intrinsic connectivity remains intact and continues to develop normally in the slice cultures. In cultures of hippocampus, Zimmer and Gähwiler (1984) have described reorganization of nerve connections similar to that observed after lesions in vivo. De Jong et al (1988) did not see obvious widespread cell death in any layers of their cultured slices, when examining them several times up to 13 days in culture; only a few individual degenerating cells were seen, especially in layer 5. On the other hand, in hamsters with the posterior thalamus lesioned at birth, a decrease in the number of cells in layer 4 of the occipital cortex has been reported (Windrem and Finlay, 1991). In my own collaborative work (Molnár et al, 1991) on the in vivo development of the cortex deprived of thalamic input by the embryonic disruption of the thalamocortical pathways, the isolated cortex develops its basic laminar organization, but there are abnormalities in the intracortical circuitry (see Fig. 9.7). However, Nissl staining alone is obviously a very crude technique to examine these questions in detail. Combined birthdating and cell filling studies would be needed to be confident about cell fate and morphology (see Katz and Callaway, 1992; Molnár et al, 1997d). Nevertheless, both in vivo and in vitro results suggest that the thalamic input might contribute to and play an important role in the formation of circuitry, at least in layer 4, and might help to sustain cells in that layer. Herrmann and Shatz (1995) described that blockade of sodium action potentials during the period of initial thalamic ingrowth to the kitten visual cortex prevents the normally occurring branching of thalamic afferents within layer 4. This experiment suggests an early role of action potential activity in the development of the layer specific stop signal. Wilkemayer and Angelides (1996) reported that blocking sodium channel-dependent activity in thalamocortical organotypic co-cultures, the thalamocortical axons develop more branches, and develop more varicosities, whereas their general pattern of axon ingrowth and the average length of collateral branches in cortex were the same as in control co-cultures. These issues deserve further study.

Target Selectivity of Thalamic Innervation in Culture

The most parsimonious interpretation of the basic result of our co-culture experiments is that a piece of occipital cortex, older than E19, is a more attractive medium for LGN axon invasion than the collagen-coated membrane that makes up the rest of the thalamic explant's environment in the culture chamber (Molnár and Blakemore, 1991). However, the unwillingness of axons to enter younger cortical explants implies that the occipital cortex develops some particularly attractive property between E19-21. To test whether this influence is specific to occipital cortex, we cultured E16 LGN with slices of neural tissue from other sites (hippocampus and cerebellum, age ranging from E18 to P8). We used the choice paradigm to compare the innervation of different structures, in the hope of revealing subtle preferences.

Figure 5.8 shows the result of culturing E16 LGN for 5 days with various combinations of brain tissue. The most dramatic result is

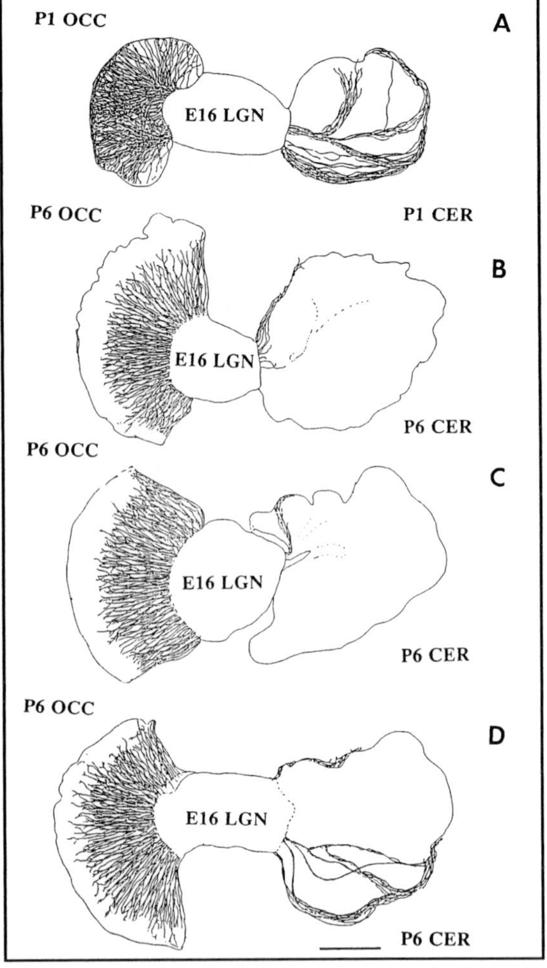

Fig. 5.8. A-D: In a choice paradigm in which each explant of E16 LGN was cultured next to a cerebellar and an occipital cortical slice the geniculate innervation was largely restricted to the slice of neocortex. In the cortex, according to its age, the fibers either run up to the pial surface (A, P1) or terminate approximately 300 μm below the surface (B-D, P6). Of the few axons that sprouted toward the cerebellar slice, most ran around one border of the slice or in fascicles across the surface of the slice to reach and course around the opposite edge. Bar: 1 mm. C is reproduced from Molnár and Blakemore with kind permission of Macmillan Magazines Limited, London (1991).

the contrast between the florid invasion of P6 occipital cortex and the virtual avoidance of P6 cerebellum (Fig. 5.8B-D). In more than 53 cultures involving LGN and cerebellum, only a tiny number of axons actually penetrated the cerebellum (Molnár, 1994). Axon growth was mainly restricted to a few fasciculated bundles running around the edge of the cerebellar slice or across its surface.

Given the choice between occipital cortex and hippocampus, LGN axons showed little preference in terms of the density of axons invading the tissue (Fig. 5.9D). Hippocampal slices were innervated almost as densely as neocortical slices, but interestingly the axons mostly continued up to the surface of the hippocampus and did not tend to terminate specifically in the middle layers, as in the occipital cortex of this age. This implies that the hippocampus is permissive to axon invasion, as is the neocortex, but has no precise equivalent to the stop signal expressed by neocortical layer 4.

Lack of Regional Selectivity for Thalamocortical Innervation in Culture

My starting question concerned the mechanisms by which thalamic axons from different nuclei find their way to their appropriate cortical fields. I have demonstrated an axon-growth-promoting influence of occipital cortex on LGN (indeed of neocortex on thalamus, in general) at E16 (and perhaps even earlier), and a growth-permissive property that appears around E19 for the occipital cortex. If these signals produced by the cortex were specific for each particular field and acted preferentially on the equivalent thalamic nucleus, they might account for the selective guidance of thalamic axons to and into the correct cortical area. Following Sperry's (1963) notion of chemoaffinity, a mere gradient of one or more molecular species from one part of the hemisphere to the opposite pole might be sufficient to establish topographic order among the inputs.

I tested these ideas by using the choice protocol, culturing explants of either LGN or some other region (more anterior or medial) of the developing thalamus, each flanked by slices of two different regions (frontal and occipital) of cortex (Fig. 5.9A-D). The result was unequivocal: the pattern of innervation was indistinguishable whatever regions of thalamus and cortex were combined. For example, the representative example in Figure 5.9C shows the very similar axonal outgrowth from an E16 LGN explant into P6 occipital cortex and P6 frontal cortex (which the LGN never innervates in vivo). Equally, explants of E16 anterior thalamus and more ventral thalamus sent axons in a similar pattern into occipital, frontal, or any other region of cortex (Fig. 5.9B).

One might imagine that regional specificity of the diffusible factor alone, evident only at an earlier age when thalamic axons are leaving the thalamus and finding their way through the intermediate zone to the correct cortical region, might be sufficient to explain their selective guidance. However, we and others saw no hint of any difference in the degree of axonal outgrowth from thalamic explants, even when cultured with slices of E19 cortex, whatever the site of origin of thalamic and cortical explants (Molnár and Blakemore, 1991; Molnár, 1994; Yamamoto et al, 1992, 1997).

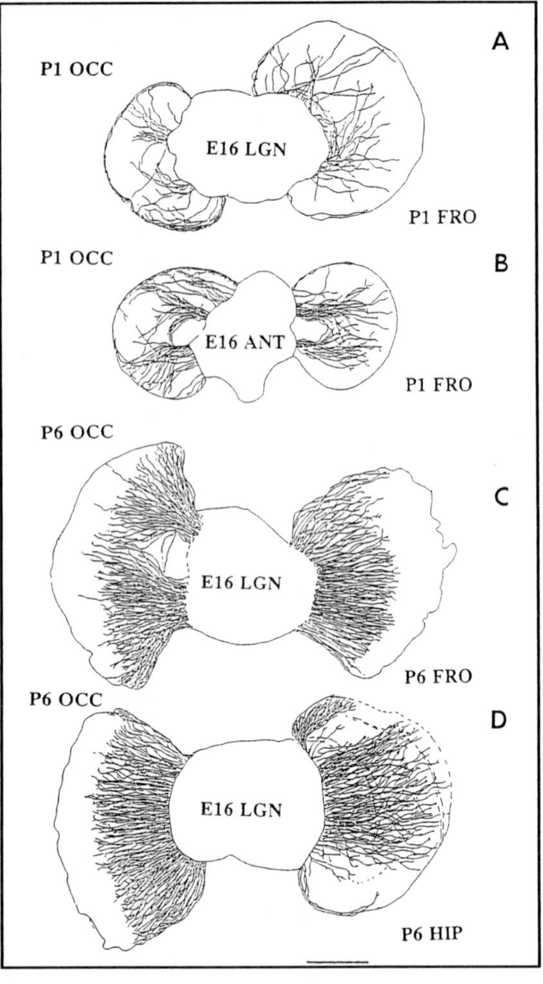

Fig. 5.9. Results of the 'choice' paradigm in which each explant of E16 LGN was cultured next to two different slices. Camera lucida drawings show DiI-labelled thalamic axons after 2.5 (A-B) or 5 (C-D) days in vitro. A-B: Confronted by P1 occipital cortex (OCC) and frontal cortex (FRO), geniculate (LGN) (A) or anterior thalamus (ANT) (B) innervation was indistinguishable. Most axons ended in growth cones, and a few already had reached the pial surface after only 2.5 days in vitro. There was a slight preference for the fibers to enter the cortical slice from the ventricular surface rather than growing around the edge and into the other layers of the cut margin. C: 'Choice' paradigm similar to A but with P6 occipital (OCC) and frontal (FRO) cortex (after 5 days in vitro). Again, there is no hint of preference for occipital cortex. Only a few axons have reached the marginal zone, and most have terminated in what appears to be layer 4, about 3-400 µm below the surface. D: Given the choice between P6 occipital cortex (OCC) and P6 hippocampus (HIP), LGN axons showed little preference in terms of density of invasion; after 4 days in vitro, hippocampal slices were innervated almost as densely as neocortical slices. However, interestingly, many axons continued up to the surface of the hippocampus and did not tend to terminate specifically in a particular layer. Bar: 1 mm. C and D reproduced from Molnár and Blakemore with kind permission of Macmillan Magazines Limited, London (1991).

These results, if they apply in vivo, make it unlikely that a remote region-specific influence of cortical areas on thalamic nuclei is responsible for the topography of thalamocortical projections. Nevertheless, Bolz and Götz (1992) have reported that the membrane-bound ingrowth permissive influence varies slightly from region to region: outgrowth from an LGN explant is somewhat more vigorous on a surface treated with the membrane fraction from P7 occipital cortex than from P7 frontal cortex. At first sight, this result seems to contradict our general conclusion that there is no obvious regional specificity in any of the observed interactions in co-culture (Molnár and Blakemore, 1991; Yamamoto et al, 1991, 1992), a result which Bolz also confirms in similar paradigms. However, it must be pointed out that Bolz agrees that the growth-permissive nature of cortex is not present at E16 (or at other younger embryonic and even early postnatal ages), when thalamic fibers in vivo have already made their way in topographic order to their cortical target areas (chapter 4; Molnár and Blakemore, 1990a; Blakemore and Molnár, 1990; Molnár et al, 1993a; Catalano et al, 1996; Juliano et al, 1996). Therefore even if there is slight regional variation in the nature of the permissiveness of the cortex at much later stages, it cannot play a part in the basic process of ordered guidance of thalamic axons to their cortical target areas.

The slight difference in the influence of membrane fractions from P7 occipital and frontal cortex on outgrowth from LGN might conceivably be due to the presence of *thalamic* axons in the explants of cortex at that late age. There might indeed be a tendency for axons from any one thalamic nucleus to prefer to grow on axons from the same nucleus rather than others. Indeed such selective adhesion and fasciculation might contribute to the maintenance of fiber order in the advancing front of thalamocortical afferents in vivo (see chapter 6). Certainly, the tendency of axons to fasciculate preferentially on existing axons is a well-described phenomenon (Cajal, 1928) and they can be selective for particular, similar axons. Description of the location of membrane bound factors during thalamocortical pathfinding is obviously important, but P7, the age used by Bolz and Götz (1992), is too late to define factors important in initial axon guidance.

Conclusions

Experiments with thalamic and cortical organotypic co-cultures, lead to the conclusion that three molecular signals are expressed by the cortex at different ages: (1) a trophic effect on the thalamus, (2) an axon-growth-permissive property from about E19; and (3) a 'stop signal', specific to layer 4, from about P3. These factors presumably influence thalamic fiber outgrowth, accumulation below the cortical plate, invasion and termination. In vitro, any region of the rat thalamus (E16) innervates any region of the cortex (E16-P11): no positional preference is apparent even when the thalamic explant is given a choice of various cortical targets. The most important question that remains, with no hint of an answer from the co-culture experiments, is: How are thalamic axons guided in an orderly fashion to their own cortical areas? That is the issue tackled in the next chapter.

CHAPTER 6

Possible Role of the Early Generated Diencephalic and Telencephalic Cells in Escorting Early Thalamic Connections

The earliest generated cells of the mammalian forebrain are in the cortical preplate, perirhinal cortex, thalamic reticular nucleus and the ventral cells of the ganglionic eminence (see Bayer and Altman, 1991). Initially, these cells form a continuous layer from the diencephalon through the primitive internal capsule to the perirhinal cortex and cortical preplate. They are very similar in many respects. They are the first to develop connections in the forebrain in carnivores and rodents (McConnell et al, 1989; Blakemore and Molnár, 1990; De Carlos and O'Leary, 1992; Mitrofanis and Guillery, 1993; Métin and Godement, 1996; Cordery et al, 1996, 1997) before they disappear through preferential cell death (Luskin and Shatz, 1995a,b; Mitrofanis, 1992a). The aim of this chapter is to review recent developments in the understanding of their early connectivity in the forebrain and to examine the possibility that early telencephalic and diencephalic axons grow and are guided along these cells and their processes.

The Regions of Transient Cell Populations in the Forebrain Coincide with the Sites of Major Rearrangements of the Thalamocortical Connectivity

The thalamocortical connectivity in the adult shows substantial rearrangement in three regions along its path:

1) The thalamic reticular nucleus (Bernardo and Woolsey, 1987).
2) The internal capsule (Mitrofanis and Guillery, 1993).
3) The cortical subplate (Nelson and Le Vay, 1985).

Development of Thalamocortical Connections,
by Zoltán Molnár. © 1998 Springer-Verlag and R.G. Landes Company.

These regions coincide with the zones where the first generated cells in the forebrain are distributed: in the thalamic reticular nucleus, ganglionic eminence and subplate (Mitrofanis, 1992a; Bayer and Altman, 1991; Luskin and Shatz, 1985a,b). The cells of these regions have numerous other similarities. They express similar transmitters (GABA, substance P, neuropeptide Y, etc.) and they are among the first to develop connections in the mammalian forebrain (McConnell et al, 1989; De Carlos and O'Leary, 1992; Mitrofanis and Baker, 1993). Since these cells and their connections are situated along the future path of thalamocortical connectivity it was suggested that they play a guiding role for later developing cells (Shatz et al, 1990; Blakemore and Molnár, 1990; Mitrofanis and Guillery, 1993). There has been great progress recently in the understanding of the various forms of interactions between these early generated cells and developing thalamocortical projections. We are beginning to understand their precise early connectivity patterns (Mitrofanis and Baker, 1993; Adams and Baker, 1995; Métin and Godement, 1996; Cordery et al, 1996; 1997). Their possible role, however, is still not fully understood. There are numerous controversial issues.

A number of workers (see McConnell et al, 1989; Shatz et al, 1990; Blakemore and Molnár, 1990; De Carlos and O'Leary, 1992; Erzurumlu and Jhaveri, 1992; Molnár and Blakemore, 1995a,b) have suggested that the axons of subplate cells pioneer the pathway from the cerebral cortex into the diencephalon. However, Clasca et al (1995) reported that carbocyanine dye in the dorsal thalamus or internal capsule of ferrets at different embryonic ages (E46-54) labelled cortical cells only in layer 5, and therefore concluded that projections from subplate and layer 6 grow towards the thalamus very slowly and are delayed for several weeks in the cerebral white matter. Therefore their role in thalamocortical guidance was questioned. Furthermore, the site of termination and/or waiting of the early corticofugal projections from the subplate is not yet established (see De Carlos and O'Leary, 1992; Miller et al, 1993).

Mitrofanis and Guillery (1993) suggested that cells in the embryonic internal capsule (the perireticular nucleus; see Mitrofanis, 1992a) might act as guidepost cells while the different classes of corticofugal projections pass through the region. Métin and Godement (1996) recently described similar cells in the medial and lateral ganglionic eminences in hamster, which might correspond to the perireticular cells. Similar transient cells were described in the human embryonic internal capsule (Letinic and Kostovic, 1996). However there are apparent differences among the different studies. In hamster the ganglionic eminence cells project to the dorsal thalamus very early, whereas in rat, according to Mitrofanis and Guillery (1993), perireticular cells do not develop their projections to thalamus until birth. In hamster, lateral ganglionic eminence cells are the first to reach the cerebral cortex (Métin and Godement, 1996) while in rat the thalamic fibers and cells from VB are the first to be backlabelled (Braisted and O'Leary, 1995).

To consider these above mentioned discrepancies further and to establish the precise early hodological relationships in the embryonic rat forebrain, I shall systematically examine these questions further in this chapter and present a review of the early connectional analysis between the various parts of the developing telencephalon and diencephalon in the rat.

Cells in the Thalamic Reticular and Perireticular Nuclei Form the Earliest Connections with Dorsal Thalamus

Carbocyanine crystal placement into the dorsal and ventral thalamus revealed a continuous chain of cells from the dorsal thalamus through the thalamic reticular nucleus (Cordery et al, 1996, 1997; see Fig. 6.1 for summary) as described by Mitrofanis and Baker (1993). We also saw blacklabelled cells within the primitive internal capsule (Cordery et al, 1996, 1997). This is almost a week earlier than previously observed in the rat (Mitrofanis and Guillery, 1993; Mitrofanis and Baker, 1993), which is similar to the findings of Métin and Godement (1996) in hamster.

The group of labelled ganglionic eminence cells seem to correspond to the cells Mitrofanis (1992a) called the perireticular nucleus. The majority of the backlabelled cells are situated in sections several hundred microns rostral to the diencephalic crystal placement sites, and their distribution follows a path similar to that of thalamocortical fibers. The number of backlabelled cells in the ganglionic eminence seems to decline from E14 to E16, and cells become more scattered. Nevertheless, some maintain their connections with the dorsal thalamus after E16, when thalamocortical fibers have already reached the cortex. We continued to observe numerous backlabelled cells among the trajectories of thalamic fibers after E17 (Cordery et al, 1996, 1997). Our optical recording study using voltage-sensitive dyes (RH482) in rat thalamocortical slice preparations

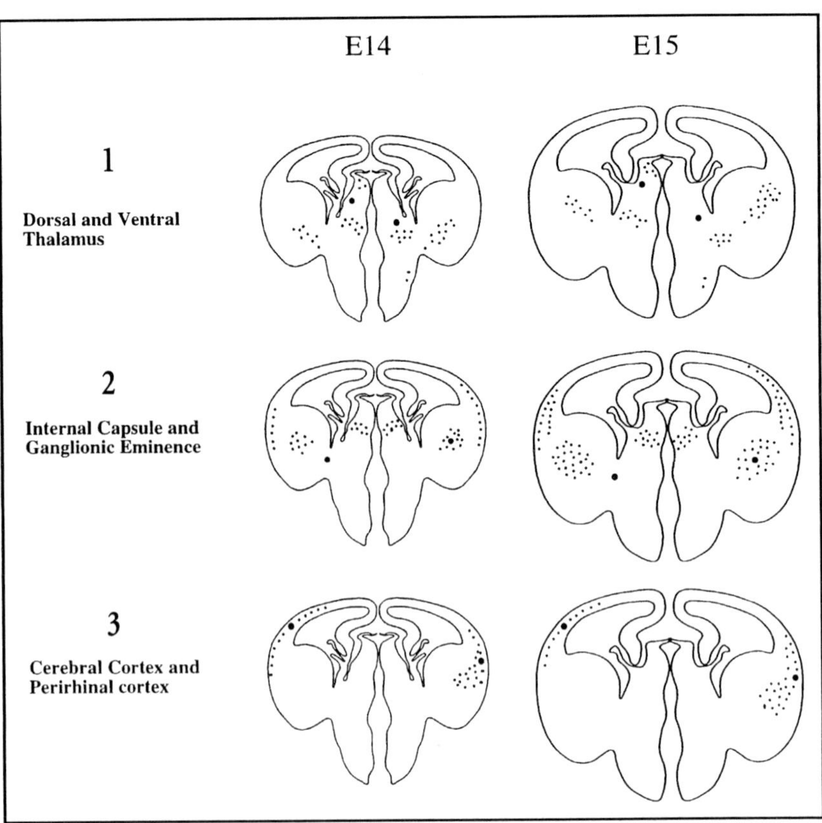

Fig. 6.1. Summary diagram of the hodology of the embryonic forebrain at E14 (left column) and E15 (right column). The upper row demonstrates the cells which have projections to the dorsal and ventral thalamus: cells in the epithalamus, TRN, hypothalamus and ganglionic eminence. The middle row gives a summary of the distribution of the backlabelled cells from the internal capsule and from the ganglionic eminence: dorsal thalamus, ganglionic eminence and peririnal cortex. The lower row demonstrates the cells labelled from the dorsal cerebral cortex and from the perirhinal cortex. Note that perirhinal cortical crystal placements labelled ganglionic eminence cells, while dorsal cortical crystal placements did not label ganglionic eminence cells.
1: The first fibers reaching the dorsal and ventral thalamus originate from cells in the internal capsule. Dye placed in the dorsal thalamus (large dot) at E14-15 backlabels cells (small dots) in the thalamic reticular nucleus, as described by Mitrofanis and Baker (1993), and the perireticular nucleus within the primitive internal capsule, but none in the cerebral wall. The patterns of backlabelled cells in the internal capsule was dependent on the exact site of the crystal placement into the dorsal thalamus.
2: Preplate projections reach the ganglionic eminence and the internal capsule and run in the region of the perireticular cells, but do not reach dorsal or ventral thalamus. At E15, after the first true cortical plate cells have migrated into the lateral cortex, some cells of the plate, as well as preplate cells (now forming the subplate and marginal zone), are backlabelled from the ganglionic eminence, or internal capsule, but still not from the dorsal thalamus, even with very long incubation periods. Only dorsal,

(Higashi et al, 1996; see chapter 9) has demonstrated that ventrobasal stimulation elicited slow, intensive depolarization within the primitive internal capsule (PIC) from E16-E18. Since these slow responses lasted more than 300 ms and were abolished by glutamate receptor antagonists, they seemed to be due to thalamocortical synaptic transmission. Insertion of a small DiI crystal into the VB of the very same slices revealed the advancing thalamocortical fibers in the PIC and SP, and backlabelled cells in the PIC (Higashi et al, 1996). This suggests that cells of the ganglionic eminence might integrate into a functioning early network with the thalamocortical fibers during development (see chapter 9 for further discussion).

In rat, crystal placement into the dorsal cortex at E14-15 labels very few or no cells within the ganglionic eminence, whereas from perirhinal cortex numerous cells are revealed. At E14, a small crystal in the primitive internal capsule labels not only so-called perireticular cells (within the lateral part of the internal capsule) but also neurons of the perirhinal cortex and the preplate of the adjacent ventrolateral cortical segment, where the cortical plate itself has not yet appeared. Dye placed in the dorsal thalamus at the same age backlabels cells in the thalamic reticular nucleus (Mitrofanis and Baker, 1993) and the perireticular nucleus within the primitive internal capsule, but none in the cerebral wall. At E15, after the first true cortical plate cells have migrated into the lateral cortex, some cells of the plate, as well as preplate cells (now forming the subplate and marginal zone), are backlabelled from the ganglionic eminence, or internal capsule, but still not from the dorsal thalamus, even with very long incubation periods. Only dorsal-thalamic (no ventral-

not ventral-thalamic cells were backlabelled from the internal capsule or ganglionic eminence.
3: Ganglionic eminence cells project to perirhinal and ventral cortex, but do not extend projections into dorsal cortex. From dorsal cortex, very few or no cells within the ganglionic eminence were labelled, whereas from perirhinal cortex numerous cells were revealed. At E14, a small crystal in the primitive internal capsule labels not only so-called perireticular cells (within the lateral part of the internal capsule) but also neurons of the perirhinal cortex and the preplate of the adjacent ventro-lateral cortical segment, where the cortical plate itself has not yet appeared. The axons of perireticular cells reach both dorsal and ventral thalamus at very early stages, and they also develop transient connections with the perirhinal and ventral cerebral cortex. Figure based on the results of Cordery et al, 1996, 1997.

thalamic) cells are backlabelled from the internal capsule or ganglionic eminence. Dye placement in the thalamus produces various patterns of backlabelled cells in the internal capsule, depending on the exact site of the crystal. The axons of perireticular cells reach both dorsal and ventral thalamus at very early stages, and they also develop transient connections with the perirhinal and ventral cerebral cortex (see Fig. 6.1 for summary).

Early Boundaries in the Developing Mammalian Pallium

Although cells in the TRN, internal capsule, ganglionic eminence and preplate have numerous similarities and form a continuous layer in the developing forebrain (Mitrofanis and Guillery, 1993), there are emerging differences in the hodological relationships suggesting boundaries between these areas. We found three interesting borders (Molnár, 1994; Cordery et al, 1996, 1997; see Fig. 6.1): 1) between the intermediate zone and the ganglionic eminence, 2) between the dorsal and the ventral thalamus and 3) between perhirinal and dorsal cortex:

1) Diencephalic crystal placements revealed a continuous chain of cells extending from the dorsal thalamus to the ganglionic eminence, but none in the cerebral wall. The backlabelled cells respect a sharp border between the striatum and the intermediate zone. In hamster, Métin and Godement (1996) described that the medial and lateral ganglionic eminences had different relationships to the diencephalon: the medial only forms connections with the diencephalon, and the lateral with the cerebral cortex. In many cases, especially in more caudal sections, we found it very difficult to make the lateral and medial ganglionic eminence distinction, but in rat and mouse (*Monodelphis domestica*) we found that the diencephalic connections are clearly not a unique property of the medial ganglionic eminence cells. Numerous cells were backlabelled both in medial and lateral ganglionic eminences, from dorsal and ventral thalamic crystal placements.
2) Labelling from the internal capsule and from the ganglionic eminence, we only saw backlabelled cells in the ganglionic eminence and the dorsal thalamus but not in the ventral thalamus. Although many of the ventral thalamic cells project to dorsal thalamus, they seemingly avoid developing projections out of the diencephalon into the internal capsule or the ganglionic

eminence. There appears to be a sharp boundary between the dorsal and ventral thalamus in this respect.

3) Perirhinal cortex and cerebral cortex differ from each other in many respects. Crystal placements into the internal capsule label perirhinal, but very few cerebral cortical cells. Very few, or no ganglionic eminence cells were labelled from the dorsal cortex, in contrast with the perirhinal cortex, where a similar crystal placement revealed numerous cells. At E15, after the first true cortical plate cells have migrated into the lateral cortex, some cells of the cortical plate, as well as preplate cells (now forming the subplate and marginal zone), are backlabelled from the internal capsule, but still not from the dorsal thalamus, even with very long incubation periods.

To correlate the above mentioned boundaries in other species and to further analyze the early connections in developing mammals, reptiles and amphibians might shed light not only on the common developmental mechanisms, but also on the homologies of different forebrain structures (see chapter 11).

Do Subplate Axons Enter the Thalamus?

In the developing cat and the ferret, Antonini and Shatz (1990) reported that subplate cells can be backlabelled by injecting tracers into the thalamus (as well as in the contralateral hemisphere and even the superior colliculus). De Carlos and O'Leary (1992) described that tracer injections into the superior colliculus labelled only layer 5 and not subplate cells in the rat (after P1.5). In the newborn rat, large injections of fast blue into the thalamus, backlabelled large numbers of subplate cells (as well as true layer 6 neurons). They concluded that 'subplate cells do not send axons into the superior colliculus, a layer 5 target, but that many do project to the thalamus, a layer 6 target.'

Our own results (Molnár, 1994; Molnár and Blakemore, 1995a,b) and those of Miller et al (1993) in rat and Sheng et al (1991) in opossum suggest an even more limited distribution of subplate axons. Miller et al (1993) observed at corresponding ages in the hamster that no subplate cells were backlabelled from thalamic carbocyanine crystal placements, and the true layer 6 cells became labelled relatively late (P6 in the hamster), but then quite suddenly and in large numbers. In the opossum the first cells to be backlabelled with horseradish peroxidase from the thalamus are in layer 6, not in the subplate

(Sheng et al, 1991). With very small carbocyanine crystal placements within the thalamus itself, we saw very few subplate cells backlabelled at any age in the rat. Such discrete thalamic placements produce clear anterograde labelling of thalamic axons all the way to the cortex (after about E16 for the LGN) but do not label substantial numbers of subplate cells or even true layer 6 pyramidal cells before birth. Slightly larger crystals, intruding into the ventrolateral diencephalon, or small crystals placed directly in the region of the primitive internal capsule, produce considerable backlabelling of subplate cell bodies (as well as of layer 6 and 5 cells, depending on age and crystal position)(Molnár, 1994).

These results suggest that very few if any subplate axons actually enter the major thalamic nuclei, and we suspect that the majority of the subplate array ends its growth somewhere in the ventrolateral diencephalon. It remains to be seen whether subplate axons actually terminate on some target structure. The hypothesis that the earliest corticofugal projections to reach the internal capsule have subplate origin has been very recently questioned by Clasca et al (1994) who found no backlabelled subplate cells after carbocyanine dye placements from thalamus and internal capsule at different ages (E46-54) in ferret. At E46, only layer 5 cells were found to be backlabelled and therefore it was concluded that the subplate and layer 6 projections grow very slowly and are delayed several weeks in the white matter. Our backlabelling studies, and most recent results from Clasca and colleagues (personal communication), however, clearly demonstrate that subplate and marginal zone projections reach the internal capsule first. There is an emerging consensus in the literature about a possible waiting period for corticothalamic fibers.

Do Corticothalamic Axons 'Wait' Outside the Thalamus Before Entering It?

For the rat, corticothalamic connections from layer 6 can be traced to the ventral thalamic nuclei by E19, according to Schreyer and Jones (1982) and Jones et al (1982). At E16 in the rat, after implanting carbocyanine dyes into the thalamic reticular nucleus (only a few hundred microns away from the dorsal thalamus), we could observe backlabelled layer 6 and subplate cells in the cortex, whereas a similar dye implantation into the dorsal thalamus (of the contralateral hemisphere of the same animal, and therefore under the same conditions) did not label any cortical cells (Fig. 6.2).

There are several possible explanations for this. The corticothalamic fibers might have reached the thalamus but not arborized extensively, and therefore not have taken up the dye sufficiently to appear labelled. Crandall and Caviness (1984a) commented on the very sparse density of corticofugal fibers within the thalamus after HRP injections into the embryonic mouse cortex at E16. We propose that the early corticofugal projections from subplate and layer 6, rather than growing slowly in the white matter (Clasca et al, 1995), reach close proximity to the thalamus quite rapidly (E15/16 in rat) but then do not enter it in substantial numbers. It is conceivable that the majority of corticofugal fibers 'wait', perhaps within the thalamic reticular nucleus, and do not enter the dorsal thalamus. It would be essential to use purely anterograde tracing from the cortex to resolve this issue.

Shatz and Rakic (1981) used ^{3}H-proline (a purely anterograde tracer) to investigate the development of corticothalamic connections in the fetal rhesus monkey. They noted that the orthogradely transported label appears in the surroundings of the LGN weeks before it is detected within the LGN itself. In the opossum, where the thalamus is more accessible for direct tracer injections in vivo, HRP injection in the thalamus revealed that thalamic fibers reach the cortex long before the first cortical cells, in layer 6 and not in the subplate, become backlabelled (Sheng et al, 1991).

It is interesting to note that in the adult there is another area within which extensive crossing of labelled fibers was observed after HRP injection to the thalamus or cortex, namely the thalamic reticular nucleus (Bernardo and Woolsey, 1987; Mitrofanis and Guillery, 1993). It was, therefore, proposed that thalamo-cortical fibers become rearranged here. Since most of these studies were performed with HRP (anterograde and retrograde tracer) we do not know whether the rearranging fibers had thalamic or cortical origin (or both). We have speculated that during development corticothalamic fibers might wait in this area (based on the work and suggestions of Shatz and Rakic, 1981; Sheng et al, 1991; Miller et al, 1993; and Clasca et al, 1995 and on our own retrograde labelling experiments (Molnár, 1994; Cordery et al, 1996, 1997). If there is indeed a waiting period for corticofugal fibers within the thalamic reticular nucleus, it might be here that the rearrangement of the corticothalamic fibers occurs, similarly to that rearrangement which occurs in thalamocortical fibers just below the cortex (Nelson and LeVay, 1985; see chapter 8). It seems to make sense for both early

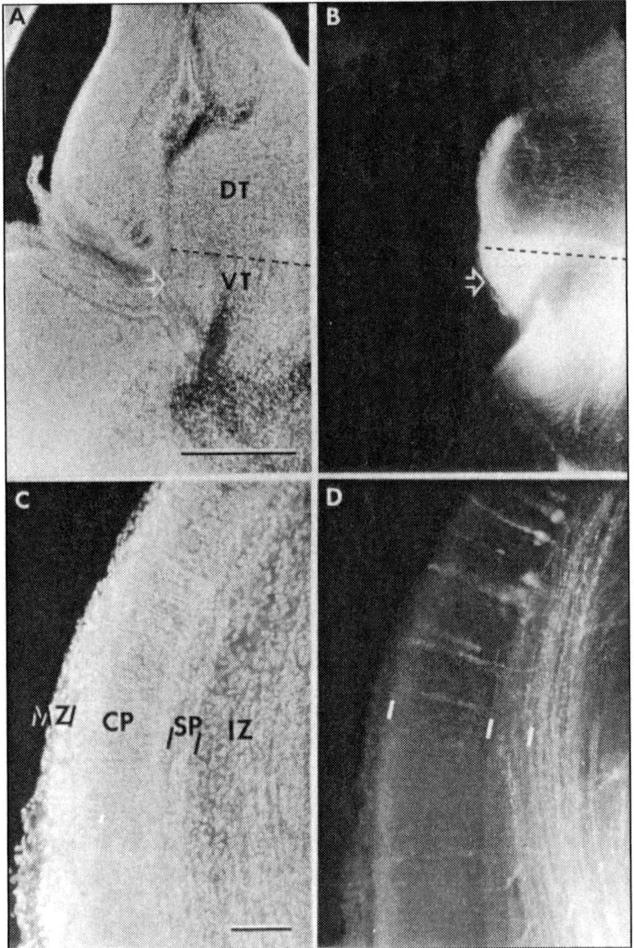

Fig. 6.2. Crystal implantation low in the ventral thalamus at E16 produces labelling very similar to that seen after dorsal thalamic placements, except that more backlabelled cells are seen in the cortex. The same incubation conditions (4 weeks at 37°C) were used for both sets of experiments. A: Same field as in B but viewed under ultraviolet illumination. B: DiI placement site (unfilled arrow) into the reticular nucleus in ventral thalamus (VT) viewed under appropriate filters for DiI. C: The bisbenzimide counterstain of the lateral cortex revealed the marginal zone (MZ), cortical plate (CP), subplate (SP) and intermediate zone (IZ). D: Same view as in C but under illumination for DiI. Back-labelled cells are seen in the lower part of the cortical plate and in the subplate. Small crystal placements within the ventral thalamus, in contrast with those in the dorsal thalamus, revealed thalamic fibers *and* backlabelled cells. Some of the cells were exclusively located in the subplate but some lay in the lowermost part of the cortical plate. If the crystal placement was strictly restricted to the body of the thalamus, then very few if any subplate cells were labelled (see Figs. 4.2, 4.4 and 4.9). This suggests that the subplate projection (and even that from layer 6 of the true cortical plate at this early age) does not extend far into the thalamus, but ends just outside it. Scale bar: 500 μm for A and B; 100 μm for C and D. From Molnár (1994).

thalamocortical and corticothalamic connectivity to be established with gross topographic order, while the distances for growth are minimal. Then the 'fine adjustment' of the thalamocortical fibers is to be performed in the subplate, and that of the corticothalamic fibers in the thalamic reticular nucleus. These two regions have numerous similarities: early generation and maturity, antigenic expression, and differential cell death (Mitrofanis and Guillery, 1993). It is plausible that they have a related evolutionary origin and might form the bridge between the reptilian and mammalian organization (see chapter 11). The crossing of fibers seen in the adult within regions corresponding to these largely transient compartments might then reveal the fine modifications in the distribution of these fibers made during the early phase of development.

Requirement for an Early Forebrain Network in the Development of Cortical Connectivity

The early generated cells in the ganglionic eminence and in the cortical preplate might be important in the initial guidance of connectivity by interacting with the growing connections as guide-post cells and through their early processes forming a three dimensional scaffold to serve as pioneers (see chapter 2). The majority of the cortical connections are formed while the early generated cells of the preplate, ganglionic eminence and thalamic reticular nucleus are already mature and have integrated into a functioning network. Initially this circuitry excludes the later generated thalamic and cortical structures; thalamic axons do not enter the cortex and corticofugal projections do not enter the dorsal thalamus, but it is conceivable that an interplay begins as soon as the two systems coexist. These functional interactions might be essential for the normal outcome of development. Lesioning parts of the immature network causes serious impairment of the development of the thalamocortical and cortico-thalamic connections: thalamic axons fail to enter the cortex (Ghosh et al, 1990; Ghosh and Shatz, 1992b), and many corticothalamic projections fail to enter the dorsal thalamus (McConnell et al, 1995). Bolz (1995) proposed that early activity might upregulate the ingrowth permissive properties in the cortical plate (Götz et al, 1992), and lesioning in the subplate after the arrival of the thalamic axons interrupts this essential process. Thalamic fibers form functioning synapses with subplate cells which might then relay activity to the cortical plate from early embryonic stages (Friauf

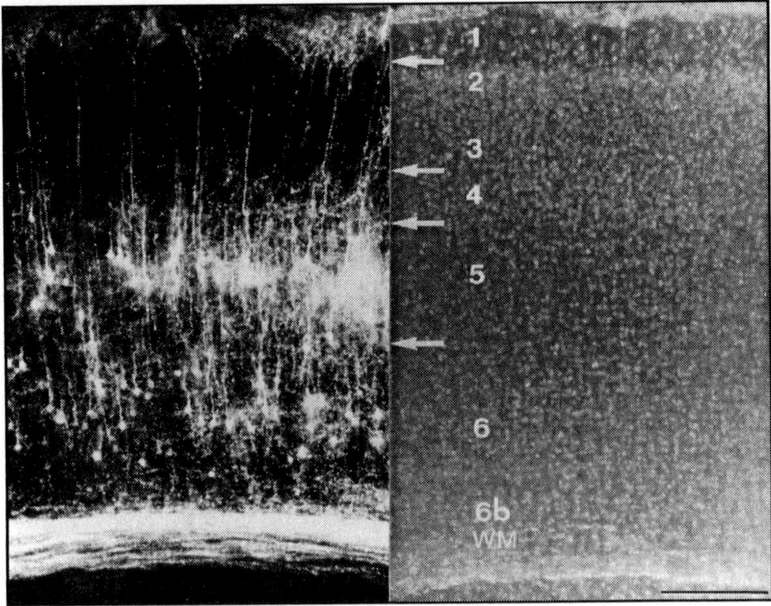

Fig. 6.3. Placement of a crystal of DiI into the lateral part of the diencephalon of a P8 brain (just lateral to the site of separation of the thalamic and peduncular fibers) backlabels cells of both layers 5 and 6. The thalamic fibers are almost completely masked by the backlabelled cell processes. The layer 6 cells project to the thalamus, while layer 5 cells project to the colliculus, pontine nuclei and spinal cord through the cerebral peduncle. The layer 5 cells possess thick apical dendrites, branching in the marginal zone. Right photomicrograph is of the same section as the left but viewed under ultraviolet illumination. The layering of the cortex (boundaries marked with white arrows) is now close to adult maturity. Scale bar: 250 µm. From Molnár (1994).

and Shatz, 1991). We know very little about the early embryonic activity patterns in the developing forebrain, but our in vitro optical recording study (Higashi et al, 1996) demonstrated that both internal capsule and subplate cells are capable of generating action potentials and are active after dorsal thalamus stimulation in embryonic whole forebrain slices (see chapter 9). If the early activation of a region of the thalamus is lacking during development (as in the anophthalmic mutant mouse), the corticofugal projections are also modified and crossmodality connections develop (Bronchti and Welker, 1995). The interplay between the early generated cells of the subplate and TRN with the developing cortical plate and dorsal thalamus might be essential to establish the precise functioning connections in the mature brain.

The Role of the Subplate and the Subplate Scaffold

The first postmitotic neurons of the mammalian cerebral cortex form a transient population, which comes to reside above and below the developing cortical plate (Marin-Padilla, 1971; Luskin and Shatz, 1985a,b). Some speculate that during their brief presence, these neurons, especially the so-called 'subplate' population, participate in numerous cell-cell interactions that are important in cortical development (Shatz et al, 1990; Blakemore and Molnár, 1990). McConnell et al (1989) drew our attention to the fact that subplate axons (as originally described by Marin-Padilla, 1971, 1972) pioneer the first axon pathways to subcortical targets, before neurons of layers 5 and 6 (which form the adult subcortical projections) extend their axons. Thus, thalamocortical axons (and conceivably later-growing corticothalamic fibers) might navigate their way to distant targets along pathways laid down by pioneering axons of subplate neurons.

To support the hypothesis that the pioneering subplate axons provide topographic guidance for thalamocortical projections, two further pieces of evidence are essential. First, the subplate projection must itself be topographically ordered, providing guidance pathways between each thalamic nucleus and the appropriate cortical area. Second, it must be shown that thalamic axons grow in intimate association with their corresponding subplate fibers. The aim of the studies reviewed in the second part of this chapter is to explore the timing of establishment and the orderliness of the first corticofugal and thalamocortical fibers, by examining the relationships among and between the two axon arrays on the way to their targets. Numerous studies (see below) have explored the possibility that the earliest-generated cells of the subplate, and their axonal projections, play a role in guiding and hence determining the topography of the thalamic connections, which in turn might influence the differentiation of a given cortical area.

Are Subplate Axons True Pioneers?

The presence of the transient subplate population and the early corticofugal scaffold that it constructs are distinctive features of the developing mammalian forebrain (McConnell et al, 1989; De Carlos and O'Leary, 1992; Blakemore and Molnár, 1990; Molnár and Blakemore, 1990a,b). McConnell et al (1989) suggested that the pioneering subplate axons could establish a topographic scaffold towards the internal capsule, which could be used to guide corticofugal and thalamic axons over their route to their target areas (Shatz et al, 1990). But is this curious cell group really needed for the establishment of the subsequent cortical connections? Does the subplate projection actually fulfil all the following conventional criteria (see chapter 2) for a pioneer projection?

Extend Axons into Territory in Which No Other Axons Exist

Certainly this is true for the subplate array. From the primordial plexiform zone through the intermediate zone to the primitive internal capsule they are the first fibers in the area (McConnell et al, 1989; Blakemore and Molnár, 1990; De Carlos and O'Leary, 1992; Erzurumlu and Jhaveri, 1992). There are numerous cells present within the primitive internal capsule (in the perireticular 'nucleus', see Mitrofanis, 1992a), but they do not yet seem to project to the cortex at this early stage (Adams N.C. et al, 1993; Molnár et al, 1994; Cordery et al, 1996, 1997; see Fig. 6.1). We have established that the subplate fibers are indeed ordered and directed towards the different regions of the primitive internal capsule as they descend, but the majority probably do not reach the thalamus itself (Molnár, 1994; Molnár and Blakemore, 1995a,b). And they most certainly do not have the chance to guide the thalamic axons over the proximal part of their journey, because the latter have passed into the telencephalon by the time the corresponding subplate axons arrive (Blakemore and Molnár, 1990; De Carlos and O'Leary, 1992; Erzurumlu and Jhaveri, 1992; Molnár and Blakemore, 1995a,b). But, until it encounters the thalamic array under the primordial striatum, the subplate does pioneer the corticofugal outgrowth.

Later Growing Axons Follow the Same Pathway

The hypothesis that the later developing ascending and descending projections follow the trajectories of the first established subplate projections is rather attractive because it could provide (or at least contribute to) the initial topography without the need for any inherent, strict, area-specific chemical matching between the interconnecting cortical and subcortical regions. The idea that thalamocortical axons require subplate axons for their growth to appropriate cortical target areas was stimulated by the observation that the first growth cones of LGN axons and occipital cortical subplate neurons are extending in opposite directions within the

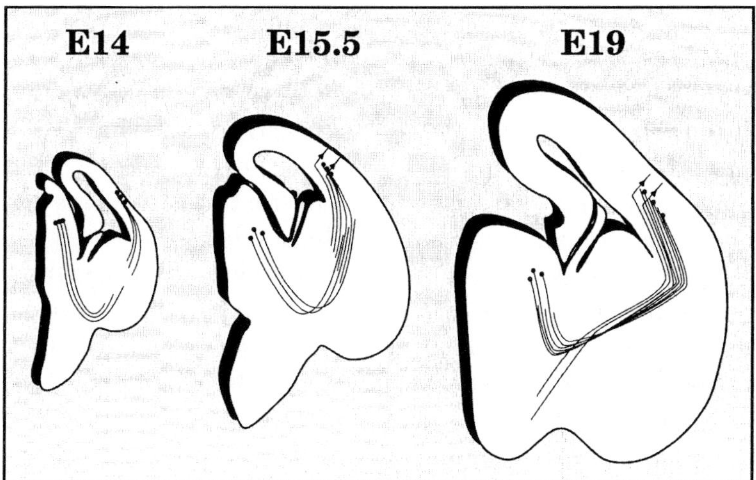

Fig. 6.4. Summary diagram from the early development of the reciprocal thalamococrtical and corticofugal projections. E14: Synchronous outgrowth of the early thalamic and corticofugal fibers. E15.5: The first encounter of the thalamo-cortical and the earliest corticofugal (preplate) projections outside the primitive internal capsule, under the striatum. The fibers from the corresponding regions of the cortex and the thalamus become closely associated with one another. E19: The thalamic fibers establish a topographic map while waiting in the subplate. The corticospinal (etc.) projections from layer 5 form the cerebral peduncle. From Molnár (1994).

internal capsule at about the same time (E30 in the cat: McConnell et al, 1989; Ghosh and Shatz, 1992a). In the rat, similar observations were made at E14-15 for the putative visual cortex (Molnár and Blakemore, 1990a,b; De Carlos and O'Leary, 1992) and for the putative somatosensory cortex (Erzurumlu and Jhaveri, 1992).

To test this idea numerous workers (see below) have attempted to label the thalamocortical and corticothalamic axons with different dyes from corresponding regions of the thalamus and cortex at the same time. Since the carbocyanine dyes travel anterogradely as well as retrogradely, this type of selective labelling is only possible before the thalamic fibers reach the cortex. However, before the thalamic fibers reach a particular cortical area it is very difficult to tell exactly where they are destined to go. Valid general predictions can be made (e.g. fibers from dorso-lateral thalamus reach the occipital cortex) but the finer matching (between a small cortical area and a small group of thalamic neurons) can be precisely defined only *after* the arrival of the thalamic fibers at the subplate layer of the corresponding cortical area. At that point (E16 for occipital cortex) they can be backlabelled from the cortex, together with the corticofugal projections, within the same bundle. But from this very moment selective labelling of thalamic and corticofugal fibers is no longer possible.

Bicknese and Pearlman (1992), in the rat, and Miller and colleagues (1991, 1993), in the hamster, have tried to discover whether ascending thalamic axons grow along the same routes as those traversed by the groups of subplate axons from corresponding cortical areas, but their double-labelling experiments were performed well after the arrival of the thalamic fibers and after the establishment of layer 6 and 5 connections. They concluded that the labelled axon bundles from the cortex and from the thalamus run in separate fascicles, with the afferent thalamocortical axons travelling more superficially, in the subplate, and the efferent corticothalamic axons deeper in the intermediate zone. This is very similar to the adult pattern in the rat (Woodward and Coull, 1984) and cat (Nelson and LeVay, 1985).

Unfortunately the location of the subplate axons in relation to the afferent and efferent pathway was not determined in these studies. In earlier stages, however, similar double labelling experiments of Bicknese and Pearlman (1992) revealed that growing thalamocortical and early corticofugal axons were intermingled within the internal capsule, as implied by the single labelling study of Molnár and Blakemore (1990a) and Blakemore and Molnár (1990).

In our own double-labelling experiments we also observed mingling within the intermediate zone in addition to the internal capsule (Fig. 6.5).

The spatially distinct pathways of corticofugal and corticopetal pathways in the *adult* does not exclude the possibility that the earliest thalamic and subplate axons do selectively fasciculate on each other. 3-D confocal microscopic reconstructions with *identified* subplate axons at *early stages* would be required to resolve this question. It is obviously extremely difficult to be sure that two dye placements at early stages are indeed in precisely corresponding positions: misalignment of the stained bundles in the experiments of Miller et al (1991, 1993) could simply reflect a failure to identify the exact relationships between thalamus and cortex in the fetal brain. Since afferent and efferent projections co-exist from E16, single crystal placements on the cortex label *both* fiber systems. The results of

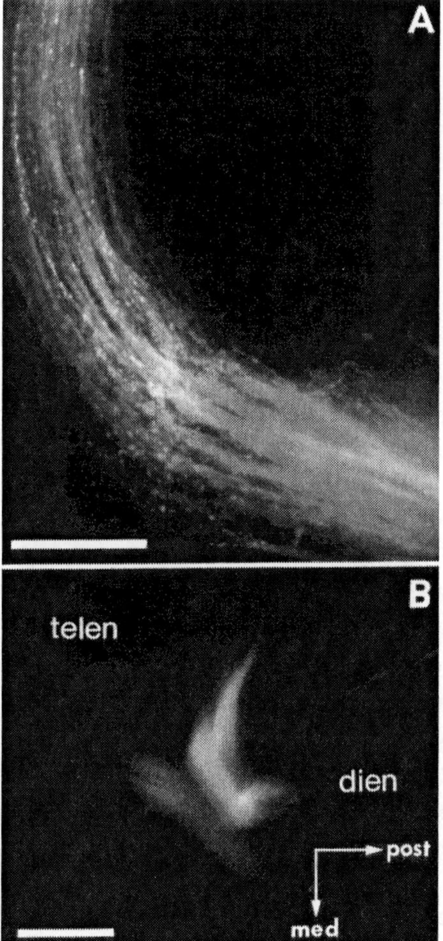

Fig. 6.5. The 'handshake' between thalamic and preplate axons and the precise topography of the early thalamocortical projection. A: In an embryonic day 15.5 (E15.5) fetal rat brain, a crystal of DiI was placed into the dorsal thalamus (putative lateral geniculate nucleus, LGN) and a DiA crystal placed into what was presumed to be the 'matching' region of the occipital cortex (putative area 17), so that both thalamocortical and corticofugal preplate fibers were labelled. At this stage, very few if any thalamic fibers have reached the occipital cortex (no cells backlabelled with DiA were observed in the thalamus), and the DiI-labelled fiber population was exclusively thalamocortical (no cortical cells were backlabelled with DiI). In this coronal section, many DiA-labelled (yellow/green) and DiI-labelled fibers (orange) are associated closely in the same plane of focus, indicating that the preplate and thalamocortical fibers become associated intimately when they meet each other. Scale bar, 100 μm. B: This fluorescence photomicrograph was taken from a 100 μm thick horizontal section of an E20 rat brain after placement of three crystals of different carbocyanine dyes, DiA, DiI and DiAsp, in a parasagittal row along the right hemisphere. Since thalamic fibers have already arrived at the cortex at E16, each placement labelled a mixed bundle of thalamocortical and corticofugal axons. Six weeks' incubation at room temperature was used to enable full anterograde and retrograde diffusion. Three distinct bundles are clearly visible passing through the primitive internal capsule without substantial mixing or crossing. Abbreviations: dien, diencephalon; med, medial; post, posterior; and telen, telencephalon. Scale bar: 500 μm. Reproduced from Molnár and Blakemore (1995a) with kind permission of Elsevier Science Ltd, Kidlington, England. See color figure in insert.

such tracing of mixed fiber bundles suggests that they do follow very similar routes, although individual fibers, especially within the depth of the intermediate zone, are difficult to resolve (Molnár and Blakemore, 1990a,b; Blakemore and Molnár, 1990).

Pioneers Actually Guide Later Axons, Providing Cues That They Follow

This point is more difficult to establish, since it would require the demonstration that thalamic fibers are unable to reach their corresponding cortical target after the deletion of the subplate cells of the corresponding cortical region. Ghosh et al (1990) have indeed lesioned the subplate in the cat by injections of kainic acid and this resulted in a bizarre behavior of the corresponding thalamic axons, which grew under and past the cortical region and appeared not to terminate in that region. However, in this experiment, the lesion was made at a late stage, almost a week after the arrival of the thalamic fibers. The crucial but technically difficult experiment is to selectively delete the subplate system *before* the growth of thalamic fibers into the intermediate zone, without changing any other properties of the region. In the grasshopper embryo the role of selective fasciculation of the anterior-growing axons on the posterior-growing longitudinal pathways was tested by specific laser deletion of the neurons of origin of the posterior-growing axons. These posterior growing pathways proved to be essential for the normal path and target finding of the anterior-growing axons (Bastiani et al, 1986).

Until similar clearcut lesion studies are performed in mammals, we have only indirect evidence, like the appearance of the growth cones (O'Leary et al, 1990; Ghosh and Shatz, 1992a), the lack of regional specificity in co-culture experiments (Molnár and Blakemore, 1991; Yamamoto et al, 1992; see chapter 5), the development of fiber ordering in *reeler* mutant mouse (Molnár and Blakemore, 1992; see chapter 7) and the expression of surface molecules in the embryonic cortex (Godfraind et al, 1988; Chun and Shatz, 1988a; Chung et al, 1991), to support the idea that the subplate scaffold might indeed be used by the thalamocortical fibers for their normal development. If the subplate scaffold does guide the distal portion of the journey of the thalamic axons, why is this necessary?

The Subplate Scaffold Compensates for the Distorting Effects of Growth

At the time of initial outgrowth of subplate and thalamic axons, the forebrain is very small and very simple in its organization. The developing cerebrum is wrapped around the developing diencephalon, the shortest route between corresponding points being a smoothly curved trajectory. But even between mid E14 (when axons first begin to leave the LGN) and E16 (when they arrive at the occipital cortex) the cerebrum expands dramatically, leading to a substantial change in the interlinking trajectories between cortex and thalamus. This inevitably distorts the pathways taken by the pioneering subplate axons. Initially, at the time when they grow through the intermediate zone, the bundle of subplate axons from a single point on the cortex follows a simple, smoothly curved route towards the primitive internal capsule. But over the following day or two, when the thalamic axons grow through the same segment of the brain, the bundle of fibers labelled from a single cortical point becomes more complex in form, with sharp bends and twists appearing in it .

A possible functional reason for the early establishment of the subplate scaffold, and the early completion of thalamocortical fiber growth (at E16 for occipital cortex in the rat) is to *preserve spatiotemporal precision* which would otherwise, because of the distorting effects of growth, become increasingly difficult to orchestrate. Interestingly the early establishment of these trajectories seems to be true for many other gyrencephalic species at the equivalent stage. I have examined human fetal brains (see Fig. 6.6) as well as rat, mouse and marsupials (*Monodelphis domestica* and *Dysaurus hallulatus*), and although the timing of the outgrowth differs enormously, the *spatial relationships* between cortex and thalamus are remarkably similar in all species: only the overall size of the brain differs sustantially. In higher mammals with large brains and gyrencephalic cortices, thalamocortical projections reach the cortex before the characteristic migration and rotation of the lateral geniculate nucleus (Le Gros Clark, 1932a,b; Rakic, 1976, 1977; see Fig. 3.6) and before the appearance of sulci and gyri.

However, inevitably, the formation of the preplate is followed by a period of rapid proliferation and migration of cortical neurons and hence thickening of the cortex and expansion of its surface area. The telencephalon grows much faster than diencephalic structures. After E14, the whole cerebrum expands rapidly, pushing backwards and downwards, around the brainstem. The relative delay in the formation of the occipital cortex (Berry and Rogers, 1965; Lund and Mustari, 1977), the closure of the internal capsule and the relative rotation of the diencephalon (Kalil, 1978) all contribute to the rapid structural complication of the trajectories between thalamus and cortex.

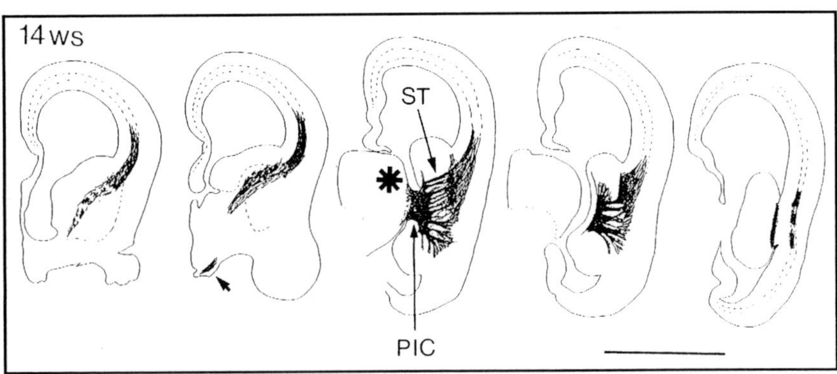

Fig. 6.6. Thalamocortical projections revealed *post mortem* with carbocyanine fluorescent dye, DiI, in a E14-week human brain (from an aborted fetus). A single crystal of DiI was placed into the dorsal thalamus (* indicates the crystal placement site) of a *post mortem* immersion-fixed brain (ethical permission for the use of human fetal material was given from the Reasearch Ethical Committee of the Albert Szent-Györgyi Medical University.) After 12 months incubation at 37°C, the whole hemisphere was coronally sectioned (200 µm) and every third section was drawn under epifluorescent illumination using a camera lucida. The serial sections are in rostrocaudal order (left is rostral, right is caudal). The other hemisphere was embedded in paraffin and 10 µm cresyl violet-stained reference sections were prepared to reveal the state of thalamic and cortical development.
The figure demonstrates that the overall topography of the thalamic projections bears a strong resemblance to that in the E15 rat (chapter 3). Some of the thalamic fibers have already left the diencephalon through the primitive internal capsule (PIC). The fibers form fascicles in the striatum (ST), but defasciculate in the intermediate zone and the subplate (area outlined with dotted lines). The fibers are not labelled throughout their full length; nevertheless the results already demonstrate that thalamic fibers reach the subplate layer of the human cortex *before* the characteristic migration of the LGN and *before* the gyrencephalic transformation of the cerebral cortex. The study suggests an analogy between the early phases of ontogeny in rodents and in primates. In these early stages, the LGN of primates is situated at the lateral aspect of the thalamus as it is in fetal adult rodents (Le Gros Clark, 1932a,b; Rakic, 1974; see Fig. 3.6). Subsequently, due to the expansive growth of the pulvinar, the LGN is simultaneously rotated upon its transverse and sagittal axes and becomes displaced to the inferior aspect of the mature thalamus (Rakic, 1977). Note that the DiI crystal also retrogradely labels the optic tract; arrowhead). Scale bar: 10 mm. From Molnár, 1994.

Thus the subplate axons establish their scaffold before the explosive growth of the cerebrum, and the structure of the pathway changes substantially by the time the thalamic axons traverse the same region. Perhaps one function of the subplate scaffold is to lay down a guidance structure at an early stage, when simple fiber-fiber ordering is sufficient to achieve topography, which can then be followed by thalamic axons as the route becomes too tortuous for simple, local interactions to preserve topographic order. The subplate array may be necessary to overcome structural complications in the telencephalic part of the pathway caused by explosive growth.

The Waiting Period

There is pressure on thalamic axons to establish their projections to the cortex as early as possible because the route between is increasing in structural complexity all the time. When LGN fibers reach the occipital cortex, only the lowest part of the true cortical plate is in place and the cortical plate is not yet permissive to axonal ingrowth (see chapters 3 and 5). Moreover, most of the main target cells of the thalamic axons, the cells of layer 4, do not take up their position in the plate until after birth in monkey (Rakic, 1976, 1977), cat (Shatz and Luskin, 1986) and rat (Lund and Mustari, 1977). Thalamic axons arrive below the correct regions of cortex before the cortical plate has fully formed and they accumulate and 'wait' over the subplate layer (for several weeks in cats and monkeys) before invading the plate proper (see chapter 4). In the rat occipital cortex, thalamic fibers arrive on E16, but do not invade the cortex significantly until around the 19th day of gestation.

The Subplate Engineers the 'Waiting Period'

Experiments with co-cultures (Bolz et al, 1992; see chapter 5) demonstrate that the cortical plate is not strongly growth-permissive for thalamic axons until about E19-20. Using an in vitro quantitative growth assay, Bolz and his collaborators demonstrated the existence of a growth-permissive, membrane-bound factor (or factors) in postnatal cortex which is not present in embryonic cortex (Götz et al, 1991; Bolz and Götz, 1992). Moreover thalamic growth cones collapse on E16 cortical membrane preparations (Hübener et al, 1992, 1995).

Interestingly, we have been unable to reproduce all the behavior characteristic of the waiting period in co-culture (Molnár and Blakemore, 1991, 1995a,b; Molnár 1994). Axons from fetal LGN cultured with occipital cortex younger than E20 simply grow haphazardly around the cortical slice and certainly do not preferentially accumulate over or terminate in the subplate layer. The only similarity to the situation in vivo is that very few axons enter the cortical plate. With cortex older than E20, they simply invade the cortical slice, without hesitating in the subplate layer. This suggests that the waiting period, which is such a distinctive feature in vivo, depends on intimate relationships having been established between the subplate scaffold and the thalamic axons *before* the latter reach their target areas. Thus, the growth of thalamic axons over the permissive subplate scaffold, in the otherwise 'hostile' embryonic extracellular environment, may be necessary to establish the waiting period. This period is itself essential to avoid the kind of unterminated ingrowth by thalamic axons that is seen in co-culture with cortex between E19 and about P2, when layer 4 is not yet expressing its 'stop signal' (Molnár and Blakemore, 1995a; see chapter 5).

If each thalamic axon literally fasciculates on a subplate axon, it may inevitably end its growth in immediate proximity to the cell body of that subplate axon, and be compelled to terminate temporarily there. Miller et al (1993) reported that occasionally a few thalamic axons grow straight up to the marginal zone through the cortical plate, while the majority are restricted to the subplate. I demonstrated (Molnár, 1994) that initially both marginal zone and subplate cells can be backlabelled from the internal capsule (at E14.5) when the thalamic axons reach the same region. One interesting possibility is that the early growth of a small proportion of thalamic fibers through the cortical plate to the marginal zone is governed by the corticofugal projections of these cells. These descending fibers, which stretch across the non-permissive territory of the cortical plate, may provide a pathway for some thalamic axons to pass through the non-permissive cortical plate. This might be based on the same mechanisms and highlights a general principle. This axonal pathfinding mechanisms is highlighted by the example of cortical malformation in the *reeler* mutant mouse, this is the subject of the next chapter.

CHAPTER 7

First Postmitotic Cells Govern the Development of Thalamocortical Projections in the *Reeler* Mutant Mouse

The first postmitotic cortical neurons form the subplate and the marginal zone (Marin-Padilla, 1971; Luskin and Shatz, 1985a,b). These neurons are thought to be generated as a single population, before the formation of cells destined for the cortical plate itself. They then migrate to the outer edge of the cerebral wall to become the primordial plexiform zone (Marin-Padilla, 1971), or preplate (Rickmann et al, 1977; Stewart and Pearlman, 1987). The subsequently migrating neurons of the cortical plate split this preplate into two laminae, the marginal zone above (the future layer 1), and the subcortical plate or subplate below the thickening cortical plate (Marin-Padilla, 1971; König et al, 1977; Luskin and Shatz, 1985a,b)(see Fig. 7.1). This mechanism is disturbed in the *reeler* mutant mouse.

The Problem of the *Reeler* Mouse

reeler is a autosomal recessive mutation in the mouse that disorganizes cortical development (Caviness and Sidman, 1973; Caviness, 1976, 1982; Caviness and Rakic, 1978; Goffinet, 1979, 1992, 1995). The gene responsible for this trait was recently cloned (D'Arcangelo et al, 1995; Bar et al, 1995; Hirotsune et al, 1995) and shown to encode a potential extracellular matrix protein named reelin, expressed in the embryonic cortex at the level of Cajal-Retzius cells. Almost simultaneously, a *reeler*-associated antigen, CR-50, was also found to be expressed by embryonic Cajal-Retzius neurons and could be an epitope belonging to reelin (Ogawa et al, 1995). D'Arcangelo et al (1997) very recently reported that CR-50 antibody

Development of Thalamocortical Connections,
by Zoltán Molnár. © 1998 Springer-Verlag and R.G. Landes Company.

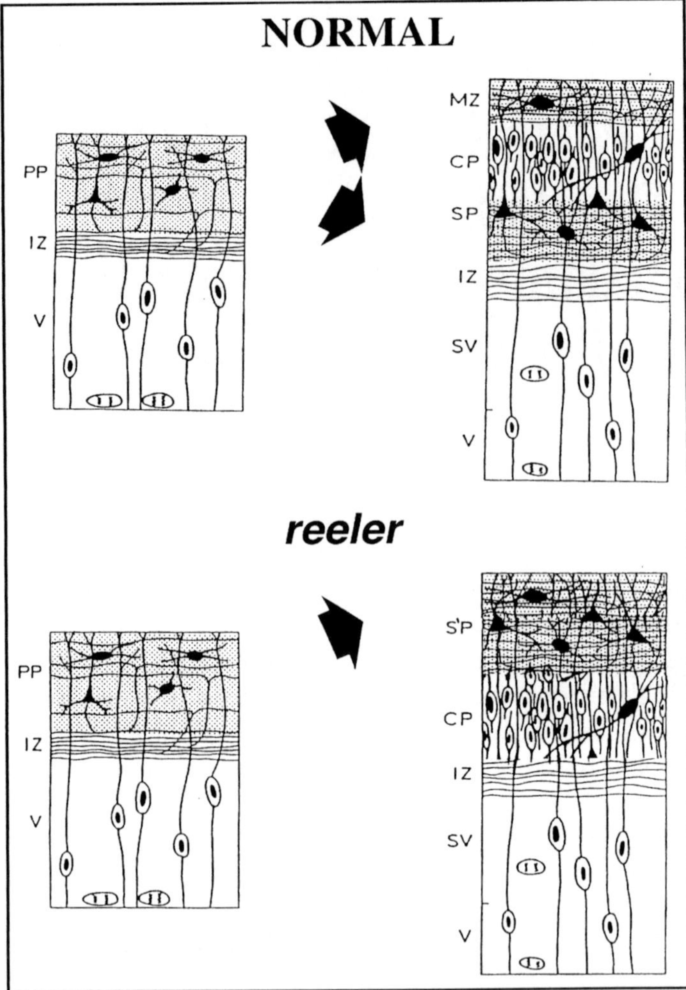

Fig. 7.1. Schematic illustration of cross sections of the cerebral wall before and after the appearance of the cortical plate in normal and *reeler* mice. The first postmitotic cells in cortical neurogenesis form the subplate and the marginal zone (Marin-Padilla, 1971; Luskin and Shatz, 1985a,b). These neurons are generated as a single population, before the formation of cells destined for the cortical plate itself, and they migrate to the outer edge of the cerebral wall to become the primordial plexiform zone (Marin-Padilla, 1971) or preplate (Rickmann et al, 1977; Stewart and Pearlman, 1987). This early phase is identical in both genotypes. In normal mice, the subsequently migrating neurons of the cortical plate split this preplate into two laminae, the marginal zone above (the future layer 1), and the subcortical plate or subplate below the thickening cortical plate (Marin-Padilla, 1971; König et al, 1977; Luskin and Shatz, 1985a,b). In *reeler*, cells of the preplate will not split and those early postmitotic cells that lie towards the bottom of

recognized the in vitro translated reelin. Reelin was found to be a secreted glycoprotein.

Del Río et al (1997) recently provided evidence that during the development of the hippocampus, reelin positive transient neurons (Cajal-Retzius cells) act as guidepost cells to direct entorhinal axons to the stratum lacunosum-moleculare and the outer molecular layer. They found that the entorhinohippocampal pathway did not form properly in vitro when the Cajal-Retzius cells had been ablated with toxins or with specific agonists. Using anti-reelin antibodies in vitro Del Rio et al (1997) described that the development of the entorhinohippocampal pathway was disturbed although not as severely as in the ablation experiments. These experiments suggests that reelin positive early generated transient cells might be involved in the development of the hippocampal connectivity.

In the cerebral cortex, reelin expression by pioneer neurons in the preplate presumably instructs the organization of the cortical plate, either directly (Goffinet, 1995) or via a trophic effect on the radial glia scaffold (Rakic and Caviness, 1995; Hunter and Hatten, 1995). Very recently another mutant mouse was described, the *scrambler*, which has a nearly identical phenotype to *reeler* (Goldowitz et al, 1996; Gonzales et al, 1996). Gilmore and Herrup (1997) proposed that the *scrambler* gene might encode the reelin receptor. Although the biochemical basis of the *reeler* trait is only beginning to be understood, the protean effects of the mutation on cortical cell migration, architectonic and hodological development have been described long ago in great detail (reviewed by Caviness, 1988).

In *reeler*, as in normal mice, the earliest postmitotic cells migrate to the outer edge of the cerebral wall to form the typical polymorphic neurons of the primordial plexiform layer or preplate (Pinto-Lord and Caviness, 1979; Goffinet and Lyon, 1979; Goffinet, 1979, 1980; Caviness, 1982). However, rather than splitting the preplate and accumulating in an inside-out sequence between marginal zone and subplate, the neurons of the cortical plate then gather, in a somewhat

the preplate, which would be destined to become subplate neurons in the normal mouse, are left stranded in the superplate of *reeler*, above the thickening cortical plate, which thus separates the arriving thalamic axons from their presumed temporary target cells. (PP, preplate; IZ, intermediate zone; V, ventricular zone; SV, subventricular zone; MZ, marginal zone; CP, cortical plate; SP, subplate; S'P, superplate). Modified from Uylings et al (1990) with kind permission of the MIT Press, Cambridge, Massachusetts.

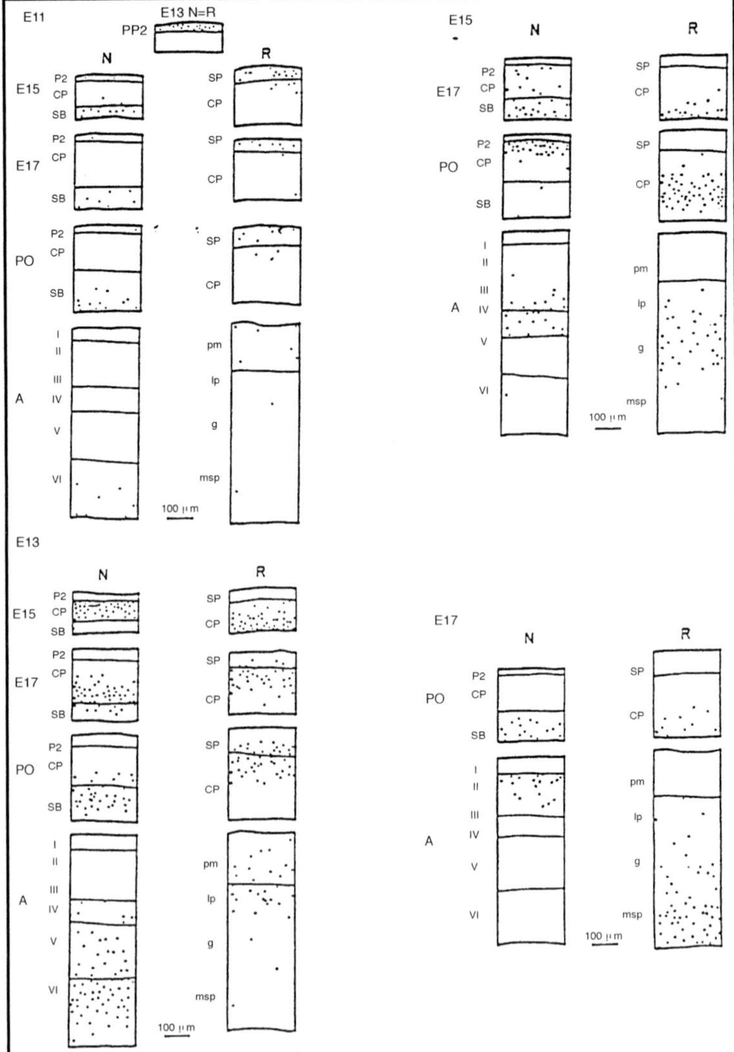

Fig. 7.2. The figure (adapted from Caviness, 1982) demonstrates the distribution of sequentially formed neurons as they achieve their final positions in the developing neocortex of normal (N) and *reeler* (R) animals. Neurons formed at 48-hourly intervals at gestational ages E11, E13, E15 and E17 were autoradiographically labelled by a single injection of [^{3}H]thymidine. Plots of the distribution of heavily labelled cortical cells are demonstrated at successive developmental stages E11-17, Po and adult in normal and *reeler* animals. ppz primordial plexiform zone, vz ventricular zone, cp cortical plate, sb subplate, sp superplate. In the adult *reeler* cortex pm, lp, g and msp indicate the zones of polymorphic, large pyramidal, granule and small and medium pyramidal cells, respectively. Reprinted from Caviness (1982) with kind permission of Elsevier Science, Amsterdam.

irregular but basically *outside-in* sequence, entirely *below* the preplate (Caviness, 1982). Caviness and his colleagues (1981) have termed the abnormal unsplit preplate of *reeler* the 'superplate'.

Despite this radical disturbance of the genesis and lamination of the cortex, thalamic axons reach the appropriate cortical areas (Steindler and Colwell, 1976; Dräger, 1976; Caviness and Frost, 1980; Simmons and Pearlman, 1982) and terminate principally in the middle of the cortical plate, corresponding to layer 4 (Frost and Caviness, 1980; Caviness et al, 1988). However, the trajectory of thalamic axons within the cortex is highly abnormal in the adult animal: instead of running radially from the white matter to layer 4, they form long loops, running up through the entire thickness of the cortex in obliquely orientated fascicles and then defasciculating and plunging down to reach layer 4 from above rather than from below (Caviness, 1976; see Fig. 7.3).

The accurate overall topography and the correct laminar termination of thalamic innervation seen in the adult *reeler* might be thought to cast doubt on the theory of thalamocortical development advanced by Shatz et al (1990) and Blakemore and Molnár (1990) described in chapter 6. If the guidance of thalamic axons and the waiting period essential for the establishment of thalamocortical connectivity depend on an intimate relationship between thalamic axons and the fibers and cells of the subplate, how can the surprisingly normal overall adult topography be produced in *reeler*, in which the cells that should constitute the subplate lie above the inverted layers of the cortical plate? And how can the strange local trajectories of thalamic axons within the cortex be explained? These questions can be answered only by a detailed study of the maturation of thalamocortical interconnections at the earliest stages of development in *reeler*. The *reeler* mouse is a remarkable 'experiment of Nature' (Caviness et al, 1988) allowing different aspects of cortical development to be studied. Inasmuch as the abnormality of cortical lamination in *reeler* mice is accomplished by normal hodological relationships, this mutant mouse provides an interesting model to address the role of pioneer neurons and fibers in corticogenesis.

Rapid Golgi, neurofilament immunohistochemical and monoaminergic histofluorescence techniques have been used to investigate some aspects of the early development of the afferent connections in *reeler* (Caviness and Korde, 1981; Yamamoto et al, 1986). Caviness et al (1988) and Yuasa et al (1994) described the analogous

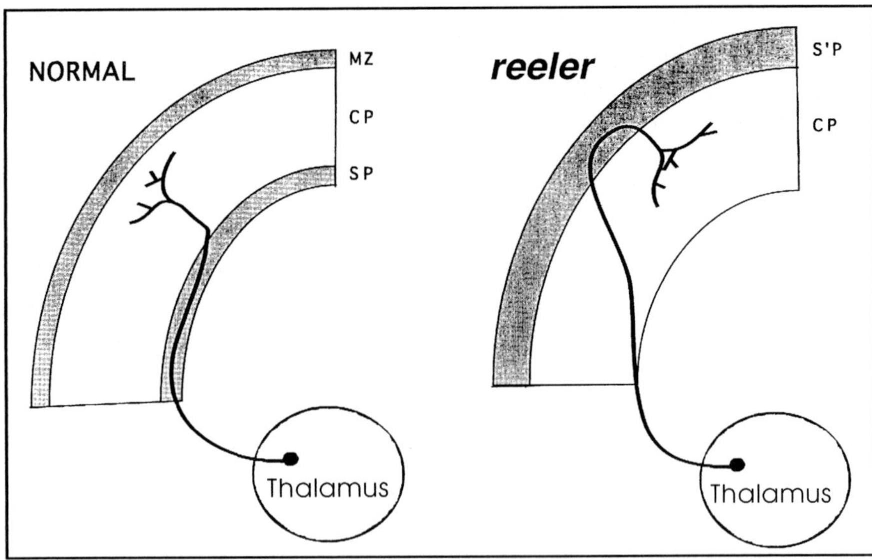

Fig. 7.3. Despite the abnormality in the position of the early postmitotic neurons and in the formation of cortical layers (Caviness, 1982), thalamic fibers reach the correct regions of cortex in *reeler*, and ultimately terminate on neurons that appear to be equivalent to layer 4 in the normal animal (Frost and Caviness, 1980). Only the local pattern of innervation within the grey matter seems highly abnormal, the thalamic fibers looping up to the top of the cortex, and reaching the middle of the cortex from above rather than from below (Caviness, 1976; Frost and Caviness, 1980; Caviness et al, 1988). (CP, cortical plate; SP, subplate; S'P, superplate; MZ, marginal zone).

targeting of the first generated cells in the subplate of normal mice and the 'superplate' in *reeler*, but the timing of the establishment of the early corticofugal pathway and its relationship to thalamocortical projections has only recently been examined (see below). The capacity of carbocyanine dyes to label several pathways in fixed tissue through anterograde and retrograde passive diffusion (Godement et al, 1987) provides a direct way to explore the development of connections. We have used axon labelling to study the interactions between thalamic axons and precocious cortical neurons in *reeler* and to examine the mechanisms by which normal regional and laminar targeting is achieved in the mutant animal despite the gross disturbance of corticogenesis (Molnár and Blakemore, 1992; 1995a,b). We wanted to understand how the correct overall topography, but local abnormality, of projections is generated during development, and to test the notion that the earliest corticofugal axons and the cells that give rise to them play a guiding role in the deployment of thalamocortical afferentation (Shatz et al, 1990; Blakemore and Molnár, 1990).

The Early Corticofugal Axon Scaffold in *Reeler*

Our results suggest that the timing and the overall topography of both the corticofugal and thalamofugal projections are essentially identical in early stages in normal and *reeler* genotypes (Molnár and Blakemore, 1992; Molnár, 1994). We could find no differences between normal and *reeler* individuals in the appearance of cortex, thalamus or their axons in normal and *reeler*. The polymorphic cells of the preplate spin out their pioneering axons, in both strains, at E13 and E13.5, *before* the arrival of any cells of the cortical plate proper, and before the arrival of thalamic or brainstem (Crandall et al, 1992) afferents (Fig. 7.4). These descending axons approach the primitive internal capsule, with a similar degree of fiber order in the two strains, at about E13.5 (for the occipital cortex).

At roughly the same stage, axons from the dorsal thalamus have grown in quite strict order through the primitive internal capsule and are approaching their corticofugal counterparts. In both wild-type and *reeler*, the thalamic fibers distribute to the appropriate cortical regions, closely intermixed with, and conceivably fasciculated upon, the pioneer corticofugal axons from those same cortical areas.

Superplate Cells as Temporary Targets in *Reeler*

The differences in cortical structure between *reeler* and normal mouse start to become evident immediately after the arrival of the first wave of migrating cortical plate cells, which stop below the preplate in *reeler* rather than penetrating and splitting it (Goffinet, 1979; Pinto-Lord and Caviness, 1979). The abnormal placement of the gathering plate cells in *reeler* forces a change in the appearance of the pre-existing early corticofugal axons, leading them to form oblique bundles, running from the superplate, through the cortical plate below (Fig. 7.5). Nevertheless, this does not change the paths of the distal parts of these axons, which have already turned through and under the developing striatum towards the primitive internal capsule, and made contact with the array of thalamic axons advancing towards the cortex, just as in the normal animal (Fig. 7.6).

Superficially, the behavior of thalamic axons as they reach their cortical field is very different in the two strains. In normal mice, as

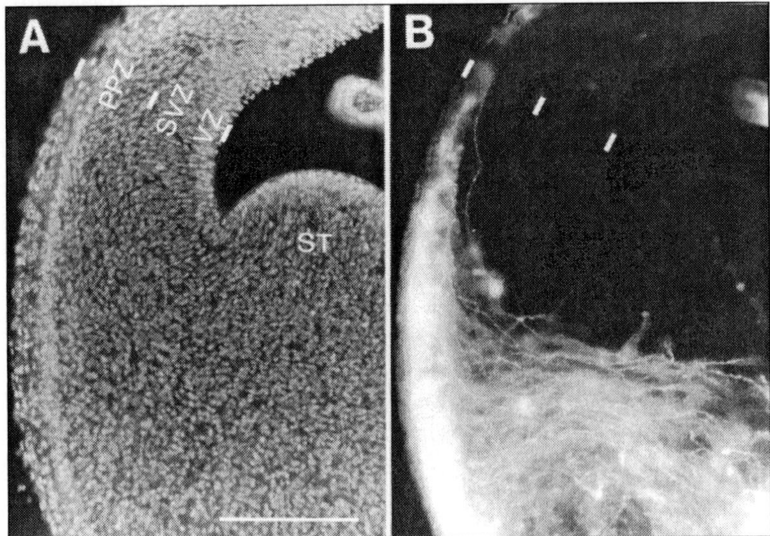

a, b

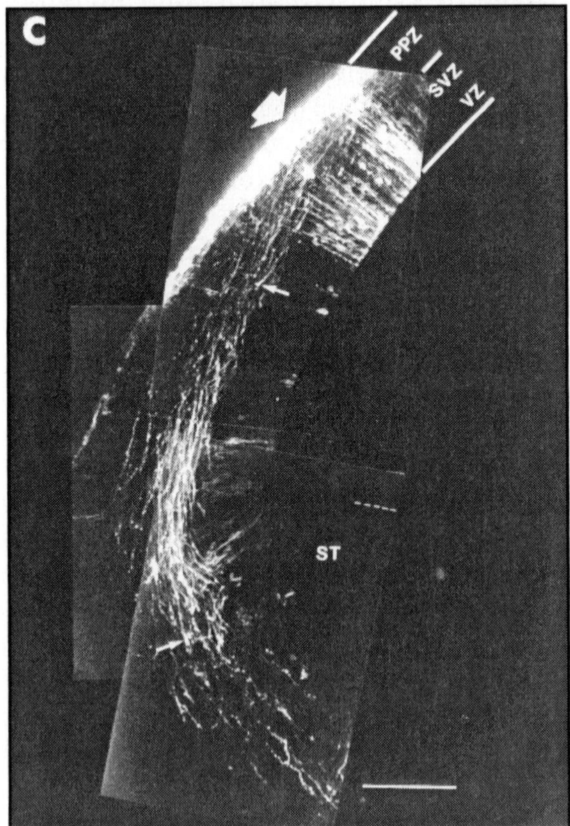

c

in rats (see Molnár and Blakemore, 1995a), with the exception of a few thalamic fibers that penetrate the cortical plate and reach the marginal zone very soon after their arrival, the vast majority of axons accumulate below the cortical plate and do not enter it for several days. But in *reeler*, the thalamic fibers immediately penetrate the cortical plate and run up diagonally to the superplate layer at the top of the cortex, where they then wait for roughly the same period of time before turning down again into the plate itself (Caviness et al, 1988; Molnár and Blakemore, 1992, 1995a,b). However, these two patterns are both compatible with and support the hypothesis that thalamic

Fig. 7.4. (Opposite) Descending projections from preplate neurons in an E13.5 mouse. Fluorescent photomicrographs were taken from a bisbenzimide stained, 100 µm-thick, coronal section of this E13.5 brain, viewed by conventional fluorescence microscopy using ultraviolet (A) and rhodamine (B) filters.
A: Bisbenzimide staining reveals that at this stage, prior to the appearance of the cortical plate, at least in the posterior part of the hemisphere, the developing cortex consists of the primordial plexiform zone (PPZ) and the subventricular and ventricular zones (SVZ and VZ respectively). Although it is rather difficult to state the genotype of the embryo with total confidence, in more anterior sections the first cells of the true cortical plate appeared to be accumulating within the preplate rather than below it, indicating that this particular embryo had normal phenotype. B: A small DiI crystal was implanted just outside the primitive internal capsule, below the striatum (ST). After 3-week diffusion the dye backlabelled cells through the full thickness of the preplate, indicating that their axons had reached the crystal placement site in the internal capsule by E13.5. Scale bar: 200 µm.
C: These assembled pictures were taken (using a confocal microscope) from a coronal section of the dorso-lateral aspect of the cortex of the left hemisphere of a 13.5 days old mouse embryo (75 µm-thick coronal section, paraformaldehyde fixed). DiI had been placed in a similar position, on the convexity of the cortex (large arrowhead), of all members of the litter (n=12) of a rl/rl male and rl/+ female (expected to be 50% *reeler*). The thickness of the optical sectioning was 2 µm, avoiding the plane of the brightest part of the crystal placement, thus allowing the visualization of spindle-shaped cells in the ventricular and subventricular zone (VZ and SVZ) outside the most intense part of the halo around the crystal site. The first postmitotic cells reside in the primordial plexiform zone (PPZ) or preplate. Their projections run down in the intermediate zone then turn under the developing striatum (ST). The pattern and the extent of the projections in *all* members of the litter (n=12, half of which should have had *reeler* genotype) were very similar, suggesting that there is no obvious difference between *reeler* and normal genotype at this stage in this respect, in the dorsolateral region of the hemisphere. Examination of the anterior ventral cortical regions showed no obvious intervention of the emerging cortical plate between marginal zone and subplate, whereas this pattern, typical of normal animals, would be seen in the ventral anterior part of the hemisphere in some other individuals in the litter (see A, B). This suggests that this embryo was of *reeler* phenotype. The ventral boundary of the lateral ventricle is demarcated with an interrupted line.
Scale bar: 100 µm.

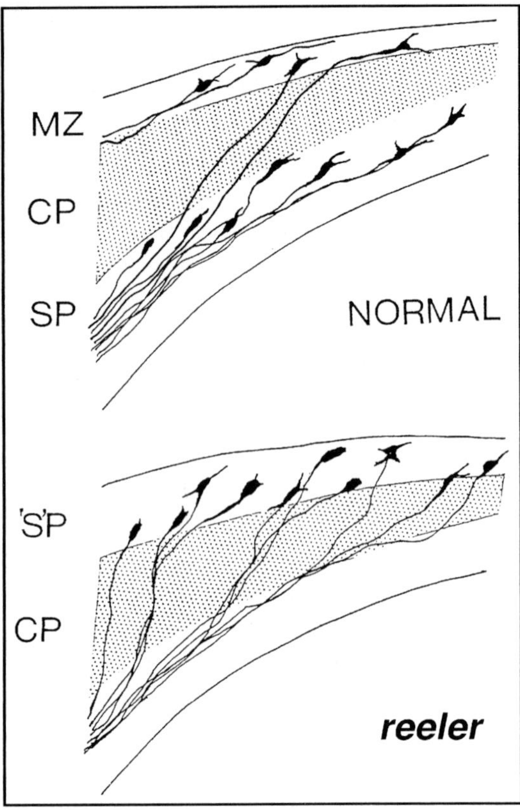

Fig. 7.5. These schematic diagrams represent equivalent sectors of the dorso-lateral aspect of the left hemisphere of the two different phenotypes at around E15, corresponding to the time of arrival of the thalamic fibers (not indicated) at the occipital pole. The upper is the normal animal (rl/+), the lower is *reeler* (rl/rl). Left is lateral and right is medial. In *reeler* the cortical plate (CP; stippling) accumulates below the entire population of first-generated cells which stays together as the superplate ('S'P). The projections of preplate cells, which were laid down *before* the appearance of the cortical plate, run through the cortical plate in fascicles in *reeler*. In the normal animals the cortical plate accumulates between the marginal zone (MZ) and subplate (SP). From Molnár (1994).

axons are specifically guided towards the preplate cells by running in close association with the axons that grew down earlier to establish the corticofugal scaffold.

Mechanisms of Guidance and Target Finding in *Reeler*

The *reeler* mouse was often presented as an argument against, rather than for, prelaid guiding trajectories and contact guidance. It was proposed that the entire guidance of thalamic axons to the cortex might be regulated by diffusible substances originating from target cells, rather than by contact cues along the trajectories of pioneer axons (Caviness and Rakic, 1978; Caviness et al, 1988). The operation of such a chemotropic mechanism was suggested to explain both the correct tangential distribution of thalamic axons throughout the cerebrum in *reeler* and their unusual local behavior.

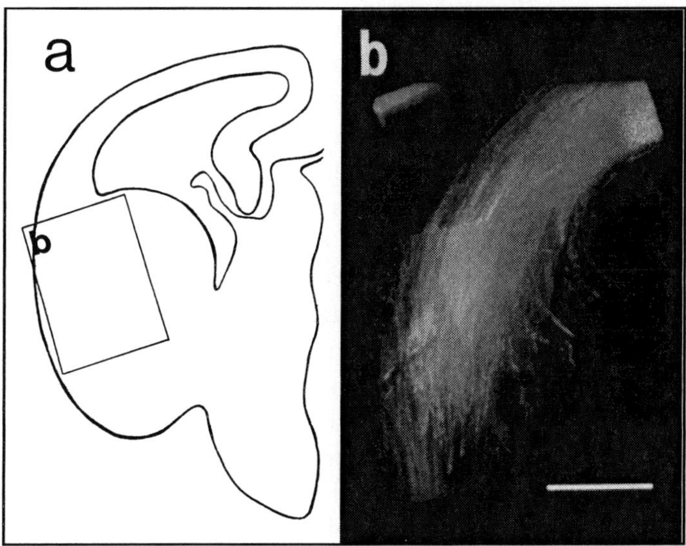

Fig. 7.6. A small crystal of DiI was placed in the convexity of the cortex and a small crystal of DiO in the dorsal thalamus of the same hemisphere of an E13.5 old mouse. The labelled fibers were imaged separately with different photomultipliers using confocal microscopy from the region illustrated with a box on the line drawing. The earliest corticofugal projections (red) fan out from the intermediate zone and loosen up as they turn under the striatum. The fibers end in large growth cones. The thalamic fibers (green) heading towards the cortex have just reached the same region. There is a considerable overlap between the differentially labelled fibers. Scale bar: 200 mm. From Molnár et al (1993a). See color figure (B) in insert.

On the other hand, results of co-culture studies (Molnár and Blakemore, 1991; Yamamoto et al, 1992; Bolz et al, 1992; Rennie et al, 1994) suggest that any *early* influences of cortex on thalamus in vitro are not regionally selective and are therefore unlikely to be directly responsible for the correct topographic distribution of thalamic axons. They also imply that the subplate is not strongly and selectively attractive to thalamic axons at any stage in culture, which one would expect if subplate cells in vivo attract and hold thalamic axons by remote trophic or tropic signals. Rather, the intimate relationship between thalamic fibers and subplate cells, and the waiting period in the subplate layer, appear to occur only if the thalamic axons approach the subplate by their normal route—over the scaffold of pioneer corticofugal fibers after interaction within the environment of the ganglionic eminence and intermediate zone.

Remember that the cortex does not become growth-permissive for thalamic axons in the rat (Götz et al, 1992; Molnár, 1994; Tuttle et al, 1995) until about E20 (presumably equivalent to about E19 in the mouse), long after the arrival of thalamic axons at the cortex in vivo. Yet, in *reeler*, thalamic axons grow through the cortical plate cell layer at a much earlier age to reach the superplate above. Interestingly, a small number of thalamocortical axons also reach the marginal zone by growing obliquely through the cortical plate, even in normal animals. The mouse does not seem to be an exception, since early ingrowth of a small fraction of thalamocortical fibers to the marginal zone has also been reported in neonatal cat (up to 2 weeks) and E14 hamster (Kato et al, 1984; Miller et al, 1993). It is conceivable that preplate cells release a diffusible agent which then attracts the thalamic fibers to the displaced preplate cells in the superplate through the cortical plate in *reeler* (Caviness and Rakic, 1978; Caviness et al, 1988): the same mechanism could account for the few fibers in normal animals that reach the marginal zone at early stages. However, axons would be expected to take the shortest available route (i.e. radial) if they were responding to the gradient of a diffusible agent, whereas in *reeler* their trajectory is clearly oblique—identical to that of the descending corticofugal bundles laid down earlier.

We suggested (Molnár 1994; Molnár and Blakemore, 1992, 1995a,b) that the characteristic pattern of thalamocortical fiber ingrowth in *reeler* might be due to their close relationship with the early corticofugal axons. The appearance of the thalamic axons, streaming in oblique, fasciculated bundles through the plate, is remarkably similar to the appearance of the outflowing axons of preplate cells seen before the thalamic axons arrive. This picture is compatible with the thalamic axons being fasciculated on the corresponding preplate pioneer axons as proposed by Shatz et al (1990) and Blakemore and Molnár (1990). Pioneer axons would then provide the attractive surface over which thalamic axons could negotiate their way to the superplate cells in *reeler*, despite their having to pass through the non-permissive cortical plate layer. Presumably, thalamic fibers selectively fasciculate on the subplate fibers at the time of their first encounter, the 'handshake' (Blakemore and Molnár, 1990).

The Same Mechanism Might Apply for Thalamocortical Fibers Targeting the Marginal Zone in Normal Mice and the Superplate in *Reeler*

In normal animals the primordial plexiform zone splits into subplate and marginal zone. Even after this splitting of the preplate, some marginal zone cells can be backlabelled from the internal capsule (together with numerous subplate cells), indicating that they too initially maintain their projections through the thickening cortical plate, similarly to the superplate cells in the *reeler*. At present we have insufficient information about how long these sparse connections from marginal zone cells exist in the rat. McConnell et al (1989) reported no backlabelled marginal zone cells from DiI crystal placements into the embryonic internal capsule of the cat. In cat these connections might disappear earlier, or perhaps they are simply more difficult to detect.

From our own birthdating studies in the rat (Molnár et al, 1991) the rough estimate for the marginal zone/subplate cell ratio is 1:4 at P5 and 1:7 at P10 (^{3}H-thymidine injection at E12). This ratio in cat, as we estimated from Figs. 3 and 5 (from Luskin and Shatz, 1985b), is 1:50 at P0 and 1:25 at P78 (^{3}H-thymidine injection at E27, E28 respectively). Differential cell death or differential labelling could explain the difference in the labelled marginal zone/subplate cell ratio. The primordial plexiform zone is itself generated in an outside-first/inside-last fashion, therefore the degree of labelling in future marginal zone and subplate might depend critically on the time of the ^{3}H-thymidine injection (Shoukimas and Hinds, 1978; Raedler and Raedler, 1978; Rickmann et al, 1977; König and Marty, 1981; Bayer and Altman, 1990), which could then affect the ratio of labelled marginal zone/subplate cells. It would be interesting to study this ratio with other techniques at an earlier time, corresponding to the arrival of the thalamic axons, and to see whether these numbers correlate with:

1) The abundance of thalamic axons reaching layer 1 versus the subplate.
2) The number of marginal zone cells backlabelled from the internal capsule in the different species.

In any case, it certainly seems possible that the sparse early corticofugal projections from layer 1 and the small number of thalamic fibers reaching layer 1 at early normal stages depend on mechanisms similar to those operating for the whole of the superplate in *reeler*.

The intimate relation of thalamic axons to preplate axons as they approach the cortex may also explain why they come to rest in the subplate (or, for *reeler*, superplate) layer and probably form synapses on the cells there (Herrmann et al, 1994). If each thalamic axon does grow along a particular preplate axon it will eventually reach the cell body of that axon (subplate and marginal zones in normal or superplate in *reeler*) and may be stimulated to terminate there, since other cell surfaces in the cortical plate are not growth-permissive at that stage. This particular trajectory of approach may be an essential prerequisite for the formation of temporary synapses and hence the establishment of a waiting period. In *reeler* mice, the thalamic afferents arrive in the superplate at 14.5/15 and gather for a period of some days, at least until E18, just as they do in the subplate of normal mice. Thus a 'waiting period', in association with the appropriate target cells, occurs in both strains, even though those target cells lie in quite different locations in the two cases.

Identical Early Thalamocortical Development in Normal and *Reeler* Mice

We have argued (Blakemore and Molnár, 1990; Molnár and Blakemore, 1995a,b; Molnár et al, 1997a,b) that the topography of both the pioneering corticofugal and thalamofugal projections is laid down at least in part by autonomous mechanisms including *chronotopic* ordering of outgrowth patterns, and maintained by fiber-fiber interaction. Within each coronal segment, the earliest generated corticofugal projections take the most superficial position in the white matter and the later generated ones (from more dorsal regions) come to lie deeper in the white matter. This type of fiber ordering could be generated by the morphogenetic gradient (for review, see Horder and Martin, 1978), the fibers becoming ordered below the subplate layer on a first-come first-served basis. The thalamic fibers grow towards the developing cerebral cortical sheet in the direction along which the cortical morphogenetic gradient has its highest differential. All these mechanisms could be available and entirely normal in *reeler* since they do not depend on the subsequent organization of the cortical plate. In particular, the early outgrowth of axons from preplate cells is achieved before the arrival of the ectopically placed cortical plate. Moreover, the scaffold established by preplate axons, and the relationship of thalamic axons to it, are indistinguishable from normal.

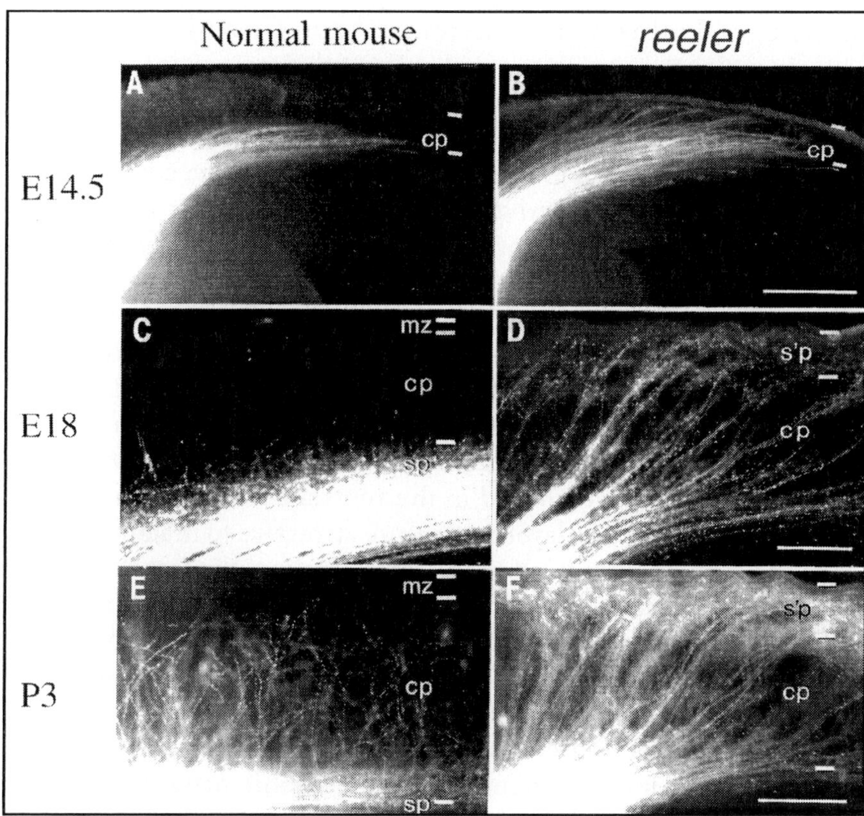

Fig. 7.7. Thalamocortical axons arrive, wait and invade in normal and *reeler* mice. These fluorescence micrographs of the lateral part of the occipital cortex (lateral is up; ventral to the left) show the sequence of events as thalamic axons approach, pause and enter the cortical plate in the normal mouse (left) and in the mutant *reeler* mouse (right), in which the cells of the preplate remain as the 'superplate' (s'p) (Caviness et al, 1988) above the arriving cells of the cortical plate (cp), rather than being split into marginal zone (mz) and subplate (sp). Thalamic axons have been labelled with a crystal of DiI placed in the dorsal thalamus. (A and B) At embryonic day 14.5 (E14.5), they are approaching the subplate in the normal mouse but are growing in oblique fascicles through the cortical plate in *reeler*. (C and D) At E18 (three days before birth), they are still concentrated densely in the subplate layer in the normal mouse but have paused in the superplate of the *reeler* mouse. Before birth in both phenotypes, they start to leave the waiting compartment and enter the cortical plate, turning up in the normal mouse and down in *reeler* (Molnár and Blakemore, 1992; Yuasa et al, 1994). (E and F) By postnatal day three (P3), most thalamic axons have terminated in layer 4 or its equivalent in the *reeler* mouse. Scale bars: 200 µm for A,B,E and F; 100 µm for C and D. Reproduced from Molnár and Blakemore (1995a) with kind permission of Elsevier Science Ltd.

Privileged Pathways for Thalamic Axon Growth

In *reeler*, the sequence of behavior of thalamic axons is entirely compatible with, and provides good evidence for, the theory of axon guidance developed in normal rodents (Blakemore and Molnár, 1990). Observations in vitro strongly suggest that the cortical plate does not become permissive for invasion of thalamic axons until several days after the time of arrival of those axons in vivo (Bolz, et al, 1992; Hübener et al, 1992, 1995; Götz et al, 1992; Molnár, 1994; Tuttle et al, 1995). This presumably explains, in part, why those axons do not extensively invade the cortical plate when they arrive but gather below it. However, at this early stage even the subplate does not offer an attractive environment for thalamic axon ingrowth in vitro, unassisted by the intimate relationship with subplate axons that appears to be established in the ventral telencephalon in vivo. In culture, the subplate layer is no more attractive to thalamic axons than the cortical plate itself in the rat. Therefore, the fact that fibers do enter and wait in the subplate layer in vivo, and even form temporary synapses on subplate cells (Friauf et al, 1990; Friauf and Shatz, 1991; Crair et al, 1993; Herrmann et al, 1994), strongly suggests that the particular, natural trajectory of approach of thalamic axons over the subplate axon scaffold is essential for this bond to be formed (rather than simply a chemotropic or trophic attractiveness of subplate or superplate cells, as previously proposed by Caviness et al, 1988, and Price, 1991). Our hypothesis does not, of course, exclude the possibility that there are cortical trophic effects on the thalamus, nor that they play some part in the process of guidance towards the cortex (Molnár and Blakemore, 1995a,b).

The subplate neurons are known to express immunoreactivity to the surface molecules L1 (Godfraind et al, 1988; Chung et al, 1991) and fibronectin (Stewart and Pearlman, 1987; Chun and Shatz, 1988a; Sheppard et al, 1991) which could provide highly attractive surfaces for the growth of thalamic axons (providing they are expressing appropriate adhesion molecules) in an otherwise relatively non-permissive environment. The selective expression of glycosaminoglycans (Derer and Nakanishi, 1983; Fukuda et al, 1997), chondroitin sulfate core proteins (Miller et al, 1992; Bicknese et al, 1994a) and specific peanut agglutinin lectin binding (Götz et al, 1992) has also been demonstrated in the subplate region and along fiber fascicles crossing the cortical plate in *reeler* (Bicknese et al, 1994b; Kurukulasuriya and Salinger, 1996) at corresponding ages. It is yet to be determined whether thalamic fibers require more than one of these extracellular cues (sequentially or simultaneously) or whether there is a specific requirement for only one or a subset of them. Fukuda et al (1997) used more sensitive visualization of neurocan in acid alcohol-fixed and paraffin-embedded sections and described that subplate and subplate projections, but not thalamic axons, express neurocan immunoreactivity in embryonic rat brain. It was previously thought that chondroitin sulfate core proteins are only expressed by subplate cells but not their projections (Bicknese et al, 1994a). Fukuda et al (1997) demonstrated that L1 immunoreactivity was specifically localized on the growing thalamic axon and using double immunostaining techniques they showed that L1-bearing thalamocortical axons were found to extend along neurocal-immunopositive regions. According to their study, L1-positive axons and neurocan immunoreactive fibers overlapped. Friedlander et al (1994) reported that neurocan can bind L1 in vitro. The heterotopic interaction between L1 positive thalamic projections and neurocan positive subplate cells might be important in thalamocortical pathfinding. It would be interesting to examine these questions in the *reeler* mutant mouse.

We have suggested (Blakemore and Molnár, 1990; Molnár, 1994; Molnár and Blakemore, 1995a,b) that thalamic axons fasciculate on their subplate axon counterparts, and hence are provided with precise pathways by which they reach the somata of the appropriate neurons (subplate cells in the normal animal) to form temporary synapses. The existence of these privileged pathways for axon growth could explain how thalamic axons in *reeler* are able to penetrate the cortical plate and stream straight up to reach the equivalent cells in the superplate, while ignoring the hostile territory of cortical plate cells around them. In the *reeler* mutant, various adhesion molecules (L1, NCAM, and fibronectin) and glycosaminoglycans were found to be *similarly* expressed by early-born cortical neurons, despite their different dispositions (Godfraind et al, 1988; Derer and Nakanishi, 1983; Bicknese et al, 1994a). There is an example for selective fasciculation between axons growing into opposite directions, 'reverse pioneer fibers', in the grasshopper embryo during the development of its longitudinal pathways (Bastiani et al, 1986). The necessity for the guiding of the anterior growing axons was demonstrated by laser ablating the posterior-growing axons. Shatz and her colleagues (Ghosh et al, 1990; Ghosh and Shatz, 1992b, 1993) have attempted to test the hypothesis that subplate neurons are crucial for the thalamic innervation of the cortex by making kainic acid lesions of the subplate in cats. However, since the earliest successful lesion was done a week *after* the arrival of thalamic axons and, given the possibility that the chemically lesioned area might undergo pathological changes other than the elimination of subplate neurons, the results are not easy to interpret. It would be interesting to perform similar experiments before the arrival of thalamic axons and to do them in *reeler* where the thalamic fibers could approach the cortical region without travelling through a damaged region of destroyed superplate cells.

Conclusions

The development of thalamocortical connectivity in *reeler* supports the hypothesis that the preplate scaffold serves a guiding role for thalamic axons. The similarity of appearance of the fascicles of preplate axons, running obliquely through the thickening cortical plate in the *reeler*, and of the subsequent bundles of thalamic axons passing upwards through it again strongly suggests that the thalamic fibers grow in intimate contact with the early corticofugal scaffold. Caviness et al (1988) suggested that the oblique fascicles of axons observed crossing the thickness of the developing cortical plate in *reeler* (labelled with the rapid Golgi technique) might consist only of thalamic fibers, but when similar preparations were stained with neurofilament immunohistochemistry, the same bundles appeared larger. Caviness et al (1988) therefore suggested that the fascicles, which they called "plexiform tributaries", include the fibers of superplate cells, running together with the thalamic fibers.

At present it is rather difficult to selectively and exclusively label both thalamocortical and early corticofugal fiber populations at the same time so close to the cortical surface. Nevertheless, over a brief period of development, this is possible. One can label the two sets of fibers with different carbocyanine dyes when the two fiber systems have already met but the thalamic fibers have not reached the cortex (over a narrow time frame, around E14.5-15.5 for occipital cortex in the rat). 3-D confocal microscopic reconstructions obtained from rat brains labelled with different carbocyanine dyes from the cortex and thalamus at this stage indeed provide evidence for fiber-fiber interaction (Molnár et al, 1993a, 1997a,b). The selective fasciculation of growing axons on pioneer axon scaffolds appears to operate, at least in this situation, in vertebrates as it does in insects (Goodman et al, 1984).

Chapter 8

The Autonomous Nature of Early Thalamocortical Organization and the Subsequent Distortions of Thalamocortical Relations

Initial Thalamocortical Organization Develops Prior to Sensory Influence

In chapter 4, I reviewed anterograde and retrograde tracing experiments in the rat which demonstrated that by E16 thalamic axons have arrived under the entire cortex in a topographically organized fashion. Back-labelling from various points on the cortex at this early stage reveals the spatially organized deployment of these thalamic fibers. The discrete fiber bundles, containing both corticofugal and thalamofugal axons, do not cross each other, and they maintain their neighborhood relationships along the entire path between thalamus and cortex. Unfortunately it is rather difficult to draw firm conclusions about the organization of individual fibers within each bundle. Nevertheless it is clear that thalamic fibers arrive in the subplate layer and wait under the cortical plate in a topographically organized fashion. The gross matching between the cortical surface and the thalamic volume is established in a rather simple fashion at embryonic ages (E16-E19 in the rat), before the thalamocortical fibers invade the cortex (Fig. 4.6-4.8; Blakemore and Molnár, 1990; Molnár and Blakemore, 1995a,b; Juliano et al, 1996). Multiple carbocyanine dye placements revealed that more anterior cortical regions are connected with more medial thalamic slabs, whereas more posterior cortical regions are connected to more lateral slab-shaped regions. The two backlabelled cell groups revealed by any two implantation

Development of Thalamocortical Connections,
by Zoltán Molnár. © 1998 Springer-Verlag and R.G. Landes Company.

sites along a coronal line seem to form a continuous slab within the thalamus. These experiments reveal that adjacent cortical fields receive their projections from adjacent groups of thalamic cells, although the arrangement of corresponding regions on the surface of the cortex and within the volume of the thalamus is rather different. The interlinking fibers (within the accuracy of the technique) preserve their neighborhood relationships along the whole length of the pathway. As thalamic fibers arrive below the cortical plate the general ordering of thalamocortical connections is established. This general mapping of the thalamocortical interconnections requires no reorganization of fibers between thalamus and cortex.

In the rat all this topography is accomplished as early as E16, before the arrival of the peripheral afferent input to the thalamus. For example, the first axons of the optic tract reach the level of the LGN at around E16, but they seem to be destined for the superior colliculus (Lund and Bunt, 1976). There is general agreement that optic tract fibers do not invade the LGN substantially until about birth in the rat (Karlsson, 1966). Similarly, in the somatosensory system, although the first trigeminal ganglion cell processes approach the brainstem at E13 in an orderly fashion, this process precedes the emergence of vibrissa rows in the periphery and the differentiation of brainstem trigeminal nuclei (Erzurumlu and Jhaveri, 1992). This suggests that the brainstem cannot relay peripheral information to the somatosensory thalamus for some time after the construction of the initial thalamocortical links (E16). The implication is that development of the early pattern of connection between thalamus and cortex is not dependent on afferent input (from sensory pathways or other subcortical structures).

The idea that the periphery does not orchestrate the *initial* layout of thalamocortical projections is not new. Cogeshall (1964) noted that "obvious fascicles of fibers can be seen joining the dorsal thalamus to the telencephalon via the internal capsule" by at least the 15th day of gestation in the rat. By contrast, ascending afferent pathways could not be traced into the dorsal thalamus until rather later. Thus, he felt he could correlate the onset of nuclear differentiation in the dorsal thalamus with the arrival of subcortical afferents: "This transforms the embryonic dorsal thalamus, a *unit that is forming connections with the telencephalon alone*, into the adult dorsal thalamus, a mosaic of units some of which connect the telencephalon synaptically with lower centres ... It seems likely that optic, tactile, and other incoming pathways plug into circuits that are, to some extent, already formed". Before the advent of fluorescent carbocyanine tracers these ideas of Cogeshall were rejected, since the general opinion was that subcortical afferents arrive in the thalamus before connections are made to the cortex (Jones, 1985). More recent results (reviewed in chapter 4) support Cogeshall's observations and hypothesis.

The notion of the autonomy of thalamocortical development is supported by the observation that anophthalmic mice (Cullen et al, 1976; Godement et al, 1979; Kaiserman-Abramoff et al, 1980), as well as bilaterally enucleated hamsters (Rhodes and Fish, 1983) and ferrets (Guillery et al, 1985), develop geniculocortical and corticogeniculate interconnections with an essentially normal topography (within the precision of tracing methods). If a monkey is binocularly enucleated at an early fetal stage, before full innervation of the LGN by the optic tract, about two-thirds of LGN cells degenerate and the total size of area 17 is proportionately reduced (Rakic, 1988). This experiment is commonly cited as evidence that cortical areas develop without thalamic afferents. Unfortunately the study documents only the adult stage. It remains to be established whether the cell loss in the LGN occurs so rapidly that the whole topography of the *initial* outgrowth of thalamocortical axons is affected (unlikely, but cannot be excluded at present), or whether the projection forms normally and is subsequently reduced in areal extent as geniculate cells die. Without this information no inferences can be made of the causal relationships.

Guillery et al (1985) suggested that the "mapped projection that grows from the retina to the lateral geniculate nucleus must in normal development conform with the geniculocortical map in order to produce the normal pattern of retinogeniculocortical projections of the adult". Guillery (1986) also considered the possibility that *early* in development, the geniculocortical and retinogeniculate systems are *initially* "two independent processes, each programmed to produce its portion of the final visual pathway". The results demonstrating the early topography at the time of thalamocortical fiber arrival at the subplate (Molnár and Blakemore, 1990a,b, 1995a,b; Catalano et al, 1996) strongly support this suggestion. The establishment of the early topographically organized geniculocortical projection, around the 16th day of gestation in the rat, cannot be dependent on the retinal input because the optic tract has not grown into the LGN at this time. This hypothesis of course does not exclude the possibility of a *later* modification. There are numerous examples where the topography between the thalamus and cortex cannot be simply explained with a simple twist of the orderly interconnections. In some areas, thalamocortical fibers have to cross. In this and the following chapters I shall argue that this modification is most likely secondary and occurs after the initial simple layout of the thalamocortical projections near the termination site. These reversals are very general in the cortex, some occur naturally during ontogeny and some could be induced experimentally. These rearrangements near the termination site, however, do not challange the simple rules developed for the initial layout of the connections.

Early Gross Specification of Thalamocortical Projections

Evidence for the early gross specification of thalamocortical topography and for the autonomy of early thalamocortical development also comes from the results of developmental manipulations that induce sensory afferents of one modality to project to central targets of a different modality (Frost and Métin, 1985; Sur et al, 1990; Pallas and Sur, 1993). In spite of such manipulations (including cortical lesions), there is very little change in the distribution of thalamocortical pathways, suggesting that these have already been specified at the time of these manipulations. After a lesion of the visual cortex, superior colliculus and LGN of a newborn ferret (corresponding in development to an E16 rat fetus) fibers of the optic tract terminate in the 'auditory' thalamus and visual information is relayed to the region of temporal cortex (the original target of the 'auditory' thalamus) that would normally become primary auditory cortex. Similarly, in the naturally blind mole rat during intermodular compensation the auditory afferents project to the LGN and, through the otherwise normal geniculocortical projection, this auditory input is relayed to what would otherwise be the visual cortex (Bronchti et al, 1989, 1991; Heil et al, 1991). In each case, the matching of a given thalamic area to its original cortical region is unchanged, indicating that some aspects of early thalamocortical development is independent of the periphery, and are based on some kind of innate automatism.

The Origin of Fiber Order: Chronotopy and Morphogenetics

In most nerve tracts the fibers are, at least to some extent, ordered with respect to both their origin and their target: the location of any fiber in the tract correlates with the relative location of its cell of origin and the location of its target (see Horder and Martin, 1978; Walsh and Guillery, 1985). One of the best known examples of fiber ordering is the ascending (*funiculus gracilis* and *funiculus cuneatus*) pathways in the dorsal columns of the spinal cord where most of the evidence for the preserved fiber ordering has derived from clinical studies (for example, see Szentágothai, 1977). Because of its importance in the differential diagnosis of spinal injury, every medical student has to know that the ascending fibers are grouped according to their origin and that their neighborhood relationships are maintained along their ascent. Fibers from the cervical regions run lateral to those from the thoracic segments, and the thoracics run lateral to those from the lumbar region, etc.

Topographically ordered projections are much more common than those that show no such ordering (for numerous examples, see Martin and Perry, 1983). Indeed topographic order may be even more widespread than it superficially appears to be, if one separately considers the independent fiber populations that share a single pathway.

The degree of fiber order in the embryonic cerebral cortex has not been studied systematically: it could be that tracts that appear relatively disorganized in the adult are much more regularly ordered at early stages and only get distorted near to their termination sites. If one is interested in the role of early fiber ordering in establishing the topography of projections then one must examine them during their establishment.

How Is the Initial Fiber Order Achieved?

Horder and Martin (1978) and Bayer and Altman (1987) pointed out that many of the major anatomical interconnections of the telencephalon can be related to the order of neurogenesis and to neurogenetic gradients of either the source neurons (entorhinal cortex and olfactory bulb mitral cells), receiving neurons (lateral septal nucleus, cortical nuclei of amygdala, anterior olfactory nucleus and primary olfactory cortex) or both (magnocellular basal nuclei and striatum) (Bayer, 1980, 1985, 1986; Bayer and Altman, 1987, 1991). The prerequisite for the translation of a temporal gradient into topographically ordered connectivity is that the outgoing fibers are organized and preserve their position as they reach their targets and occupy a position on a first-come first-served basis. However, this has not been established for most of these systems. The possibility that fiber ordering is dependent on the relative timing of their establishment, i.e. the translation of temporal events into spatial distributions, is termed *chronotopy*. Let us return to our original example in the spinal cord (ascending fibers of the dorsal columns) for illustration. The development of the spinal cord follows a rostrocaudal gradient. The first established cervical fibers run deeper in the funiculus posterior (*funiculus cuneatus* and *gracilis*) than the later developing thoracal and lumbal fibers. The fibers originating from the sacral segment take the most superficial position. This arrangement is called Kahler's Rule (see Szentágothai, 1977). The *tractus spinothalamicus* has a very similar arrangement. This type of fiber arrangement reflects the timing of its establishment and has chronotopic fiber ordering.

I would like to propose that the arrangement of descending (subplate) and possibly of ascending (thalamic) axons in the fetus is determined, at least in part, by the *chronology* of their outgrowth. The idea is that a spatiotemporal wave of cell generation and differentiation occurs across corresponding parts of the cortex and thalamus, and that this wave is converted into two arrays of axons.

Our own study clearly demonstrates that both the subplate projection and the thalamic outgrowth form ordered arrays (Molnár, 1994; Molnár and Blakemore, 1990a,b, 1995a,b). This is revealed by the experiments in which several different dye crystals were placed in parasagittal or coronal rows along one hemisphere, showing that the labelled bundles containing both corticofugal and thalamocortical fibers are topographically organized. The *timing of outgrowth* of the pioneering subplate projection from different cortical regions follows the overall pattern of maturation of the cerebral hemispheres (see Fig. 3.4). Fibers leave the subplate of the rostral cortex 1.5-2 days before they grow out of the occipital cortex. They all stream towards the primitive internal capsule, creating a fan of axon bundles, preserving the front-to-back lateral to medial topography of their cortical origins. Thus the spatial arrangement of axons is partly determined by the timing of their outgrowth.

However, time is unidimensional and the cortical surface is essentially two-dimensional: clearly chronotopy cannot account for everything. While the arrangement of the subplate scaffold along the anteroposterior axis certainly correlates with the time of outgrowth of the axons, I believe that chronotopy is much more important for the organization of the *mediolateral* axis of the subplate scaffold (and of the thalamocortical axon array within the telencephalon). Subplate fibers from regions at different anteroposterior levels do not share the same segment of the intermediate zone and therefore inevitably stay distinctly separate from each other if they simply grow towards the primitive internal capsule. But fibers from all points along any strip of cortex radiating out from the primitive internal capsule (coronal strips in the middle of the hemisphere) share the shallow depth of the intermediate zone and therefore have ample opportunity to mix and lose their order. This seems to be prevented by strict arrangement in the white matter according to the time of outgrowth of the axons. The one-dimensional time axis essentially determines the mapping of the mediolateral axis of the axon array. At any coronal level, axons first stream out from the more ventral and lateral parts, and fibers are subsequently extruded into the intermediate zone from progressively more medial cortical locations. The results of multiple labelling with dye crystals placed in mediolateral rows clearly establish that the later generated subplate axons, from more dorsal and medial cortex, lie *deeper* in the white matter. Thus, the depth profile of subplate fibers in the intermediate zone of the ventral portion of the hemisphere is a quite precise reflection of their time of outgrowth, which itself depends on the mediolateral positions of their cells of origin across the cortex. Woodward et al (1990) reported that in the adult rat, fibers arising from lateral cortical neurons are superficial to those from more medially located cells. My own observations on the distribution of pyramidal cell axons of layer 6 and 5 in the white matter, labelled from the internal capsule or cerebral peduncle after E16, also showed remarkable fiber ordering, following the same lateral-superficial/medial-deep sequence (see Fig. 8.1). Chronotopic ordering of corticofugal fibers in the white matter might be a general principle.

The array of thalamic axons is also influenced and partly determined by the timing of their outgrowth. In particular, thalamic fibers destined for more ventral areas of cortex leave the thalamus before those that reach more dorsal parts at the same coronal level. This is reflected in their times of arrival under the cortical plate (ventral anterior areas first) along trajectories parallel to the highest

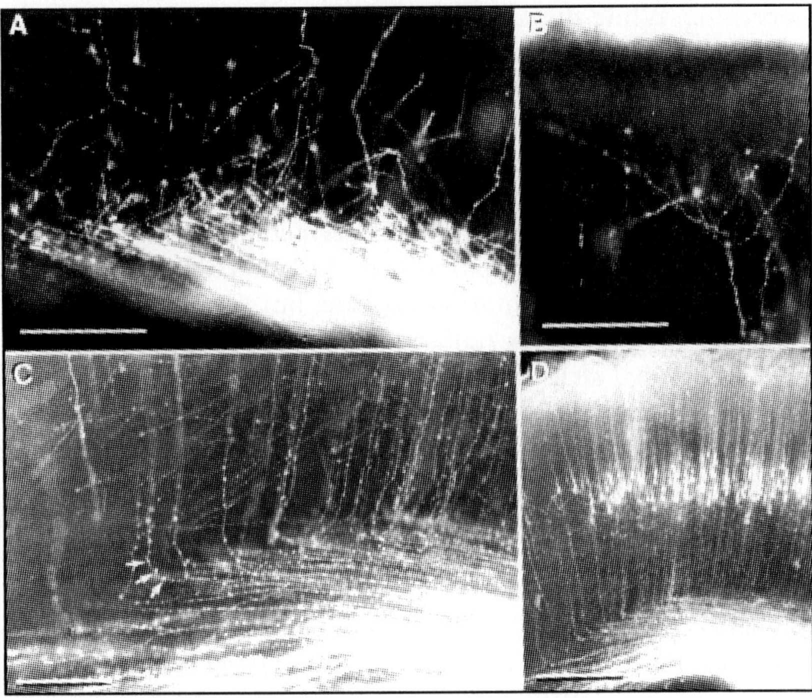

Fig. 8.1. Photomicrographs of the occipital cortex of a P2 rat in which a DiI crystal had been placed in the dorsal thalamus (A, B) or the cerebral peduncle (C, D) to illustrate possible chronotopic fiber arrangements of thalamocortical and layer 5 projections in white matter. A: The thalamic fibers are parallel to each other until they enter the cortex. There they follow more irregular patterns as they ascend to the middle layers, where they arborize. B: The photomicrograph was taken from the same section as A, but focusing on a single arbor extending over a large area, some 200 µm below the pial surface. C and D: Inserting a single crystal of DiI into the cerebral peduncle leads to the labelling of layer 5 pyramidal cells (D). Their axons are highly ordered in the white matter (C). In both the thalamocortical and the corticospinal projections, the fibers are so arranged that the fibers belonging to ventral areas run superficial to fibers destined for, or originating from, more dorsal areas. Since the earlier fibers enter the more ventral cortical areas first, and since these areas develop relatively earlier, this type of fiber ordering reflects the timing of its establishment. There is *chronotopic* fiber ordering within the white matter. By following individual axons of the labelled layer 5 cells within the white matter and in layer 6, varicosities of the axons could be observed, which might correspond to en passant synapses on subplate and to layer 6 cells. By electron microscopy Shering and Lowenstein (1994) recently demonstrated in the cat that axons of layer 5 and 6 neurons form asymmetrical synapses on subplate cells. The varicosities seen in the photomicrograph in C (arrowheads) suggest a similar arrangement in the rat. Scale bars: A,B: 200 µm; C: 100 µm; D: 500 µm.

differential of the neurogenetic and morphogenetic gradient (see chapter 3; Catalano et al, 1991, 1996; Molnár and Blakemore, 1990, 1995a,b). This arrangement of thalamocortical trajectories remains preserved in the adult (Caviness and Frost, 1980; Frost and Caviness, 1980).

I proposed (Molnár, 1994, 1995a) that both thalamocortical and early corticothalamic subplate fibers are generated in a precise spatiotemporal sequence and they simply maintain their fiber-fiber relationships from thalamus to subplate and from subplate to near the region of the thalamic reticular nucleus respectively. Both projections then accumulate and change their topography, the thalamocortical fibers doing so in subplate and the corticothalamic fibers near the thalamic reticular nucleus. But within most parts of the pathway, the timing of outgrowth will certainly determine which particular fibers reach the 'zone of interaction' below the corpus striatum at any particular time. If the timing of initial outgrowth from thalamus and cortex is roughly synchronized and the distance from each to the zone of interaction is roughly the same, then when the first thalamic fibers reach the interaction zone, they presumably meet only the first-formed subplate axons (from rostromedial cortex). Then, over the following 1.5 days or so (in the rat), successively generated thalamic fibers might reach the interaction zone just as their cortical subplate counterparts arrive. This sequential interaction of arriving fibers, on a first-come-first-served basis, might reduce the possibility of aberrant interaction between the two arrays.

Contact Guidance (Attraction and Repulsion): The Morphology of Subplate Growth Cones

O'Leary et al (1990) in the rat, and McConnell et al (1989) and Kim et al (1991) in the cat and the ferret, suggested that the growth cones of subplate axons growing down in the nascent intermediate zone are unusually large (as much as 10 μm across), with complex morphology (having excessive numbers of florid filopodia). As the scaffold extends close to the primitive internal capsule, the growth cones become smaller and their morphology much simpler. Mason and Godement (1992), using time-lapse techniques to observe optic axons in the optic chiasm, suggested that the morphology of the growth cones (which were also large, with many filopodia) could reflect a stage of 'decision-making' (i.e. route selection) by the growing axons. Brown (1984), examining the formation of the neuromuscular junction, observed similar phenomena. As the developing axons of motor neurons grow rapidly, the structure of their growth cones is simple, but, at choice points, their morphology changes, gaining a characteristic 'exploratory' appearance. In an in vitro assay, Stuermer (1992) observed the growth of axons into a 'dead-end' stripe consisting of a growth-permissive surface surrounded by a growth-inhibiting membrane preparation. When the axons reached the end of the stripe, time-lapse recording showed spectacular changes in the morphology of their growth cones. Numerous filopodia were extended over the entire distal 200-300 μm of the fiber, before the axon pulled back or died.

This all implies that subplate axons are in an 'exploratory' mode of growth as they pioneer the pathway through the intermediate zone. They are presumably using some form of 'contact guidance' to determine their trajectories, i.e. they are influenced in their patterns of growth by the molecular (and/or mechanical) characteristics of the surfaces with which they come into contact. It is interesting to consider what surfaces are available at this stage with which the developing axons could interact. One obvious candidate is the pre-existing processes of other cells. Mitrofanis (1992a) recently described a transient cell population in the internal capsule, in the gateway to the thalamic nuclei. These so called perireticular cells are in a strategic position; they could potentially interact with both corticofugal and thalamocortical projections; they reach the thalamus very early (Métin and Godement, 1996; Cordery et al, 1996, 1997) but they do not seem to develop projections to the cortex in large numbers until E16 (Adams N. et al, 1993; Cordery et al, 1996, 1997). Recently Ramcharan and Guillery (1997) described that the dendrites of the transient perireticular cells extend their dendrites across the path of cortical and thalamic fibers in the internal capsule at E17 and early membrane contacts might indicate their involvement in guidance between E14 and E19, when both thalamocortical and corticofugal projections pass through the region. We are far from understanding the significance of these interactions. Radial glial cells, which are widely considered to be primarily responsible for neuronal migration, are generally distributed orthogonally to the routes taken by subplate axons. They might also influence their growth, with their roles in neuronal migration and axonal pathfinding possibly depending on entirely *different* surface molecules, thus avoiding 'crosstalk' between the two functions. Recently Richards et al (1997) reported that axon outgrowth from explants of E15 rat neocortex was directed towards the cocultured internal capsule in collagen gels, suggesting that the internal capsule releases a chemoattractant for cortical axons.

Possible Inhibitory Factors in and Around the Internal Capsule

Not all influences on growing axons are attractive. Some molecules (membrane bound or soluble) inhibit or discourage axon growth (Caroni and Schwab, 1988a,b; Schwab and Caroni, 1988; Pini, 1993) and it may be that the pioneering subplate axons are constrained within the routes that they take by the existence of hostile surfaces outside those routes. The behavior of the advancing front of the axon bundle as it reaches the ventral telencephalon (late on E14 for the occipital cortex) is interesting in this respect. The leading axons diverge over a much wider territory than that occupied by the bundle behind them. Could these very first axons be searching laterally and finding hostile characteristics that force them back into the correct, tight trajectory? Recently Skaliora et al (1996) described an interesting expression of semaphorin G in the ganglionic eminence at E15 in the rat. The pattern seems to correspond to the region of the perireticular nucleus (Mitrofanis, 1992a). It is conceivable that the internal capsule is a hostile environment for axon growth. I have proposed this based on the fasciculation pattern of thalamic axons in the region (Molnár, 1994; see Fig. 4.2).

Soluble chemosupressant molecules, in addition to the above mentioned surface bound molecules, might also be involved in the guidance of these early connections. Richards et al (1997) recently observed that dorsal telencephalon suppresses the outgrowth from E15 rat neocortical explants in collagen gels and suggested that the dorsal telencephalon releases a soluble chemosupressant which in addition to the soluble chemoattractant released by the internal capsule might direct all initial corticofugal projections towards the internal capsule.

On the other hand, one has to keep in mind the fact that the extracellular space surrounding the route taken by one group of subplate axons already is (on the superficial side) or will soon be (on the deep side) occupied by subplate axons from neighboring regions of cortex. Thus, if growth-inhibiting factors do exist outside the preferred route of any subplate bundle, either they must be specific for each particular bundle (unlikely, because it would demand a myriad of different molecules to regulate every bundle), or they could exist on the surfaces of those neighboring axons. Perhaps the first descending axons diverge from the advancing bundle until they contact neighboring subplate axons. One principle underlying the maintenance of the ordered array could simply be that subplate axons tend to deposit themselves on pre-existing axons but cannot cross them.

Expression of Cell Adhesion Molecules in the Embryonic Telencephalon

Godfraind et al (1988) and Chung et al (1991) have demonstrated complex and rapidly changing patterns of different antigen expression in the embryonic mouse cortex (from E12 on). Many of the observed antigens were associated with the subplate or subplate fibers. For instance, L1 was specifically expressed along the path of the first corticofugal projections. Chun and Shatz (1988a) described fibronectin immunoreactivity in cat subplate, demonstrating that the subplate might play a role in shaping the extracellular matrix. The subplate might establish a sheet or a three-dimensional compartment which is preferred only by ingrowing thalamic fibers, but not by cortical plate efferents or callosal axons. There is some evidence that specific cortical pathways can be marked with different molecules. Two condroitin sulphate proteoglycan core proteins are apparently restricted to the afferent thalamocortical pathway and subplate but not the corticothalamic pathway (Bicknese et al, 1991; Sheppard et al, 1991; Miller et al, 1992). In the *reeler* mutant mouse, chondroitin sulfate positive streaks cross the cortex (Bicknese et al, 1994b). Henke-Fahle et al (1996) recently described a carbohydrate epitope of a glycoprotein which is selectively expressed in thalamocortical axons but not in the corticofugal pathway. In vitro assays suggest that the antibody against this carbohydrate epitope (mAb 10) has opposite effects on the growth of thalamic and cortical axons on postnatal cortical membranes. It reduces the growth rate of thalamic, but increases the growth rate of cortical, axons.

The various fiber tracts traversing the intermediate zone are segregated from each other (radial glia, callosum, thalamic fibers), so each might conceivably have its own preferred growth-promoting surface expressed within the intermediate zone. Alternatively, since they are laid down successively, each underneath its predecessor, they might grow preferentially on the surfaces of the pre-existing fiber bundles, which might, at that stage, be expressing growth-promoting surface molecules appropriate for the next set of axons. Obviously the whole system might operate with few classes of surface molecules along the organized trajectories of prelaid fibers and radial glia. Contact guidance need not necessarily be limited to mechanical factors but might involve a range of different cell-cell interactions. The orderliness of the fiber paths may indicate the existence of either: 1) precise chemoaffinity cues which are followed by the growing fibers along the entire pathway; or 2) local cues at each point along the route. These two mechanisms are not mutually exclusive—the first fiber path might be guided entirely by chemoaffinity cues along the extracellular pathway (e.g. chemoattractants emanating from the internal capsule and chemosuppressants emanating from medial cortex for corticofugal fibers, whereas thalamocortical projections might be attracted by cues from the cortex (Métin and Godement, 1996; Richards et al, 1997; Molnár and Blakemore, 1991; Rennie et al, 1994). The rest might be 'blindly' led by contact guidance over the surface of the pioneering fibers. This two-phase mechanism has been suggested to explain the development of the retinotectal system (Horder and Martin, 1978) and the interhemispheric projections (Hankin and Silver, 1986).

Chronotopy in Other White-Matter Systems

From examination of the *corticospinal* projection from layer 5 it is apparent that its fibers are also arranged chronotopically, the fibers originating from ventral areas running superficial to the fibers from more dorsal areas. The more recently outgrowing fibers have to undercut and run deeper than the older ones.

Hankin and Silver (1986), studying the chronology of *callosal* development (in the rat), showed that callosal fibers are generated in a spatiotemporal sequence, those from more medial cortical areas (cingulum) crossing first, with those from more lateral areas reaching the midline later, coming to lie deeper in the callosum. The initial gross matching of the two hemispheres on to each other might be based on the timing of the establishment of these connections. It is interesting to note that, in the rat, subplate axons pioneer the interhemispheric route, but only from the cingulate cortex (Koester and O'Leary, 1991). These connections develop slightly later than the ones growing towards the internal capsule because of the neurogenetic gradient (see chapter 3) and a soluble chemosuppressant released from the medial cortex at early stages (Richards et al, 1997). The initial layout of the callosal fibers is very clear in embryonic life, but their distribution across the hemisphere is substantially altered with subsequent development. Interestingly the order of fibers in the callosum itself preserves the simple juvenile pattern (i.e. the original chronotopic order) as demonstrated by Nakamura and Kanaseki (1989) in the adult cat. It is interesting to speculate whether the rule that earlier fibers lie in a superficial position to the later ones in the white matter can be extended to the relationship between different fiber systems. Table 8.1 gives a summary of the timing of the establishment of the different fiber systems in rat. Callosal projections (of the very same cortical segment) lie under the thalamocortical ones in the white matter. It has previously been shown in the adult rat (Woodward and Coull, 1984) and cat (Nelson and LeVay, 1985), that geniculocortical fibers are localized most superficially in the white matter, whereas later developing corticofugal pathways occupy successively deeper strata. These studies were performed in the adult and do not reveal the earliest corticofugal projections from the subplate. Nevertheless they seem to support the notion that the earliest generated fibers occupy a more superficial position (closer to the cortical plate) than the later developing ones.

Observations of the arrangement of developing fiber bundles of different origin in the cortical white matter suggest general principles operating during the development of the extrinsic connectivity of the cortex: early connections maintain their neighborhood relationships during their growth; they show gross ordering from the time of their establishment, and they are chronotopically arranged. Individual fibers of each system and different fiber populations appear to occupy positions in the cortical white matter according to their time of establishment. During this developmental sequence, the chronology of the appearance of the different tracts is preserved in the ordering of the fibers in the adult white matter. In other words, *for two different (thalamo-cortical, subplate, corticospinal or callosal) fiber systems, the first developed will usually occupy a more superficial position under the grey matter.*

Subsequent Distortions of Thalamocortical Relations

So far, I have argued only that topographic ordering in the early thalamocortical projection establishes the *gross* distribution of input to cortical areas as early as the beginning of the waiting period

Table 8.1 Timing of the establishment of cortical connections in the white matter

E14-15	subplate projections reach the internal capsule	
		De Carlos O'Leary, 1992
		Erzurumlu and Jhavery, 1992
		Molnár and Blakemore, 1990a
E15.5-16	thalamic fibers reach cortex	
		Catalano et al, 1991
		De Carlos and O'Leary, 1992
		Molnár and Blakemore, 1990a
E16	corticothalamic fibers reach TRN	
		Molnár, 1994
		(see chapter 6)
E19	corticospinal fibers reach cerebral peduncle	
		Mitrofanis and Guillery, 1993
		Molnár, 1994
E19-20	callosal projections reach contralateral hemisphere	
		O'Leary et al, 1990

(chapter 4). This might even determine most of the general areal mapping of the thalamocortical interconnections (Scannell et al, 1997). I have argued that this initial map requires no fiber reorganization between the thalamus and cortex; a simple twist of the fiber bundles can explain the relationships between the two representations (see Figs. 4.6-4.8). However, it is tempting to extend this argument to the proposal that the ordering also creates the internal maps that are such a vivid feature of sensory and motor areas of the cortex. This attractive notion, that cortical mapping is initially determined by simple ordered projection from thalamus to cortex, has been challenged on the basis of the topography of the visual field representations in LGN and striate cortex. Conolly and Van Essen (1984), analyzing the visual field representation in the dLGN and the visual cortex of the macaque monkey, found that the map is mediolaterally but not anteroposteriorly reversed between LGN and cortex. Since this map transformation cannot be explained with a single twist, they concluded that the geniculocortical fibers in the white matter must reverse positions along one major axis but not the orthogonal one.

Nelson and LeVay (1985), using pairs of tracer injections to reveal fiber order in the adult cat, did indeed report crossing of thalamic

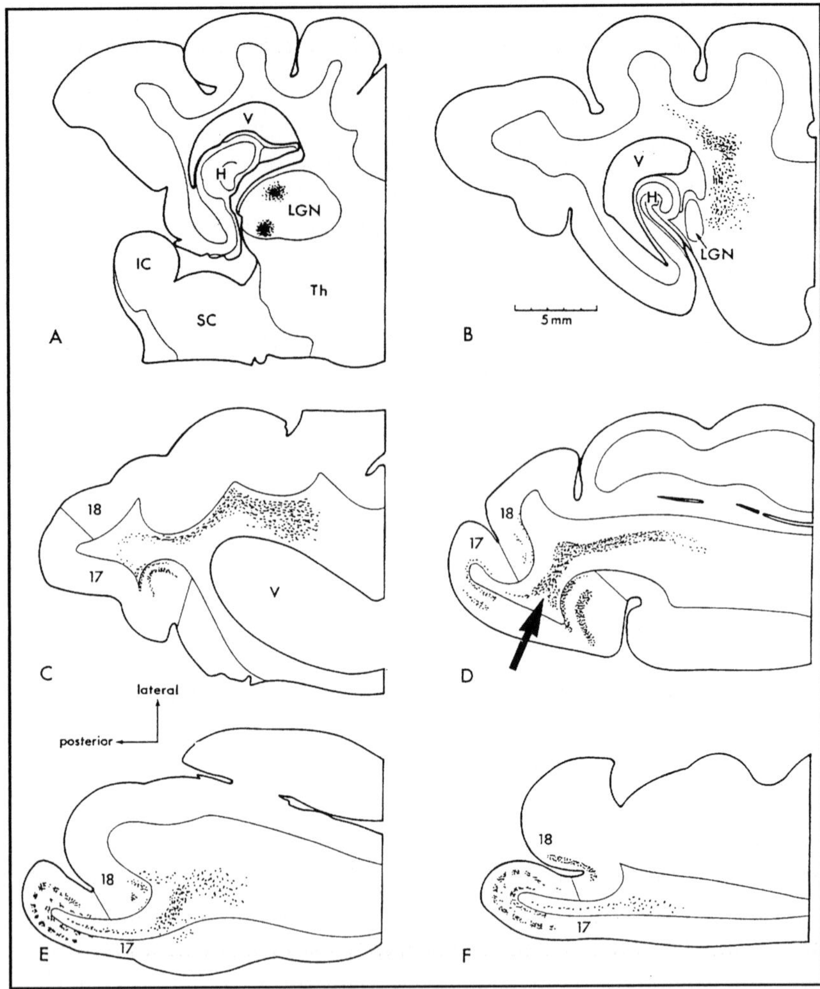

Fig. 8.2. Reconstruction of optic fibers labelled from a mediolateral pair of injections of peroxidase-conjugated wheat-germ agglutinin (WGA)(red) and [^{3}H]proline (green) into the lateral geniculate nucleus. Horizontal sections oriented so that lateral is up, posterior is to the left, and presented ventral to dorsal (A-F). The crossing of the two sets of fibers occurs shortly before reaching area 17, shown in D (arrow). Optic fibers showed no such reversal when labelled from anteroposterior pairs of injections. Figure reproduced from Nelson and LeVay (1985) with kind permission of John Wiley & Sons, Inc, New York. See color figure in insert.

axons in the mediolateral but not in the rostrocaudal axis of the optic radiation. This anisotropic decussation occurred, however, only a short distance (2-500 µm) below the visual cortex, at a depth that might well have been in the transient subplate during development.

For instance, in albino mutants and in monocularly enucleated animals, abnormal retinogeniculate input can cause secondary changes in the geniculate and in the optic radiation (for reviews, see Guillery, 1986; Trevelyan and Thompson, 1992; chapter 10). It would be interesting to follow the initial phases of development in such animals. From what stage and how will the periphery influence thalamocortical topography? Does the initial layout differ in these animals? Do these reversals require early cortical activation patterns? At present we know very little about the nature and mechanisms of these secondary changes.

Perhaps the decussation pattern seen in the adult represents a modification of the array of thalamocortical fibers that occurs during or at the end of the waiting period (Molnár, 1994, 1995b). Perhaps the transient side branches in the subplate described by Naegele et al (1988) and Ghosh and Shatz (1992) form the anatomical substrate for this reversal. It would be interesting to follow up this question by examining the early ordering of thalamic fibers in the occipital cortex of normal, albino and early monocularly enucleated animals (see chapter 10).

Recently Adams et al (1997) proposed that the thalamocortical fiber rearrangements which require fiber crossing are very common in the cortex. Sensory maps of the primary and secondary visual (e.g. Allman and Kaas, 1971) and somatosensory areas (for example, see McCasland and Woolsey, 1988; Catania and Kaas, 1995) are reversed to each other. This implies that only one of these maps can be established with a simple transformation, while the other requires the rearrangement and crossing of the thalamocortical connectivity at some point along the pathway. Interestingly, the secondary areas seem to correspond to the initial overall layout of thalamocortical connectivity. At present we do not know the development of these mirror reversals (see chapter 11).

I suspect that these transformations and map reversals might even operate within the same cortical field (Molnár, 1994, 1995b). The body surface is represented as a sensory homunculus in the primary somatosensory cortex. In humans this map is fragmented in an interesting way, differently from the motor homonculus (see Szentágothai, 1977). It would be interesting to examine the fiber ordering all the way from the somatosensory thalamus, as Nelson and LeVay (1985) did for the visual projection. The initial layout of thalamic axons might be very clear and regular in embryonic life, but

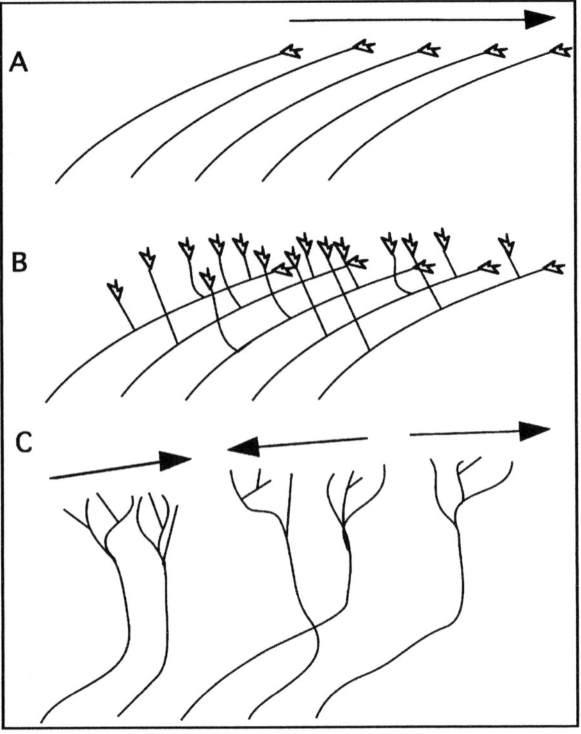

Fig. 8.3. Dynamic interactions between thalamocortical fibers and subplate could change the initial topography during the waiting period. A: Initially the growing thalamic fibers each consist of a single parent axon tipped with a growth cone. B: Naegele et al (1988) in the hamster (see chapter 4), and more recently Ghosh and Shatz (1992a) in the cat and Catalano et al (1996) in the rat have shown that during a limited period some thalamic fibers send transient side branches towards inappropriate regions of cortex on their way to their correct target area. Some of the first fibers entering the cortex are side branches of principal axons. The thalamic fibers arriving beneath the non-permissive cortical plate (chapter 5) might activate the appearance of lateral filopodia, from which these transient side branches can develop, as in the experiments of Stuermer (1990) where the growing axons found themselves surrounded by hostile membrane preparations in an artificial 'dead-end' street. Williams et al (1993) demonstrated that filopodia-like outgrowth can be induced along a developing neurite shaft in vitro by a locally applied electrical field. The development and extension of side branches might conceivably be caused by electrical fields set up by electrical activity in thalamic axons within the subplate region. Shatz and her colleagues (Friauf et al, 1990; Friauf and Shatz, 1991; Herrmann et al, 1994) have provided considerable evidence that at least some thalamic axons in the cat actually terminate and form functional synapses on subplate neurons and these connections might relay activity into the subplate sheet, which is in a strategic position (see Fig. 9.1) to influence further development of thalamocortical and intracortical circuitry. C: Some of the side branches enter the cortical plate and form arbors; the others disappear. The final topography can become substantially altered near the termination site, but the fiber ordering preserves the initial juvenile topography along the rest of the pathway.

becomes substantially altered with further development near *the termination site*. However, if the fiber ordering at the *rest of the pathway* preserves the simple juvenile topography, the ideas developed above are not challenged. Direct injection of tracers at the internal capsule in thalamus would be necessary to resolve this question.

Initial Fiber Ordering of the Connections Might Be Preserved Along the Pathway, but Substantially Modified at the Target Fields

Thalamic connections seem to be laid out in an orderly fashion on the basis of simple spatial and partly chronological algorithms (see chapter 4; Blakemore and Molnár, 1990; Shatz, 1992a,b). The topography of the initial intercortical connectivity is also rather simple and may depend on very similar mechanisms (Hankin and Silver, 1986). Fiber ordering preserves the chronological order of its establishment. Since the final representation is dependent on and complicated by many other factors, these initial clear-cut basic principles have gone unnoticed.

If we could examine the fiber ordering in the adult internal capsule or in the adult corpus callosum (away from the target area) the fibers might still show their basic topographic arrangement. Fortunately, there has been such study on the adult callosum in the cat (but not yet on the internal capsule). Nakamura and Tanakesi (1989) made small, discrete injections of HRP into different parts of the callosum of adult cats. Although there was no correlation between functional areas, there was a strict sectorial topographical correspondence between the corpus callosum and the cortical areas from which the cells were sending their projections (Fig. 8.4). Small injections into the adult cat presplenium labelled a zone of cortex very similar to that resulting from a neonatal injection into areas 17 and 18. Kind and Innocenti (1990) note that the topography of the juvenile callosal projection neurons may be related to that of their axons in the corpus callosum. I would rather say that the initial fiber ordering of the intercortical connections is *preserved* in the callosum, but substantially modified at the target fields (Figs. 8.4 and 8.5).

Such experiments, examining fiber arrangements in tracts distant from origins and targets, would be extremely interesting in developing and adult albinos or after early enucleation. I suspect that the initial connectivity between the different cortical areas might also be explained with a few very simple principles (see chapter 10).

Thalamocortical and Corticofugal Topographies Become Modified Near Their Termination Sites

It is interesting to note that in the adult there is another area within which there is extensive crossing of fibers, namely the thalamic reticular nucleus (Bernardo and Woolsey, 1987; Mitrofanis and

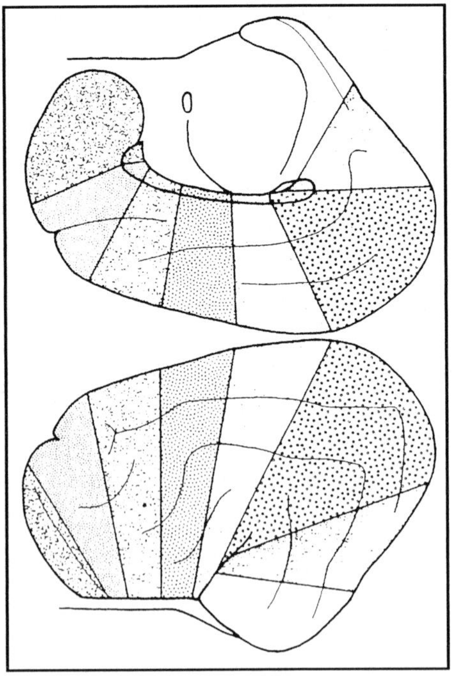

Fig. 8.4. A diagram showing the sectorial topographical correspondence between the corpus callosum and the cortex in the cat. The lower drawing shows the lateral view, the upper one shows a medial view (inverted). The rostral side of the hemisphere is to the left; the dorsal side is at the junction of the two drawings. Reprinted from Nakamura and Tanaseki (1989) with kind permission of Elsevier Science, Amsterdam.

Guillery, 1993). Since the studies demonstrating this were performed with HRP (an anterograde and retrograde tracer) we do not know whether these fibers had thalamic or cortical origin. I have speculated (see chapter 6) that corticothalamic fibers might wait in this area (based on my own retrograde labelling experiments at E16, and on the work and suggestions of Shatz and Rakic, 1981; Sheng et al, 1991; and Miller et al, 1993). If there is indeed a waiting period for corticofugal fibers within the thalamic reticular nucleus, then it might be important in the rearrangement of the fibers. Lozsádi et al (1996) provided evidence for corticofugal fiber rearrangement close to the diencephalon. It seems to make sense for early connectivity to be established with gross topographic order, while the distances for growth are minimal, and then the 'fine adjustment' of the thalamocortical fibers to be performed in the subplate, and that of the corticothalamic fibers in the thalamic reticular nucleus. These two regions have numerous similarities (Mitrofanis and Guillery, 1993). They contain cells which are among the earliest generated cells in the mammalian pallium; they contain similar markers (e.g. GABA, calretinin and parvalbumin) and form the first connections within and outside the cortex before they presumably undergo preferential

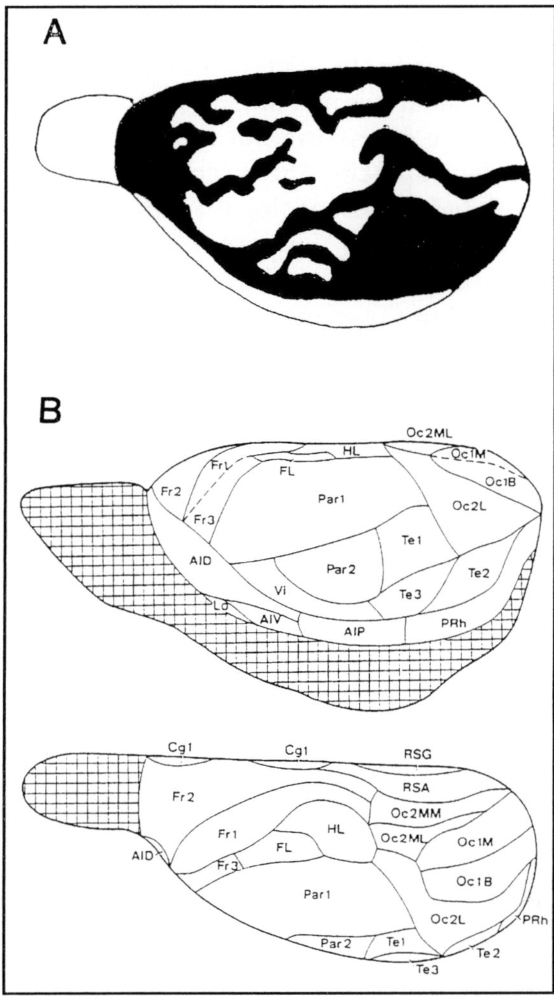

Fig. 8.5. A: Summary of the organization of the callosal connections, on a dorso-lateral view of the adult rat brain. The filled regions indicate areas of callosal projection revealed by degeneration studies (Ackers and Killackey, 1978). B: Comparison of the above diagram with cyto- and myelo-architectonical maps (lateral, dorsal and medial views respectively) suggests that adult callosal connectivity does not include primary visual cortex or certain regions of the motor and somatosensory cortex. Fr1-Fr3-frontal areas; Par1-Par2, Fl, HL-parietal areas; Te1-Te3-temporal areas; Oc1B, Oc1M, Oc2L, Oc2MM, Oc2ML-occipital areas; Cg-Cg3-medial prefrontal cortex; IL-infralimbic area; MO, LO, VO-orbital cortex; AID, AIP, AIV-insular cortex; PRh-perirhinal cortex; RSA, RSG-retrosplenial cortex. The allocortex and parts of the periallocortex are cross-hatched (Zilles et al, 1980). A: reproduced from Ackers and Killackey (1978) with kind permission of John Wiley & Sons, Inc., New York. B: reproduced from Zilles et al (1980) with kind permission of the MIT Press, Cambridge, Massachusetts.

cell death (Shatz et al, 1990; Mitrofanis and Guillery, 1991, 1993). The crossing of fibers seen in the adult within regions corresponding to these largely transient compartments might then reveal the fine modifications of the distribution of these fibers made during the early phase of development. These modifications require that the thalamocortical and the corticofugal axons begin to interact with their target regions during the period of their accumulation and change their own distribution and topography. The next chapter will examine the nature of these interactions.

CHAPTER 9

Can Early Thalamocortical Afferents Govern Their Own Topography in Cortical Plate? Activity Patterns During Early Cortical Circuit Formation

Functional Early Thalamic Synapses in Subplate

In mammals thalamic fibers arrive at the appropriate cortical regions before their ultimate target cells are born (Lund and Mustari, 1977; Rakic, 1976; Shatz and Luskin, 1986) and therefore they have to "wait" before they can establish their final pattern of innervation within the cortical plate. The subplate contains pre-mature neurons with a characteristic morphology, and with early expression of various transmitters and receptors together with numerous synapses (Marin-Padilla, 1971; Molliver et al, 1973; Valverde and Facal-Valverde, 1988; Valverde et al, 1989; Innocenti and Clarke, 1984; Shatz et al, 1988; Kostovic and Rakic, 1990). Anatomical studies suggested that some of the thalamic fibers form synapses with subplate cells (Kostovic and Rakic, 1990; Herrmann et al, 1994) and at least some of these connections were shown to be functional in cat (Friauf et al, 1990; Friauf and Shatz, 1991) and rat (Higashi et al, 1996). Since the development in rodent is more accelerated, the existence of the waiting period was questioned (see Catalano et al, 1996). Optical recordings and current source density analysis in embryonic whole forebrain slices (Higashi et al, 1996; Molnár et al, 1996b); however, support the notion that there is a short period after the arrival of the thalamic fibers, between E17-19, when thalamic fibers change their mode

Development of Thalamocortical Connections,
by Zoltán Molnár. © 1998 Springer-Verlag and R.G. Landes Company.

of growth, form functional synapses in the subplate and the lowermost segment of the cortical plate but not within the rest of the immature cortical plate. This pattern does not seem to change between E17-19, whereafter the thalamic fibers substantially invade the cortical plate and the entire cortical plate becomes depolarized from synapses distributed at layer 6 and in the dense cortical plate (Higashi et al, 1996).

Dynamic Interactions During the Waiting Period

The fact that thalamic axons wait in the subplate does not mean that thalamic axons do not engage in a highly dynamic interaction with the developing cortex during this period. They form numerous transient side branches (Naegele et al, 1988; Ghosh and Shatz, 1992a; Catalano et al, 1996) and then may start to transmit early patterns of activity. It has been demonstrated that the peripheral neurons already generate spontaneous activity patterns in corresponding ages (Galli and Maffei, 1988; Meister et al, 1991; Wong, 1993) when the sensory afferents begin to reach the thalamus (Lund and Bunt, 1976). These activity patterns could elicit EPSPs on thalamic projecting neurons (Shatz and Kirkwood, 1984; Mooney et al, 1996), and thus the activity patterns may alter the forming terminals within the subplate and cortical plate. In the rat, thalamic fibers are capable of conducting action potentials from the time they reach the cortex (E16) and our optical recording and current source density analysis experiments in thalamocortical slices demonstrated functioning thalamocortical synapses from embryonic day 17/18 in the subplate and lower cortical plate. The functioning connections might elicit activation patterns, which perhaps act on the side branch formation of the thalamic axons determining the site of ingrowth and final arbor formation. These naturally cannot modify the construction of the initial thalamocortical links, since by the time the spontaneous early sensory input reaches the thalamus from the sensory afferents, the primary deployment of the thalamic fibers to the planar sheet of the cortex has been completed in an autonomous fashion (E16 for rats, see Molnár and Blakemore, 1995b; Catalano et al, 1996).

Nevertheless, this initial layout of the thalamic fibers might then be confronted with slightly different activity patterns, which now are beginning to arrive from the periphery. Indeed, Nelson and LeVay (1985) demonstrated in the adult cat that thalamic axons cross each other in the mediolateral but not in the rostrocaudal axis of the optic radiation only a short distance (200-500 μm) below the visual cortex, which could very well correspond to the subplate zone during development (see Fig. 8.2 in chapter 8). It is possible that the crossing of the two sets of fibers seen in the adult represents a modification of the array of thalamocortical fibers that occurred during, or at the end of, the waiting period. The side branch formation might be regulated by electric fields of the activity patterns, as it was demonstrated in vitro that filopodial-like outgrowth can be induced along a developing neurite shaft by locally applied electric field (Williams et al, 1993). In contrast with the initial gross deployment of thalamocortical and corticothalamic connections, which might be due to a cascade of a few, relatively simple mechanisms (Shatz et al, 1990; Blakemore and Molnár, 1990), the remodelling of cortical circuitry during thalamic fiber invasion is probably a more complex process in which patterns of afferent and local activity, expression of surface molecules and growth factors, and cell death all play crucial roles. In turn the remodelling leads to the formation of new synapses and therefore to new distributions of activity and this might influence the morphological and physiological differentiation of cortical neurons (see Katz and Shatz, 1996).

Cortical Circuitry Changes Coincide with Thalamic Invasion

It is rather difficult to get access to cortical neurons in order to monitor their activity in situ during early phases of cortical development, but with the development of various new in vitro recording and imaging techniques, our understanding of the pattern of development of the functional intra- and extra-cortical connections has dramatically increased (see Gilbert, 1983; Toyama et al, 1991; Katz, 1993; Fox, 1995; Katz and Shatz, 1996). These studies demonstrated that the early cortical circuitry has rather different properties than the adult. There are transient connections and cells involved, and the mode of their interactions can be quite different, not relying on conventional synapses, but using gap junctions (LoTurco and Krigstein, 1991; Yuste et al, 1992) and non-conventional transmitters like nitric oxide (see Vincent and Hope, 1992).

In most mammals a substantial remodelling of cortical circuitry occurs around birth, when the accumulating thalamic fibers leave the subplate and enter the cortex (see Fig. 9.1) (Shatz et al, 1990; Friauf and Shatz, 1991; Allendoerfer and Shatz, 1994). Some subplate

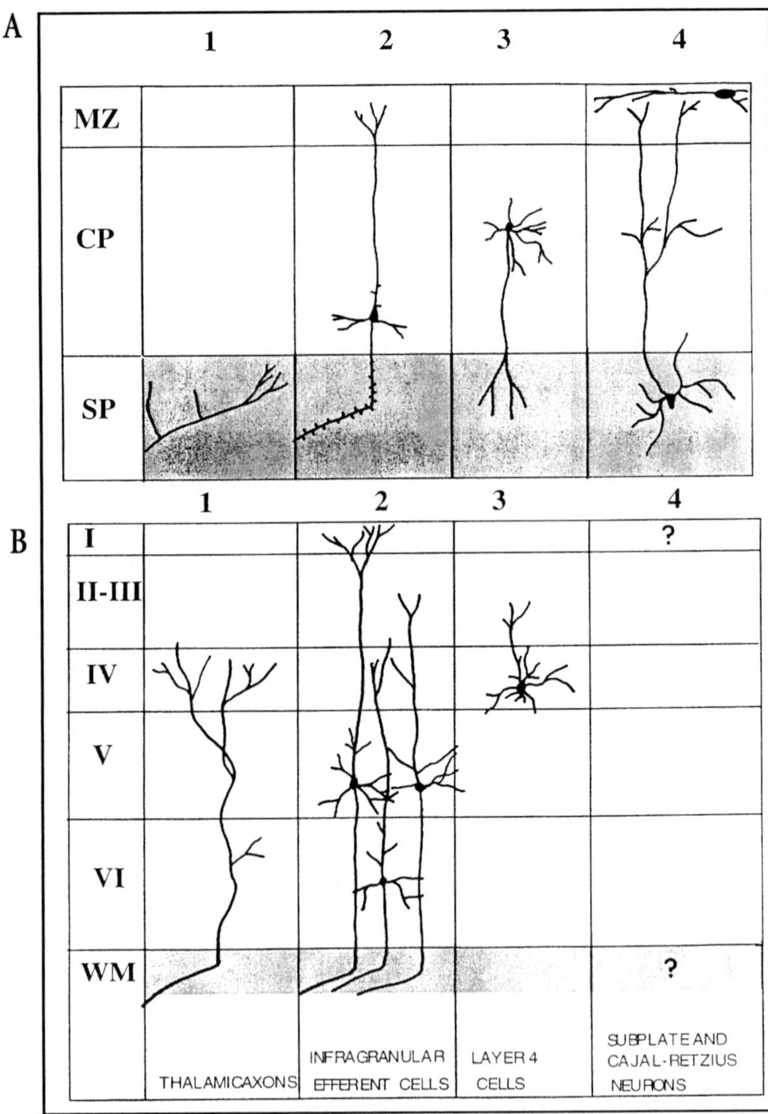

Fig. 9.1. Summary diagram showing the changes in thalamocortical fiber arrangement and the morphology of granular and infragranular cells as well as the subplate during the re-assembly of thalamocortical circuitry. The vertical panels represent: 1: Thalamic axons, 2: Infragranular efferent cells, 3: Layer 4 neurons, 4: Subplate and Cajal-Retzius cells. A: Thalamic fibers accumulate below the cortical plate (CP) and form synapses in the subplate (SP) (panel 1) (Rakic, 1976; Herrmann et al, 1994). In carnivores some subplate cells send their axons to the marginal zone (panel 4) and on the way the ascending axons collateralize in the middle of the cortical plate (Friauf et al, 1991). In kittens corticofugal fibers from layers 5 and 6 form boutons en passant on subplate cells (panel 2) (Lowenstein and Shering,

cells receive terminals *en passant* from collaterals of the corticofugal axons of layer 5 and 6 cells in newborn cat (Lowenstein and Shering, 1991). Many subplate cells in the cat also send an axon into the developing cortical plate, which terminates in the marginal zone (probably on transient Cajal-Retzius cells) and gives collaterals into the middle of the cortical plate (Friauf et al, 1990). In return, spiny stellate cells of layer 4 (in the cat) send their axons into the subplate at an early postnatal age (Callaway and Katz, 1992). At this phase, then, cortical microcircuitry is very different from that of the adult and might involve the subplate, marginal zone and neurons of the cortical plate (Friauf and Shatz, 1991). Synchronously with the establishment of the ultimate thalamic innervation pattern, there are major rearrangements (see Fig. 9.1B). Subplate cells, and most probably Cajal-Retzius cells as well, stop participating functionally in the circuitry (Friauf and Shatz, 1991) and many of them start to die (Luskin and Shatz, 1985b; Derer and Derer, 1990, 1992; Anderson et al, 1992). Within this period some of the contralaterally projecting pyramidal cells transform into stellate cells in the kitten (Vercelli et al, 1992) and some contralaterally projecting layer 5 cells in the rodent lose their apical tufts (Koester and O'Leary, 1992; Kasper et al, 1994). Unfortunately we only have fragments of knowledge from various species; much more is needed to understand the logic and the causal relationships between these changes. For a better understanding a more detailed analysis of cell morphologies and locations of synapses is required, and is not available at present. It is not yet established in rodent whether the subplate cells form functioning synapses within the cortical plate and marginal zone as has been proposed for carnivores (Friauf and Shatz, 1991).

1994). At this stage most, if not all cells have pyramidal morphology in layer 5 and 4 (Vercelli et al, 1992; Koester and O'Leary 1992; Kasper et al, 1994). In carnivores spiny stellate cells of layer 4 send axons that form contacts with subplate cells in this early period (panel 3) (Callaway and Katz, 1992). In the rat this period would correspond to E16-20. B: Just before birth (E20-21 in the rat) thalamic fibers (panel 1) leave the subplate and (in addition to layer 6) arrive principally in layers 4 and, to a lesser extent, layer 1 (Kato et al, 1984), which also represented the major targets of subplate cells. Some of the pyramidal cells in layer 4 and 5 loose their apical tuft (Vercielli et al, 1992; Koester and O'Leary, 1992; Kasper et al, 1994). Speculations about the possible activity patterns in the developing cortex would have to take into account other projections (including the corticothalamic, corticotectal and callosal) which also have contacts with the subplate (Innocenti and Clarke, 1984).

In Vitro Approaches to Study Early Functional Thalamocortical Development

To investigate certain aspects of the formation of functional thalamocortical and cortico-cortical connectivity, in vitro models are desirable and offer many advantages. Recordings from organotypic co-cultures of thalamus and cortex (Yamamoto et al, 1989, 1992; Bolz et al, 1990; 1992; Molnár and Blakemore, 1991, 1995a) and from cortical slices which received input from transplanted thalamus (Hamasaki et al, 1987; Kurotani et al, 1993) have certain advantages over isolated cortical slice preparations (e.g. Friauf and Shatz, 1991) because the thalamus and thalamic input are kept together with the cortex during recordings, whereas in conventional acute cortical preparations these connections are cut and it is difficult to implement selective stimulation paradigms. The disadvantage of these models is, however, that the thalamocortical connections develop artificially and therefore they might not be representative for all aspects of natural development. Therefore, there is a considerable interest to develop slice preparations where distant interconnected structures are kept intact in numerous systems (see Halloran and Kalil, 1994, for callosum; Bastmayer and O'Leary, 1996, for corticospinal projections; Tóth et al, 1997 for septo-hippocampal connections), and to use them for physiological and imaging experiments. Due to the unique anatomy of the somatosensory thalamocortical fibers, it is possible to prepare a brain slice which contains both somatosensory thalamus and primary somatosensory cortex in addition to their interlinking fiber pathways (Bernardo and Woolsey, 1987; Agmon and Connors, 1991; Behan et al, 1991; Gil and Amitai, 1996). The direct, selective stimulation of the ventrobasal thalamus in slices of rodent forebrain that also include somatosensory cortex enables the investigation of the early functional interactions between the thalamic fibers and the developing cerebral cortex during the time of thalamic fiber arrival, their cortical invasion and synapse formation. This preparation was used by Toyama's group to image the spatio-temporal pattern of cortical activation using voltage- and Ca^{2+}- sensitive dyes, intracellular recording and to perform current source density analysis of field potentials after selective thalamic stimulation (Higashi et al, 1993, 1996; Crair et al, 1993, 1997; Molnár et al, 1995a, 1996b; Kurotani et al, 1996; see Fig. 9.2).

The above techniques are complementary and reveal different aspects of the forming circuitry. The sites of developing thala-

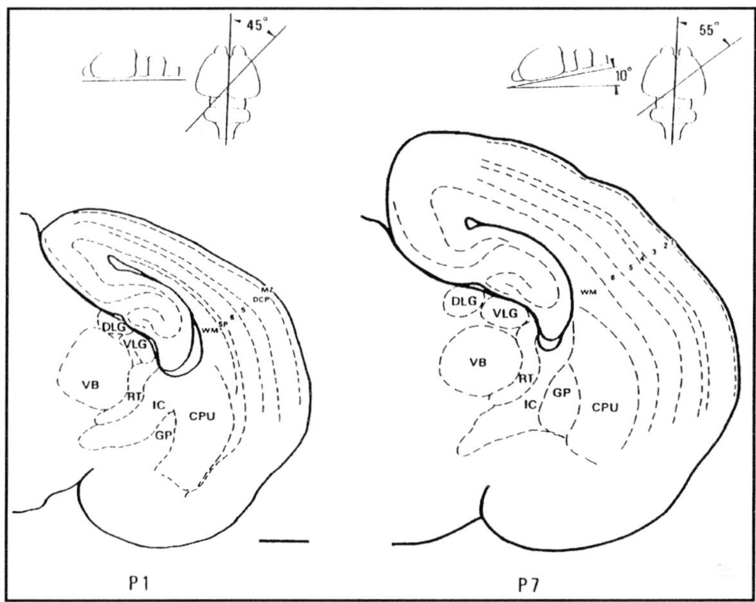

Fig. 9.2. Preparation of whole forebrain slices containing the somatosensory cortex and thalamus with intact connections at various ages. The upper schematic drawings explain the different angles used for the sectioning at different stages (45° at P1 and 55° at P7). The camera lucida drawings demonstrate two examples of thalamocortical slices used for our study. They were drawn from 50 µm thick, Cresyl violet stained section (results obtained from the P1 slice are shown in Fig. 9.3). CPU: caudate putamen; DLG: dorsal lateral geniculate nucleus; DCP: dense cortical plate; GP: globus pallidus; IC: internal capsule; MZ: marginal zone; RT: thalamic reticular nucleus; SP: subplate; VB: ventro-basal complex; VLG: ventral lateral geniculate nucleus; WM: white matter. Scale bar: 1 mm. From Molnár et al (1996b).

mocortical synapses can be determined with current source density analysis (CSD), while imaging techniques give valuable information on the site of depolarization (voltage-sensitive dyes) and the location of the activated cells (Ca^{2+} sensitive dyes).

Current Source Density Analysis (CSD)

Extracellular recording reveals the local dominant ionic currents which flow through cell membranes during synaptic activations near the recording site. These currents can be inward currents (current sinks) or outward currents (current sources). It was demonstrated that mostly EPSPs are responsible for the extracellular current sinks (Mitzdorf and Singer, 1978) and, thus, CSD can detect

EPSPs, even those which do not necessarily lead to postsynaptic depolarization. The radial arrangement of the adult and developing cerebral cortex, where the current sources and sinks are directly proportional to the second spatial derivative of voltage along its depth, makes the cerebral cortex a suitable system to employ CSD analysis (Nicholson and Freeman, 1975). CSD analysis of field potentials thus promises a valuable complementary technique to intracellular and optical recordings to determine the location of the forming synapses. It has been used to study major transmembrane currents in the cerebral cortex of adult (Mitzdorf and Singer, 1978; Bode-Greuel et al, 1987) and developing cats (Friauf and Shatz, 1991), and in various in vitro thalamocortical model systems: in organotypic co-cultures (Yamamoto et al, 1989, 1992) and in slices of visual cortex and transplanted lateral geniculate nucleus (Hamasaki et al, 1987; Kurotani et al, 1993). Agmon and Connors (1991) examined the adult pattern of CSDs in thalamocortical slices of the mouse. To examine the location of the thalamocortical synapses during embryonic and early postnatal development, we (Molnár et al, 1996b) performed CSD analysis after stimulation of the ventrobasal thalamus selectively in slices of rat forebrain (Bernardo and Woolsey, 1987; Agmon and Connors, 1991) that also included somatosensory cortex at various ages (E17-P14).

At E18 to birth a short-latency current sink is observed in the subplate and layer 6, indicating that thalamic axons can conduct action potentials and some have already formed functional synapses there. Between birth and P3, when thalamic axons are completing their upward growth, a sink gradually appears more superficially, in the dense cortical plate, perhaps corresponding to future layer 4 or the proximal segments of apical dendrites of layer 5 cells. Synchronously a current source arises in layer 5. Both sinks and sources disappear if synaptic transmission is blocked.

The adult-like distribution of CSDs becomes apparent after P7. The antidromic component, which results from the backfiring of layer 6 cells through their thalamic projections, can no longer be blocked completely after this age.

In rat, the proposed transient circuit of Friauf and Shatz (1991), which might be essential in further cortical development in carnivores (Ghosh and Shatz, 1992b), is not apparent. In our CSD study, the direct thalamic stimulation at embryonic ages (before the thalamic fiber ingrowth to the cortex) did not elicit the long latency

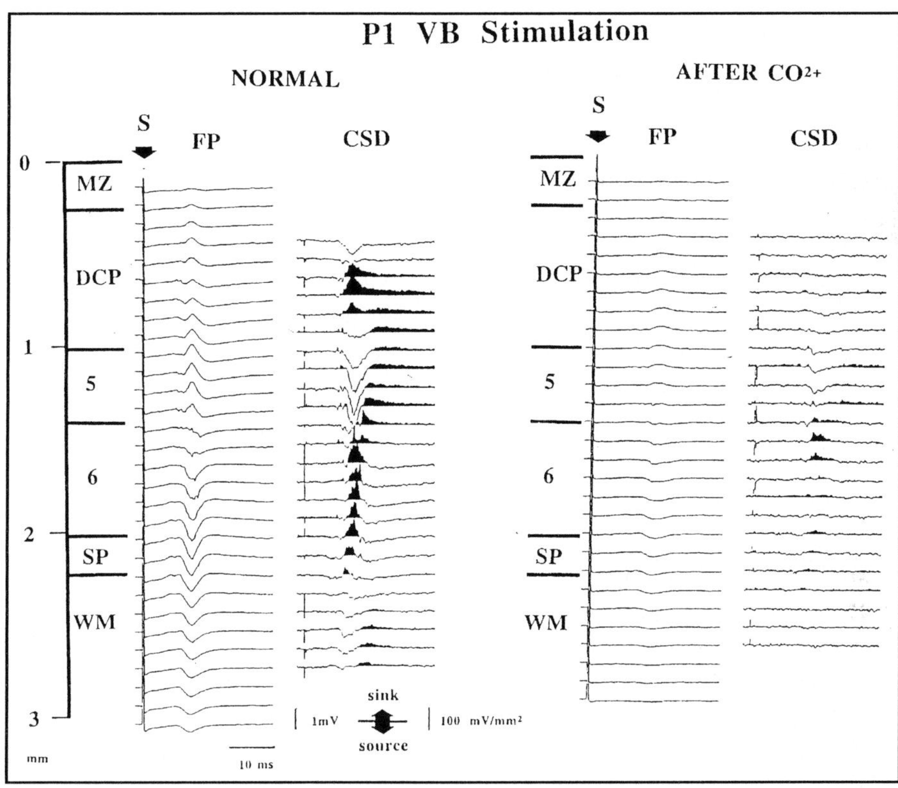

Fig. 9.3. Field potentials (FP) and corresponding current source density profiles (CSD) from a P1 thalamocortical slice elicited with direct VB stimulation before and after the application of Co^{2+}. The *camera lucida* drawing and a cresyl violet stained section of this slice is shown in Fig. 9.2. Submaximal VB stimulation (stimulus intensity: 3mA; threshold: 1.2 mA; maximum: 3.4 mA) elicited field potentials (FPs) in almost all cortical layers with 10 ms latency. CSDs calculated from the field potentials (collected at every 100 µm) demonstrate strong current sources in layer 6, in the subplate and in the dense cortical plate; synchronously intensive current sinks appeared in layer 5. CSDs lasted for more than 25 ms in layer 5 and DCP. The right two columns demonstrate the FPs and the CSDs after 30-minute perfusion with modified Ringer solution where the 5 mM Co^{2+} substituted Ca^{2+} to block synaptic transmission. Both current sources and sinks were almost completely abolished by Co^{2+}. From Molnár et al (1996b).

CSDs in the putative layer 4 of the cortical plate (Molnár et al, 1996b; Molnár, Kurotani, Higashi and Toyama, unpublished observation). The difference might be due to the greater antidromic component elicited during white matter stimulation (Friauf and Shatz, 1991), where the recurrent cortical plate collaterals might activate the cortical plate itself. Blocking the synaptic transmission eliminates most of the responses in the upper CP, but this does not necessarily indicate that the activation was antidromic, through thalamic afferents.

Our study in rat, in contrast with the cat (Friauf and Shatz, 1991), did not demonstrate apparent late current sources in the upper cortical plate elicited after direct thalamic, or even after distal white matter, stimulation. This indicates that thalamic activation cannot elicit cortical activation either directly through thalamic afferents or through subplate activation. This discrepancy might be due to species differences or the more selective direct VB stimulation paradigms employed in our study. Perhaps subplate neurons in the rodent have no time to engage in a more complex interplay with cortex and marginal zone as in carnivores (Shatz et al, 1990; Friauf et al, 1990) because the cortical plate matures more rapidly. CSD analysis in combination with direct thalamic stimulation in embryonic carnivores, an experiment rather difficult to perform, could help to resolve this issue.

Antidromic Activation in Thalamocortical Slices

The antidromic stimulation of corticofugal projections presents a great problem in the interpretation of the CSD analysis or other electrophysiological and imaging results. Nonsynaptic antidromic activation of subplate neurons via their long-distance axons or antidromic activation of recurrent collaterals of subplate, layers 5, 6 and other corticofugal supragranular neurons could contaminate and complicate the findings when stimulating from the optic radiation or from the white matter. These collaterals might elicit EPSPs through their synaptic contacts in various other neurones in the subplate and cortical plate. By demonstrating that the vast majority of these responses can be abolished by blocking synaptic transmission does not necessarily mean that they do not contribute to the CSDs (see Fig. 9.4).

In thalamocortical slices a more selective stimulation paradigm can be achieved than from the white matter, by excluding some of the antidromic components, but unfortunately it still does not eliminate the above described possibilities completely, especially in late postnatal ages when the layer 6 corticofugal projections innervate the specific sensory nuclei of the thalamus. Since the thalamocortical and layer 6 corticofugal projections take slightly different paths (Woodward et al, 1990; Agmon et al, 1993), it was suggested that thalamocortical slices preferentially preserve the thalamocortical but not the corticofugal projections (see Agmon et al, 1993). In our experience this does not apply in all of our cases using 450-500 μm

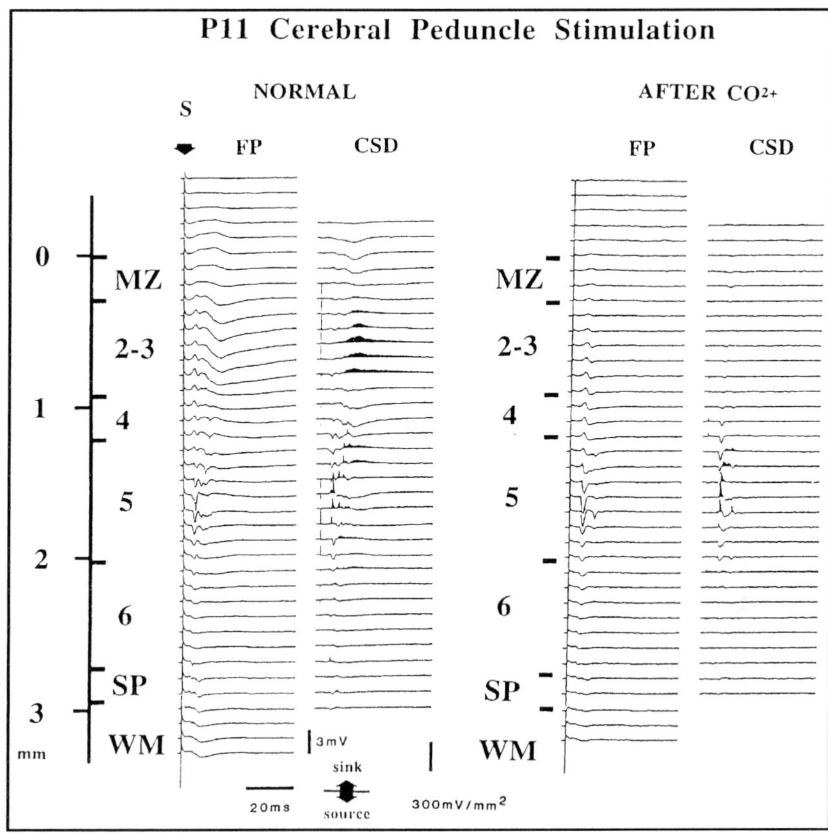

Fig. 9.4. Current source-density analysis in a P11 thalamocortical slice obtained after selective cerebral peduncle stimulation during perfusion with normal ACSF (left column) and after perfusion with 5mM Co^{2+} ACSF (right column). The FPs and CSDs remained unchanged within layer 5 after Co^{2+} application, but the FPs and CSDs disappeared from layer 2-3. From Molnár et al (1996b).

thick whole forebrain slices, where we could observe antidromic activation of layer 6 from P7 (Molnár et al, 1996b). Our labelling study on the very same thalamocortical slices which we used for imaging studies demonstrated that layer 6 cells can be backlabelled from the stimulation site in VB at postnatal ages (Higashi et al, 1993; Crair et al, 1993) but not in embryonic ages (Higashi et al, 1996). Indeed anatomical studies raised the possibility in various species that corticofugal projections from layer 6 and subplate are delayed in their growth towards the thalamus (see sections 6.4 and 6.5 in chapter 6).

Optical Recording

In Professor Toyama's laboratory (Kyoto Prefectural University of Medicine) we have been using optical recording of voltage sensitive dyes (Grinvald, 1984) with Fuji HR Deltaron 1700 differential image acquisition system to reveal the pattern of cortical activation elicited by selective thalamic stimulation in thalamocortical slices (Fig. 9.2). In our study we determined the site of monosynaptic activation and, using pharmacological manipulations, the nature of thalamic transmission at different embryonic and postnatal ages (Higashi et al, 1993, 1996; Crair et al, 1993; Kurotani et al, 1996). At E17, ventrobasal nucleus stimulation elicited spike-like responses which propagated from the thalamus through the primitive internal capsule to the border between the cortical plate and subplate. These fast activations seemed to represent fiber responses, since the application of DNQX (50 μM) and APV (50 μM) (NMDA and AMPA receptor antagonist mixture) had no effect on them. Prolonged responses lasting more than 300 ms and sensitive to the glutamate receptor antagonists started to appear first in the primitive internal capsule at E17/18, then in SP and the layer 6 at E19, but no responses were found in IZ and in the upper layer of CP. At E21-birth, the slow depolarization became detectable in the entire CP and SP. These results indicate that thalamic afferents are capable of conducting action potentials as early as E16, but synaptic transmission starts to work at E18/19 in SP and layer 6, one or two days after the arrival of the thalamic afferents to the developing cortex. Insertion of a small DiI crystal into the VB into the stimulation site, in the very same slices from which we recorded revealed the advancing thalamocortical fibers in the internal capsule and subplate and backlabelled cells in the internal capsule (Higashi et al, 1996). No retrogradely labelled cell appeared in the subplate or cortical plate in any of the slices, consistent with our tracing study (see chapters 4 and 6), indicating that at these prenatal and early postnatal ages corticothalamic projections had not invaded the thalamus yet. It also suggests that cells in the internal capsule (perireticular nucleus) are capable of synaptic transmission from early stages.

Shortly after birth (P0-P4) the evoked cortical response is rather diffuse and involves several or all layers in a columnar fashion. The zone of monosynaptic activation is initially restricted to the lower layers (subplate and layer 6) and gradually moves up to layer 4 within an activated column by P4. Although we did not see slow activation of cortical plate before the entry of thalamic fibers into the cortical plate (perhaps one would in carnivores, see Friauf and Shatz, 1991), it is still a rather interesting possibility that subplate cells are involved in the formation of an early, transient, functional column (in which all cells are activated polysynaptically) from the beginning of the thalamic fiber arrival and accumulation, while the monosynaptic activation gradually moves up through the infragranular layers.

After P5 the response is restricted to layers 4 and 6 and subplate is no longer activated after thalamic stimulation. These interesting changes of spatial distribution of activation are accompanied by a dramatic change in the timecourse of the activation. The time taken to peak for the responses is 86.9 +/- 11.2 ms at birth and it is reduced to 23.6 +/- 8.9 ms between postnatal day 2 and 5; it is only 6.7 +/- 0.6 ms at postnatal day 10-13. This dramatic reduction in the length of the depolarization, combined with other changes, might be responsible for the decreased capacity of thalamocortical projections to elicit long term potentiation (Crair and Malenka, 1994; Fox, 1995). Isaac et al (1997) recently suggested that during the early postnatal period (postnatal days 2-5) a significant proportion of the thalamocortical synapses are 'silent' and they can be converted to functioning ones during long term potentiation (LTP). They explained the loss of the susceptibility of the thalamocortical connections to LTP with the disappearence of the 'silent' synapses by postnatal day 8-9. It seems likely that of the different glutamate receptors the NMDA receptors have a special role in developmental plasticity (Bear et al, 1990) because they have their highest level at the peak of the plastic period and then decline dramatically. NMDA receptors are the only glutamate receptors whose expression can be prolonged by dark rearing (Gordon et al, 1997). NMDA receptors have a more prominent role during development; nevertheless some adult thalamocortical transmission remains to be mediated through them (Gil and Amitai, 1996).

Ca^{2+} *Imaging in Thalamocortical Slices After Selective Thalamic Stimulation*

With optical recording using voltage-sensitive dyes we could determine the site of depolarization, but unfortunately we could not resolve cells, and the technique cannot show the exact location of the cell bodies which depolarize. We used Ca^{2+} imaging with confocal microscopy to be able to resolve individual cells and to further

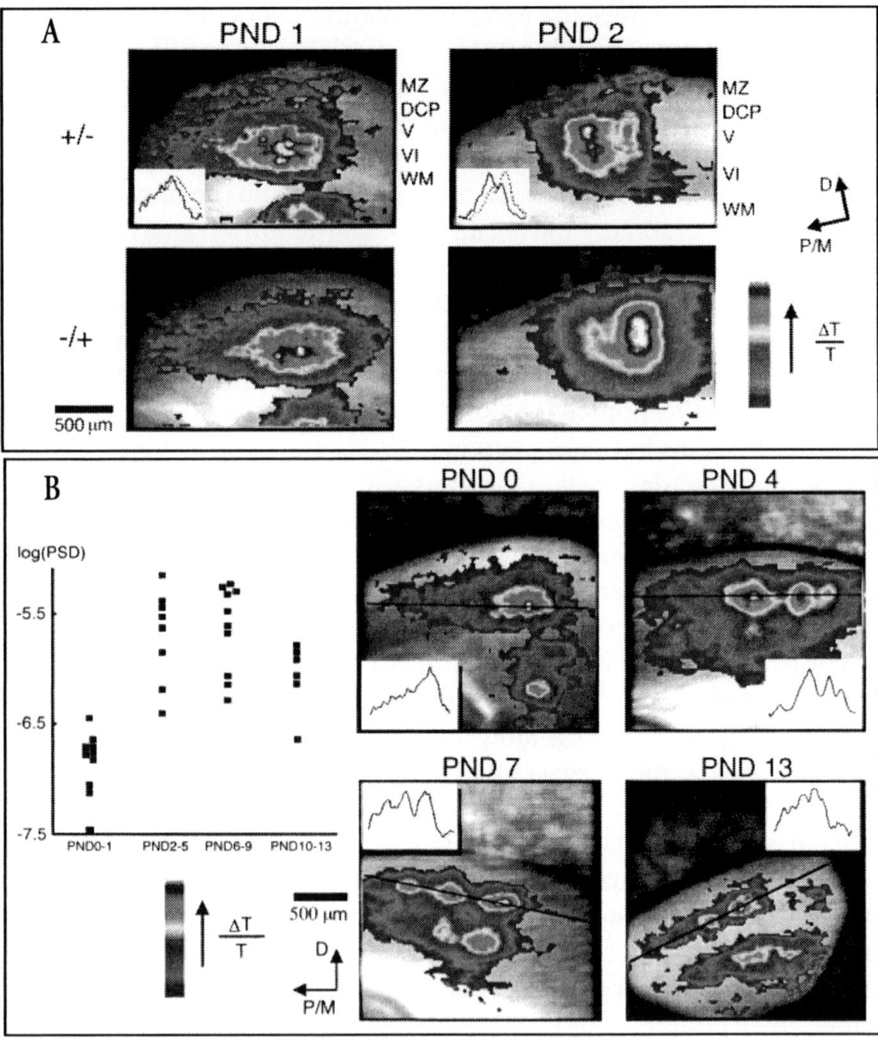

Fig. 9.5. A: Development of the columnar activation pattern from a broad activation pattern at postnatal day 2 studied with optical recording. Thalamocortical slices (400 µm thick) were stained with voltage sensitive dye (RH-482 NKK; 20 minutes with 0.05-0.1 mg/ml) and placed in a warmed (32°C) interphase-type recording chamber on the stage of an inverted microscope. Stimulus induced changes in the intensity of transmitted light (700 +/- 30 nm) were measured with a 128 x 128 pixel array of photosensors every 0.6 ms (Fujifilm HR Deltaron 1700) for up to 300 ms at 8 bits resolution (all 8 bits are devoted to measure changes in transmitted light intensity relative to a pre-stimulus reference image). Pseudocolor figures show relative changes in transmitted light intensity ΔT/T (where T is the intensity of transmitted light) with lower cutoff at twice the standard deviation of the pre-stimulus noise and superimposed on the gray scale reference image of the slice. The response to a single 100 µm stimulus of the thalamus using a bipolar stimulating electrode was averaged 4 to 16 times; intervals between stimuli were at least 30 seconds. During the optical recording extracellular electrodes were used to continuously monitor field

examine the development of the thalamocortical and intracortical circuitry in thalamocortical slice preparations (Molnár et al, 1995a). Scanning at intervals of 170 ms, occasional spontaneous activity was detected in just a few cells. Brief electrical stimulation (1-5, 100 μs pulses at 100 Hz), delivered with a bipolar electrode positioned into the ventrobasal complex of the thalamus, elicited a rapid and sustained elevation of Ca^{2+} within cells of the putative somatosensory cortex indicating the existence of functional connections between thalamus and cortex from the earliest ages studied (P0). The vast majority of the responses were observed in the upper cortical plate, corresponding to layer 4, the lower dense cortical plate and some in layer 1. Quantitative analysis of single cell responses revealed a heterogeneous population according to their reaction onset and duration of Ca^{2+} elevation. A subpopulation of cells responded rapidly with onset of less then 60 ms. Some of these responded with a single,

potentials. Stimulation of the thalamus at birth or at postnatal day 1 (P1) results in broad activation of the cortex, without clusters or columns of response. In contrast, at P2, the depolarizing response in cortex is arrayed in columns extending from the dense cortical plate (DCP) through layer 6, with individual columns of response separated by sparse responding inter-columnar regions. On the left frame (frame A and B) is the response from a P1 slice, with stimulation from slightly different positions, using opposite polarities of the same bipolar electrode. The black and white inset shows the spatial profile of the response across layer 6 for each stimulus polarity superimposed. There is only a single peak, which shifts slightly when switching polarity of the stimulation, reflecting the topographic mapping of thalamocortical afferents. In contrast, at P2 (frames C and D), there are two distinct columns of response, each of which can be preferentially activated depending upon the polarity used for the thalamic stimulation, indicating topographic order in the thalamocortical connections. The inset shows the superimposed laminar profile for the two responses (2.25 mm in width), revealing the separable peaks for each column. (D is dorsal; P/M is posterior/medial; MZ is marginal zone; DCP is dense cortical plate; WM is white matter). B: Development of barrel modules. Examples for cortical activation patterns at four ages between P0 and P13. Black and white insets in each color picture depict the response profile along the line through layer 6 of the cortical plate (indicated with black on each panel). The images were taken with the same magnification, all 2.25 mm in width. At birth (PND 0), there is a broad single peak. In contrast, on postnatal day (PND) 4, 7 and 13, there are barrel clusters seen as repetitive peaks and troughs in the laminar profile along layer IV. For each slice, the degree of barrel patchiness was quantified using one-dimensional fourier techniques (results shown in left diagram). The amplitude of the power spectral density is significantly lower at PND 0 and 1 than at all other ages tested. Reproduced from Kurotani et al (1996) with kind permission of Kyoritsu Shuppan Co. Ltd. Publishers, Tokyo. See color figures in insert.

short and rapid Ca^{2+} transient, others with long, sustained, elevation lasting as long as 2.5 seconds, but in most cells the onset of sustained response was delayed with respect to the stimulus by hundreds of milliseconds. These experiments show that even at early stages of cortical development cells respond differentially to thalamic stimulation which might be correlated with their intra- and extra-cortical connectivity. It would be interesting to correlate the imaging results with different backlabelling techniques. Perhaps cells with different projections (thalamus, contralateral hemisphere or superior colliculus) react differentially to thalamic stimulation in a very specific fashion.

Our observations on rodent forebrain slices using various techniques (voltage- and Ca^{2+}-sensitive dyes: Crair et al, 1993; Higashi et al, 1993, 1996; and current source density analysis: Molnár et al, 1996b) demonstrate that cells of the lowest layers of the cortex receive functional thalamic input before birth and that thalamocortical axons form synapses with more superficial cells as they grow into the cortical plate.

Invasion of the Cortical Plate and Death of the Subplate

The wholesale entry of thalamic fibers into the cortical plate occurs around the time that the cells of the subplate start to die (Ferrer et al, 1992; Price et al, 1997). This raises the possibility that there is causal relationship between the two events. Is the subplate death triggered by the cortical invasion of thalamic afferents or do thalamic fibers leave the subplate because they are beginning to die? Perhaps the death of subplate neurons is *necessary* for the release of the axons and their invasion of the cortex. If so, the subplate would play the additional role of determining when thalamic axons are able to invade the cortex. The obvious next question is: what causes subplate cell death? Is it simply innately 'programmed' to occur at the appropriate time in development, as is much of the apoptotic cell death that occurs in the nervous system of both invertebrates and vertebrates (Driscoll and Chalfie, 1992)? Or is it triggered and regulated by other events (see Purves, 1988; Oppenheim, 1989)?

There is considerable recent interest in the 'social' control of cell death patterns during development (Raff, 1992). One attractive possibility is that spontaneous impulse activity might start in thalamic axons before and around the time of birth, releasing glutamate from their terminals, which kills the temporary subplate target cells

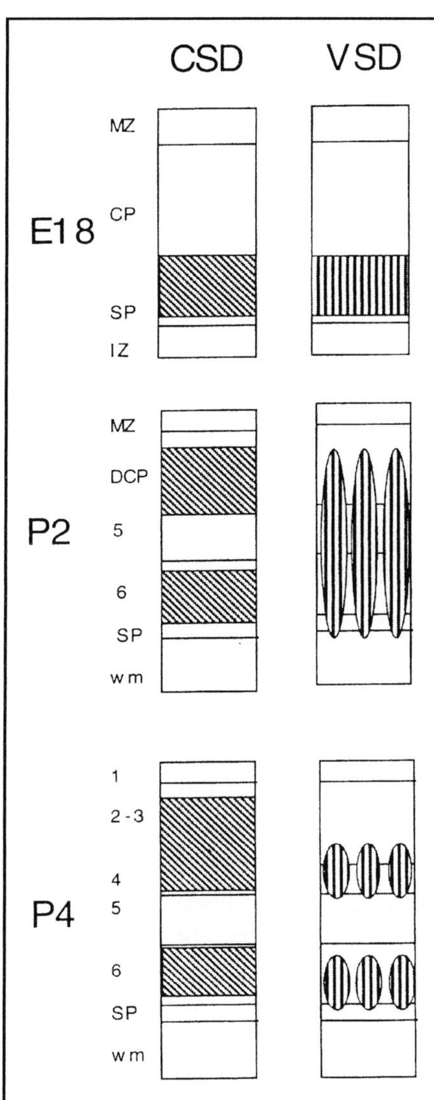

Fig. 9.6. Comparison of our current source analysis results with our optical recording findings in thalamocortical slices using direct VB stimulation (Higashi et al, 1993, 1996; Crair et al, 1993). The site of the major short latency current sinks in general corresponded well to the sites of initial, short latency cortical depolarizations following direct thalamic stimulation. The only exception, perhaps, is layer 5, which shows strong signals with optical recording. Layer 5 might get depolarized via thalamic input into its apical dendrite rather than its cell body.

through excitotoxic effects. Certainly subplate cells express NMDA receptors (Herrmann et al, 1992) and are particularly vulnerable to excitotoxins: injection of kainate into the cortex of the newborn cat selectively kills subplate cells and leaves the cortical plate apparently intact (Ghosh et al, 1990). My colleagues (Molnár et al, 1991) and I have tried to test this hypothesis by producing lesions of the primitive internal capsule at around E15/16, so as to interrupt the thalamic innervation of the posterior part of the hemisphere (and the subcortical projections from that region of cortex). We have preliminary evidence that subplate neurons change their projection pattern and are indeed to some extent 'rescued' in cortical regions where the connections through the internal capsule were lesioned at embryonic life. We birthdated subplate cells with [^{3}H]thymidine at E12 and lesioned the internal capsule unilaterally at E16 in rat embryos. The heavily labelled cells were counted at both sides and the ratio of labelled cells was determined between the lesioned and control side. At P10 this ratio was between 1.31 and 1.53, indicating that more heavily labelled cells remained at the isolated side. Interestingly, DiI placement into cortex at the lesioned side revealed many backlabelled cells in the subplate (not seen in normal animals), indicating that subplate cells developed or maintained connections with distant cortical plate within the isolated cortex (Fig. 9.7).

Similar abnormal connections were recently described in the cerebral cortex of prematurely born human infants who suffered white matter necrosis because of perinatal asphyxia (Marin-Padilla, 1996). Such perinatal asphyxia selectively destroys the afferent and efferent axons, but leaves the cortical gray matter relatively intact because of its intact blood supply through the leptomeningeal perforating vessels. Perinatal white matter infarction deprives the overlying developing cortex of its external connections, which might be needed for its normal maturation. Due to the white matter necrosis the patients subsequently suffer from acquired neocortical dysplasia (Marin-Padilla, 1996). The disturbed development of a relatively isolated cortical region could lead to local cortical dysfunction responsible for the pathogenesis of neurological disorders including childhood epilepsy (see Squire et al, 1997).

Further work is needed to establish causal relationships between thalamic innervation and subplate cell morphology and cell death. Shatz and colleagues reported that, for isolated cultures of ferret cortex, a larger fraction of subplate cells survive than in vivo (Hohn et al, 1993). This may support the view that external factors, including thalamic innervation, play a role in the initiation or regulation of subplate death in vivo, since in the ferret cortex the majority of subplate cells disappear by adulthood (Shatz et al, 1990). Price and Lotto (1996) suggested the contrary: that the presence of thalamus in organotypic cocultures is rescuing subplate cells from their preferential death, as they observed in single cortical cultures. However,

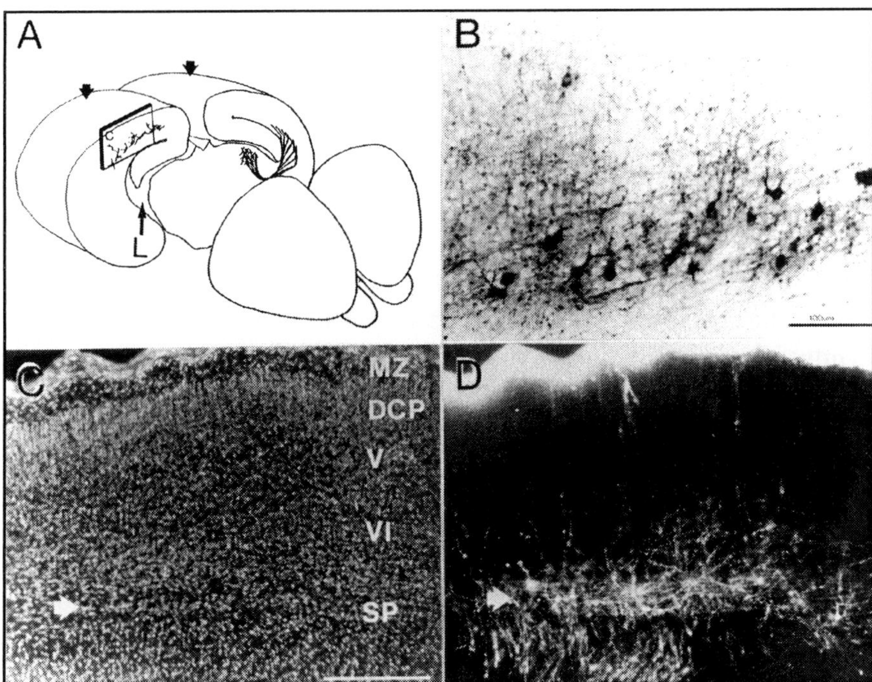

Fig. 9.7. Subplate cells change their projection patterns within the hemisphere where the internal capsule was lesioned at embryonic day 16. DiI placement on the isolated cortex at P2 revealed many cells backlabelled in the subplate, not seen on the non-lesioned side or in normal animals, indicating that subplate cells developed or maintained extensive projections into the cortical plate. A: Schematic drawing illustrating the lesion (L) and the crystal placement sites (arrows at the convexity of the occipital cortex). C: Bisbenzimide stained coronal section anterior to the crystal placement demonstrating the layering of the isolated cortex. D: Same field viewed under DiI filter to demonstrate the numerous cells labelled in subplate. (White arrow are in identical position in C and D). B: High power confocal microscopic image from the same section to further illustrate the morphology of the labelled subplate cells. Scale bar: 100 μm for B; 200 μm for C and D. From Molnár, Yee, Lund and Blakemore, 1991.

more precise quantification will be needed to settle this point. Using a sensitive, quantitative assay for dying cells involving the confocal microscope, we saw a distinct layer of dead or dying subplate (and marginal zone) cells in postnatal rat solitary cortical cultures (Adams R et al, 1993), but did not examine the effect of thalamus systematically. The issue clearly deserves further study.

Growth and Plasticity of Thalamic Axons in the Cortical Plate

The precise mechanisms regulating the entry of thalamic fibers into the cortical plate are not understood and deserve further study. In particular, the substantial transitions or partial reversal of parts of thalamocortical maps (see Guillery, 1986) might depend on the fate of the transient side-branches described by Naegele et al (1988) and Ghosh and Shatz (1992a)(see chapter 8).

After thalamic fibers detach from the subplate layer, they turn upwards to enter the cortical plate, presumably responding to the local trophic and growth-permissive influences of the cortex revealed in co-culture experiments (Molnár and Blakemore, 1991; Bolz et al, 1992; Rennie et al, 1994; chapter 5). By comparison with the remarkably strict parallel arrangement of thalamic fibers in the white matter at early stages, they become more topographically *disordered* as they enter the cortical plate. Many axons grow up more or less radially but others grow more obliquely within the plate, following varied courses through the deeper layers. It almost looks as if the high degree of order, which may be created by fiber-fiber interactions during growth is 'deliberately' slightly disturbed as it is transformed into the topography of projection into the cortex itself. Perhaps such local mixing of the projection is advantageous in that it allows postsynaptic cells to have access to a scattered group of axons, and hence to have the opportunity to employ correlation detection mechanisms to detect local correlation of activity. In the rat, by P0-2, some axons appear to have terminated in layer 6 and others to have run all the way up to the marginal zone; but the majority of the fibers grow upwards, then branch and lose their growth cones, about 2-300 μm below the pial surface. Each arbor extends laterally over about 300 μm diameter, and the zone of termination is not necessarily centered above the point of entry of the axon in visual cortex, in contrast with the primary somatosensory cortex (Agmon et al, 1993), where the arbors form above the region where they enter the upper layers. In the primary visual cortex of the hamster Naegele et al (1988) demonstrated that there is a slight reduction with age in the size of individual thalamic arbors relative to the total size of area 17. They suggested that changes in arbor size due to collateral formation and withdrawal contribute to fine map adjustment. This, however, might be different in primary somatosensory cortex, where Agmon et al (1993, 1995) described precise thalamocortical fiber targeting and arbor formation using anterograde and retrograde labelling techniques. We repeated the experiments of Agmon et al (1995) in whole brains (rather than in thalamocortical slices as it was originally performed) and this made it possible to establish comparisons between the initial thalamocortical targeting in the somatosensory and visual cortical plate (Molnár et al, 1996a). We found two interesting differences at P2. From identical pairs of localized carbocyanine dye placements in putative primary visual cortex we observed more scatter and 3-5 times smaller numbers of backlabelled thalamic cells in LGN than from primary somatosensory area in VB. This might suggest that cortical areas differ in the initial order and innervation density of their thalamic afferents (see Fig. 10.4).

Although a rough thalamocortical map can form without a peripheral afferent projection or input (Kaiserman-Abramoff et al, 1980; Rhodes and Fish, 1983; Guillery et al, 1985), there is a wealth of evidence that the physiological input from the periphery influences the local connectivity of thalamocortical axons in sensory areas of the cortex (for review, see Shatz, 1992b). In some areas, such as the barrel field of the mouse (van der Loos and Woolsey, 1973) and layer 4 of the monkey and cat striate cortex (LeVay et al, 1980; Swindale et al, 1981; Stryker and Harris, 1986), these activity-dependent changes are particularly pronounced and rapid during a well-defined 'sensitive period' early in postnatal life, but, even in adults, maps can apparently rapidly shift locally (within 300 μm or so in a few days) in response to deafferentation or even local over- or under-stimulation (Merzenich et al, 1983; Gilbert and Wiesel, 1992). At least some of the remapping in these kinds of experiments is likely to be occurring at the cortical level, but it is important to note that, for the somatic sensory system, rapid changes in response to deafferentation have been described at the level of the thalamus, the dorsal column nuclei and even the spinal cord (see Wall and Egger, 1972; Merrill and Wall, 1978). After peripheral nerve injury in the adult monkey, Garraghty and Kaas (1991) reported that functional reorganization occurred not only in the somatosensory cortex but also, to some extent, in the ventrobasal thalamus.

The organization of immature cortical neuronal circuits is very different from that in the adult and involves the subplate as well as neurons of the cortical plate (Friauf and Shatz, 1991; see Fig. 9.1). Obviously this immature circuitry should be taken into account if we are to understand the changes in cortical representation that can occur during development and to contrast them with superficially similar plasticity in the adult.

It appears likely that the fine grain of cortical topography is adjusted locally on the basis of correlations in the patterns of firing of thalamic axons and the extent of such modification is relatively large in late embryonic and early postnatal life (Jensen and Killackey, 1987b; Guillery, 1986). The modification of arborization patterns of individual thalamocortical axons within the cortical plate has been demonstrated as a result of early sensory deprivation (e.g. Garraghty et al, 1986; Jensen and Killackey, 1987a,b; LeVay and Stryker, 1979; Blakemore et al, 1981; Swindale et al, 1981; Friedlander et al, 1991; Hata and Stryker, 1994). Although such redistributions of thalamic axons are clearly correlated with physiological changes, especially clear for the ocular dominance columns of the primary visual cortex (LeVay et al, 1980), it is not at all clear whether all arbors that survive are functional: some may have few or no synaptic connections. Antonini and Stryker (1993) demonstrated that there was very little difference between the total length of arborization and the total number of branch points of individual geniculocortical arbors after long term (4 weeks) and short term (6 to 7 days) deprivation during the critical period in the cat, but in both cases there was a dramatic reduction of complexity compared to the arbors from LGN cells driven by the non-deprived eye. This implies that both functional and anatomical changes are rather prompt and at least approximately matched in extent during this period of development.

It is conceivable that some of the dendritic modifications of cortical neurons are also influenced by the patterned activity delivered to the cortex. Kossel et al (1995) reported that visual deprivation can change the dendritic morphology of spiny stellate cells in cat visual cortex. Peinado and Katz (1990) demonstrated that infraorbital nerve cut can change the retraction of the transient apical dendrites of spiny stellate cells in rat barrel cortex. We have recently shown that long term blockade of NMDA receptors during the first postnatal week in the rat somatosensory cortex causes a significant fraction of callosally projecting layer 5 cells to fail to retract their apical tuft (Molnár et al, 1997d). This might be due to direct or indirect effect. Although in rat visual cortex, layer 5 neurons express NMDA receptors before and after (P3 and P11) the differentiation of their apical dendrite (T. Doubell and Z. Molnár, 1997, unpublished observations), perhaps their apical dendritic differen-

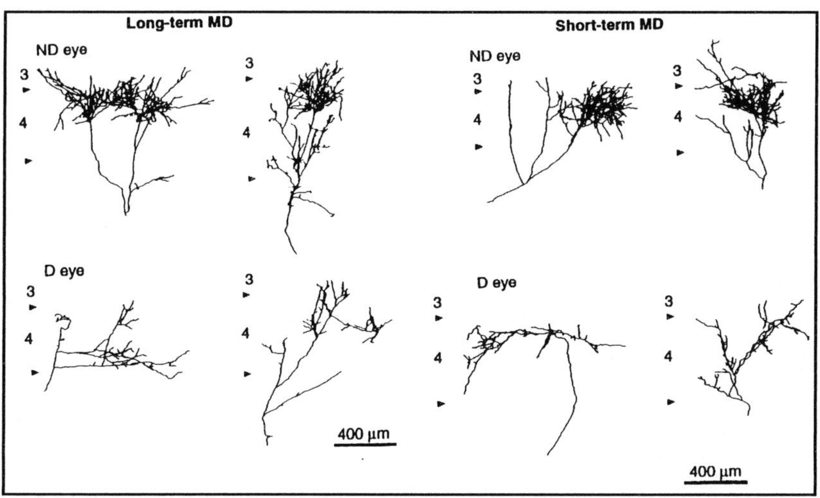

Fig. 9.8. Antonini and Stryker (1993) demonstrated a rapid and dramatic reduction of complexity of individual geniculocortical arbors compared to the arbors from LGN cells driven by the non-deprived eye after long term (4 weeks, left two columns) and short term (6 to 7 days, right two columns) deprivation. Upper row of the reconstructed arbors (all coronal view) is from the nondeprived eye (ND), lower row is from the deprived eye. Cortical layers 3 and 4 are indicated by arrowheads. This implies that both functional and anatomical changes are rather prompt and at least approximately matched in extent during this period of development. Reproduced from Antonini and Stryker (1993) with kind permission of the American Association for the Advancement of Science, Washington, D.C.

tiation is indirectly influenced by the activity of marginal zone cells. We are currently investigating the relative number and morphology of Cajal-Retzius and subplate cells in the very same experimental paradigm. It will be a great challenge of further research to understand the causal relationships involved in cortical circuit formation.

Possible Role of Non-Conventional Cell-Cell Interactions During Cortical Circuit Formation Involving Gap Junctions and Nitric Oxide

There is considerable interest in the mechanisms of neuronal communication before the establishment of synapses. Before cortical cells establish their final connections they might synchronize their activity and begin to communicate through gap junctions and through transmitters which do not require conventional release and uptake mechanisms.

During embryonic and early postnatal life, cortical neurons are locally linked into coupled domains via gap junctions (LoTurco and Krigstein, 1991; Yuste et al, 1992; Dermietzel and Spray, 1993; Nadarajah et al, 1997). Intracellular injections of neurobiotin, a derivative of biotin, revealed that neurons in the cortical plate of neonatal rats are extensively coupled and that this coupling allows the neurons to establish coordinated activity patterns, revealed by Ca2^{+}-imaging in acute cortical slices, before the establishment of synaptic connectivity (Katz, 1993). Ca^{2+} acts ubiquitously as an intracellular messenger and molecular trigger. It plays a part in the generation of the action potential and in transmitter release; it activates a variety of intracellular pathways, including nitric oxide production; it could play a part in growth regulation through its influence on the cytoskeleton; its entry via the NMDA receptor has been implicated in activity-dependent long term potentiation (Kandel, 1981; Eccles, 1983; Lynch and Baudry, 1984). The interplay and transition between the two phases of intercellular communication (between the coupled and connected phases) is uncharted territory. There might be additional cell-cell interaction between cortical cells via nitric oxide (NO). In view of the potential power of NO to regulate the growth of cellular processes, the conductivity of gap junctions and cell death, and both to control synaptogenesis and to modulate existing synapses (see Hölscher, 1997), I shall briefly review the possibility that NO is involved in cortical development. The most exciting potential role of the NO system (which has yet to be demonstrated) is to act during development before the formation of synapses. I shall examine the possibility that NO-producing neurons might play an important role in regulating the formation of cortical circuitry, neuronal excitability and local differentiation of cortical fields.

The Nitric Oxide System

Thomas and Pearse (1961, 1964) first described a sparse population of 'solitary active cells' in the central nervous system, especially the cerebral cortex and striatum, containing the oxidative enzyme *NADPH-diaphorase* (NADPH-d). This enzyme transfers electrons from nicotinamide adenine dinucleotide phosphate to electron acceptors; the histochemical reaction used by Thomas and Pearse involves the NADPH-dependent reduction of a tetrazolium salt, generating a blue formazan reaction product. Since Scherer-Singler et al (1983) introduced a modification of the method to allow its use in aldehyde-fixed tissue, the distribution of NADPH-d-positive cells has been widely studied. They remained a functional mystery until Hope et al (1991) discovered that NADPH-d is a Ca^{2+}-activated *nitric oxide synthase* (NO synthase): it catalyzes the production of NO from L-arginine via a Ca^{2+}/calmodulin-dependent process.

Possible Functions of Nitric Oxide

NO is thought to act as an intercellular messenger (see Vincent and Hope, 1992). It is produced in cells that contain NADPH-d when intracellular Ca^{2+} rises (as a result of activation of ligand-gated or voltage-gated Ca^{2+} channels, or the mobilization of intracellular Ca^{2+} stores). It was demonstrated in the rat brain that the nitric oxide synthase (NOS) neurons express NMDA receptor mRNA (Price et al, 1993). NO is lipid soluble, and therefore crosses the cell membrane. Although it has a very short half-life, only 0.3 sec, its small size allows it to diffuse very rapidly through aqueous and lipid phases to penetrate other neurons (no surface receptor system is needed) within a 3-4 cell radius (Gally et al, 1990). It can then act by stimulating the heme-containing heterodimer, *soluble guanylyl cyclase*, to produce cyclic GMP (Arnold et al, 1977; Miki et al, 1977; Ignarro, 1990), which then regulates protein kinases, phosphodiesterases and ion channels (see Vincent and Hope, 1992).

Soluble guanylyl cyclase can be revealed immunohistochemically (Ariano et al, 1982). Significantly, it occurs in different neuronal types from those containing NO synthase. For instance, soluble guanylyl cyclase is localized in the efferent neurons of many structures (pyramidal cells of cerebral cortex, Purkinje cells of the cerebellum, medium spiny neurons of the striatum), while NO synthase occurs in smaller interneurons (see Vincent and Hope, 1992). Indeed, the very different optimal Ca^{2+} concentrations required for the two enzymes makes it unlikely that they can act in the same cell. This implies that there are separate donor and recipient cells in the NO system.

The most exciting possible role of NO in the central nervous system is in interneuronal signalling. The stimulation of cGMP production and hence of protein kinases and phosphodiesterase could enable NO to act during development and in the mature animal to regulate a variety of interactions between neurons (Vincent and Hope, 1992), including the control of their growth and differentiation (Mattson, 1988).

Cell Types Producing Nitric Oxide During the Development of the Cortex

In all cases in which they have been studied, NADPH-d-positive cells, presumed to be generating NO, are a distinctive subpopulation of neurons. They are usually small interneurons rather than larger principal cells, and their axons distribute locally.

In the adult monkey, individual intensely labelled NADPH-d cells are seen in all layers of striate cortex and in the white matter (Sandell, 1986). They are most abundant (though they constitute only a small fraction of the entire population) in layers 2, 3 and 6, and least common in layers 4 and 5. The white matter is also especially rich in NADPH-d cells (Sandell, 1986; Mizukawa et al, 1988). Vincent and Kimura (1992) and Druga and Syka (1993) obtained similar results in adult rats, except that the NADPH-d cells were distributed with equal density from the white matter to layer 2, a finding which we have confirmed (Molnár et al, 1993).

We have shown in the rat that NADPH-d-reactive cells, which synthesize NO (Hope et al, 1991), first appear around the time of birth (P0) in the subplate and the lower layers of the cortical plate, but not in the remainder of the first-generated cell population located in the marginal zone, suggesting that subplate and marginal cells are phenotypically distinct (Molnár et al, 1993). With increased maturity, NADPH-d-positive bipolar and bistellate cells gradually appear in increasing numbers in the upper layers, while NADPH-d polymorphous cells remain present in the lower layers, including the subplate, and in the white matter (beginning around P9). With further development, the density of all these cells generally declines (starting about P13), reaching approximate adult levels about three weeks postnatal. The processes of these cells establish a ubiquitous, dense network, which infiltrates all areas and all layers of the developing cortex.

Our results show that NADPH-d-positive cells in the rat cortex are of the same morphological classes throughout development: in other words, there does not appear to be transient expression of NADPH-d in an unusual population in which it is not found in the adult. Only certain bipolar, bistellate and multiform neurons ever express NADPH-d. No pyramidal cell, identified by morphological criteria, was labelled at any of the stages examined.

Labelled cells deep in the white matter, whose individual processes could be followed more easily than those of cells in the grey

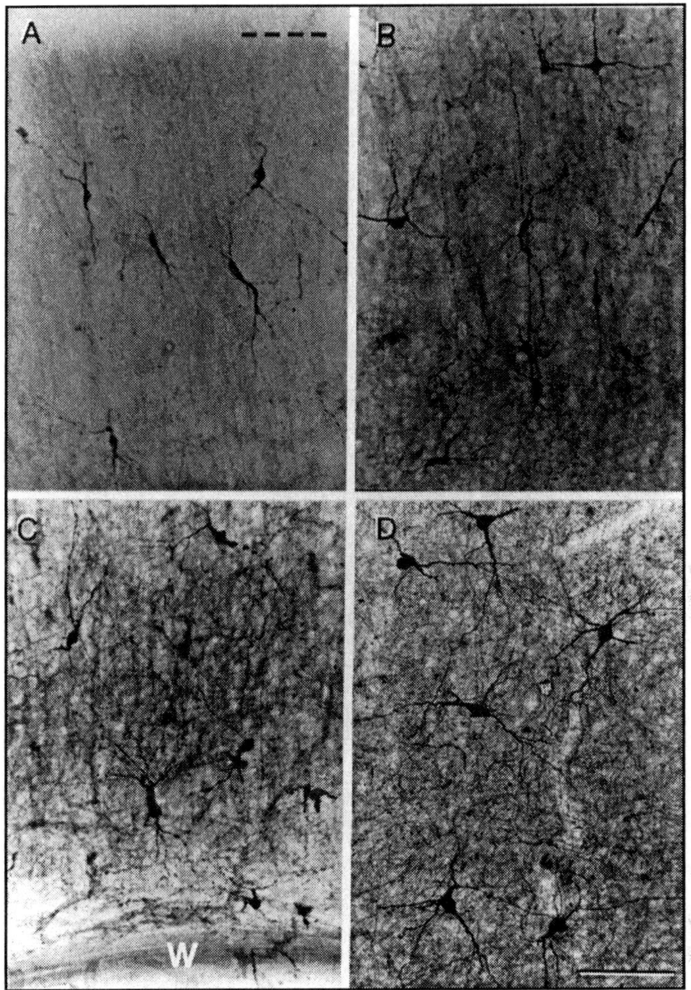

Fig. 9.9. Photomicrographs of NADPH-d-reactive cells in coronal sections from P7 (A and B), P10 (C), and in P24 (D) somatosensory cortex. Scale bar: 100 μm for all. In general, the cells show highly variable morphology, do not have dendritic spines (they are smooth), and are most probably interneurons.
A: This shows the upper layers of the P7 cortex, where most of the labelled neurons are smooth bipolar or bitufted, oriented vertically. The staining in this particular preparation is less intense, and the nuclei appear unstained. The layer 1/2 border is indicated with an interrupted line. Note the beaded (but not spiny) appearance of some of the dendrites. B: The lower part of the P7 cortex (including sublayer 6b, presumed vestige of part of the subplate) contains bipolar and multipolar cells. C: The white matter (W) and layer 5 and 6 of P10 somatosensory cortex. Polymorphous cells are very intensely labelled, revealing Golgi-like staining. D: Polymorphous cells at the layer 5/6 border of P24 cortex. The cells now show clearer processes, which can be traced into the dense, elaborate meshwork between the labelled cell bodies. From Molnár et al (1993b).

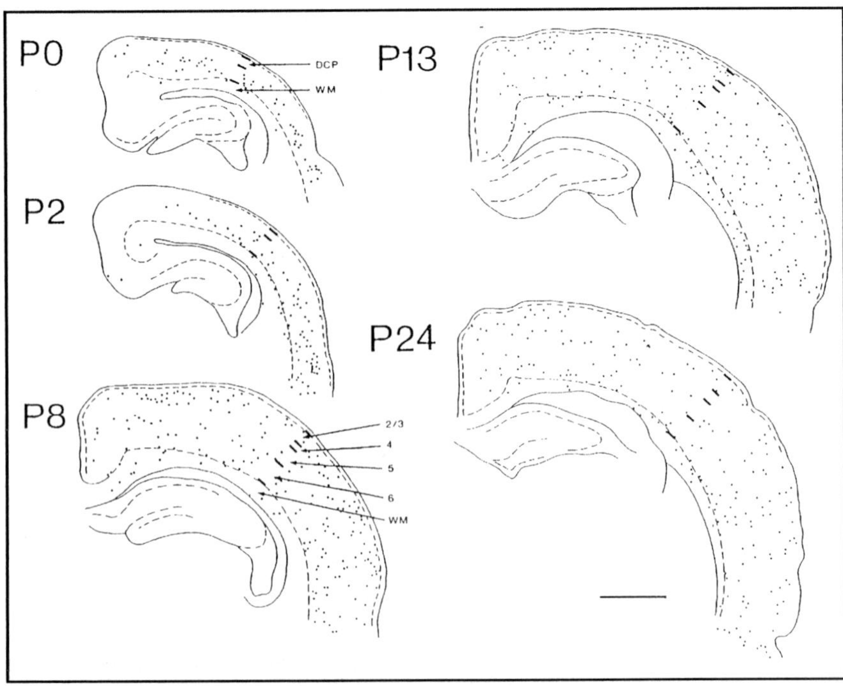

Fig. 9.10. The areal and laminar distribution of NADPH-d cells at different ages (P0, P2, P8, P13, P24) illustrated in camera lucida drawings of coronal sections through comparable sectors of the somatosensory cortex. Each dot represents a stained cell body (but labelled cells in the hippocampus were not drawn). Over the first week, NADPH-d cells gradually appear in all layers (following the general cortical neurogenetic and morphogenetic inside-out gradient) except in layer 1. In the second and third week there is a slight decline in the cell number per unit area. The white matter also contained labelled cells. At P0 and P2, the two superficial bars indicate the borders of the dense cortical plate and the lower two the white matter: at all the other ages, all layers are indicated, the superficial pair of bars indicating the borders of layers 2 and 3 combined. (DCP: dense cortical plate, WM: white matter). Scale bar: 1 mm. From Molnár et al (1993b).

matter, have a tremendous lateral spread, as described by Sandell (1986) and Mizukawa et al (1988). In the cortical layers, the arbors of the major dendritic branches can usually be followed and they are clearly very extensive. It was usually impossible to follow the fine axons of reactive cells far from the cell body, but it seems likely that they also contribute enormously to the widespread dense meshwork of fine NADPH-d-positive processes that covers the entire cortex. It has been previously reported that in the adult rat cerebral cortex SOM, APP and NPY are co-localized with NADPH-d in the same cortical neurons (Vincent et al, 1983). It is not surprising, then, that the majority of NADPH-d reactive cells, at all ages, have morphological appearance very similar to that described for interneurons containing the neuropeptides SOM and NPY (Parnavelas et al, 1988; Hendrickson, 1983; Hendry et al, 1984a,b). Indeed, the timing and laminar sequence of expression of NADPH-d reactivity closely resemble those described for the appearance of NPY and SOM in the rat cortex (Parnavelas et al, 1988, Hendrickson, 1983; Hendry et al, 1984a). Recent research in human cortical developmental disorders (schizophrenia) demonstrated specific alterations in the number of NADPH-d positive neurons (Jones, 1995).

The Fibrous NO Meshwork

One of the most interesting features of NADPH-d reactivity revealed by studies on the cortex is its distribution in a dense, ubiquitous plexus of fine processes (see Fig. 9.9 from Molnár et al, 1993). Although, even at the peak of expression, around P10, only a tiny proportion of cortical neurons stain for NADPH-d, the stained processes, assumed to be the dendrites and intrinsic axons of these cells, form a very widespread meshwork. This plexus extends from the white matter to the pial surface in mature animals, but during the first week of life, there is clearly more fiber staining in the infragranular layers, surrounding the labelled cell bodies.

Since the vast majority of cortical neurons in the adult are immunoreactive for soluble guanylyl cyclase (in the cell body and the projections), they are presumably capable of responding to NO (Ariano et al, 1982; Nakane et al, 1983). This means that the system of NO-synthesizing neurons, though relatively small in number, is, through its extensive fiber meshwork, potentially capable of influencing all those cells that contain guananyl cyclase. It would be interesting to know how the distribution of guanylyl cyclase matures in the developing cortex and to see whether its appearance correlates with the distribution of NADPH-d-reactive cells and their processes.

Possible Functions of the NO System During Cortical Circuit Formation

Ca^{2+} influx (perhaps particularly through the ionophore of the NMDA receptor) is needed for NO production from arginine in the presence of calmodulin and NO synthase. NO needs no release mechanism or surface receptor system: it is membrane soluble and diffuses rapidly in both lipid and aqueous domains. So, although its half life is only 0.3 secs, it is capable of influencing several cells surrounding its site of production. Since the fiber meshwork is very densely distributed, it is likely that all cortical neurons are effectively within reach of a site of NO production.

NO appears to act by stimulating soluble guanylyl cyclase (Arnold et al, 1977; Miki et al, 1977) and may directly activate a cytoplasmic protein ADP-ribosyltransferase (Brune and Lapetina, 1989). It may also play a part in regulating cell death (Wallis et al, 1992). The cGMP produced by activated guanylyl cyclase could have a variety of intracellular effects via its influence on protein kinases and phosphodiesterase. These include influences on the cytoskeleton (and hence growth) and the regulation of gap junction conductivity. The most exciting possible role for the NO system is as an intercellular messenger mediating synaptic modification (see Vincent and Hope, 1992; Izumi and Zorumski, 1993; Chapman et al, 1992). It seems likely that NO released by donor interneurons in the cerebellum triggers long term depression of parallel fiber/Purkinje cell synapses (Ito and Karachot, 1990; Shibuki and Okada, 1991). In the hippocampus, too, there is evidence that NO, again produced by interneurons, might be responsible for mediating changes in transmitter release contributing to long term potentiation in the hippocampus (see Vincent and Hope, 1992). Since similar long term changes in synaptic efficacy are known to occur in certain cortical areas, it is possible that the NO system in the developing cerebral cortex plays a crucial role in such synaptic plasticity. Although plasticity seems to persist, at least in some cortical areas, throughout life, there is no doubt that cortical circuitry is generally especially modifiable during an early postnatal period. It may then be significant that the density of the NO system in the rat appears to reach a peak just after the first postnatal week and then to decline somewhat to adult values (Molnár et al, 1993).

As well as its quite likely role in mediating synaptic modification, NO could contribute specifically to many other processes during development. It is interesting to note that there is a similar age-related pattern of NADPH-d reactivity in the rat retina, where NADPH-d cells reside in the inner nuclear layer (INL) and their diffusely labelled dendrites spread in the inner plexiform layer (IPL). The soma sizes and dendritic fields of NADPH-d cells are largest and most prominent between P11 and P15, and reduce with further development (Mitrofanis, 1989). The period of their maximum size coincides closely with the first appearance of amacrine and bipolar cell synapses in the IPL (Horsburgh and Sefton, 1986) and with the beginning of electrical activity (Weidman and Kuwabara, 1968). It seems possible that NO plays a role in synaptogenesis in the inner plexiform layer. The possible modulatory effect transmitted by NO might be responsible for the synchronization of activity patterns during development (Wong et al, 1991).

The demonstration of NADPH-d-positive interneurons in the developing cerebral cortex raises the possibility that NO diffusion could supplement gap junction and synaptic transmission as a form of cell-cell communication during development. NADPH-d cells appear in the cortex when coupling of the immature neurons through gap junctions is rapidly declining (Yuste et al, 1992). The NADPH-d fiber plexus could synchronize activity patterns at a stage when the majority of the cortico-cortical synapses have not yet formed, but the gap-junction coupling of developing neurons has declined. Its unique characteristics suggest that NO could play an essential role in the establishment of the precise cortical microcircuitry. The 'functional crystallization' of cortical circuitry might need the local 'binding' between neighboring neurons that NO could provide, before the establishment of most cortico-cortical synapses. In spite of all the initial excitement surrounding the NO system no study demonstrated its direct role in cortical plasticity. On the contrary, Gillespie et al (1993) demonstrated that a significant reduction of nitric oxide synthase activity after the chronic administration of inhibitor had no effect on the ocular dominance shifts that normally result from ocular dominance deprivation in kittens. Nevertheless it is too early to discard its role as a signalling molecule in the early activity dependent phases of cortical circuit formation.

Self-Organizing Capacity of the Cerebral Cortex and Early Thalamic Innervation: Conclusions

We can consider, as a first approximation, that the basic initial topography of the thalamic projection is established during embryonic life. From different thalamic regions, fibers arrive to their target cortical areas at different times along roughly parallel trajectories. Fiber bundles preserve their neighborhood relationships as thalamic regions interlink with their appropriate cortical fields. But then different mechanisms presumably operate subsequently:

1) to form branches within the subplate;
2) to permit individual axons to cross each other and disorganize slightly as they grow into the cortical plate;
3) to control the local sculpting of axon arbors within the 200-300 mm field of their initial arborization; and
4) to enable individual synapses to be adjusted in strength within their anatomical distribution. Even prenatally, *multiple* mechanisms are involved in determining thalamic innervation, but additional, activity-dependent processes appear to take over after the basic innervation is in place and can substantially modify the topography near the termination site.

Chapter 10

Anophthalmic, Albino and *Barrelless* Mutants Help Us Understand the Multiple Mechanisms Involved in Thalamocortical Development

The initial topography of thalamocortical connectivity is accomplished before the arrival of the peripheral afferent input to the thalamus (chapter 8), but then this topography is substantially altered as the fibers accumulate and enter the cortical plate. This latter phase is no longer autonomous; it is orchestrated by the early input from the periphery (chapter 9). These ideas lead to the suggestion of two phases of thalamocortical development: one which is independent from the periphery and a second which is controlled by the input from the sensory periphery. Depending on the relative dominance of the two phases, the peripheral input can substantially modify the pattern of final projections. These ideas are not new. Cogeshall (1964) noted that: "It seems likely that optic, tactile, and other incoming pathways plug into circuits that are, to some extent, already formed". Guillery et al (1985) suggested that the "mapped projection that grows from the retina to the lateral geniculate nucleus must in normal development conform with the geniculocortical map in order to produce the normal pattern of retinogeniculocortical projections of the adult". Guillery (1986) also considered the possibility that, *early* in development, the geniculocortical and retinogeniculate systems are *initially* "two independent processes, each programmed to produce its portion of the final visual pathway". In this chapter I shall examine some mutant animals, each of which could be considered as a clever experiment of nature, which

Development of Thalamocortical Connections,
by Zoltán Molnár. © 1998 Springer-Verlag and R.G. Landes Company.

give insight into the interplay between the autonomous and activity-dependent processes and help us to understand further the development and plasticity of thalamocortical connections. For the sake of clarity I shall discuss them in two major groups:

1) In the first group of mutants the abnormalities have indirect effects on the thalamocortical development. The initial thalamocortical development is not disturbed, but then the changed peripheral input to thalamus (e.g. due to the lack or altered development of sensory organs and their connections to thalamus) is confronting the early autonomously developed thalamocortical connectivity with information which is in conflict with the one which developed autonomously. This can lead to the reduction of the primary cortical sensory area (anophtalmic mutants or binocular enucleations) or to the altering of the cortical topography (reversal, map compression in albinism or monocular enucleation).
2) In the second group of mutants I shall discuss the NMDA receptor knockout mouse and various *barrelless* mutants where mechanisms involved in the normal thalamocortical interaction are disturbed due to the lack of a receptor, or ligand of unknown origin. These problems can lead to abnormal arbor formation and/or stabilization and a failure in cortical pattern formation. Studies on these mutants help us to understand more of the multiple mechanisms involved in various phases of thalamocortical development.

Thalamocortical Topography in Anophthalmic Mutants and After Early Binocular Enucleation

Anophthalmic mice (Cullen et al, 1976; Godement et al, 1979; Kaiserman-Abramoff, 1980), as well as bilaterally enucleated hamsters (Rhodes and Fish, 1983) and ferrets (Guillery et al, 1985), develop geniculocortical and corticogeniculate interconnections with an essentially normal topography (within the precision of tracing methods). If a monkey is binocularly enucleated at an early fetal stage, before full innervation of the LGN by the optic tract, about two-thirds of the LGN cells degenerate and the total size of area 17 is proportionately reduced (Rakic, 1988). This experiment is commonly cited as evidence that cortical areas develop without thalamic afferents. Unfortunately the study documents only the adult stage. It remains to be established whether the cell loss in the LGN occurs so rapidly that the whole topography of the *initial* outgrowth of thalamocortical axons is affected (unlikely, but cannot be excluded at present), or whether the projection forms normally and is subsequently reduced in areal extent as geniculate cells die. Without this no inferences can be made of the causal relationships.

In the rat the early topography is assumed by the time of thalamocortical fiber arrival to the subplate (Molnár and Blakemore, 1990a, 1995a,b; Catalano et al, 1996). The establishment of this early topographically organized thalamocortical projection, around the 16th day of gestation in the rat, cannot be dependent on peripheral input because the optic tract has not grown into the LGN and the ventrobasal nucleus has not received peripheral input from the whiskers at this time. This, of course, does not exclude the possibility of a *later* modification.

Evidence for the early gross specification of thalamocortical topography and for the autonomy of early thalamocortical development also comes from the results of developmental manipulations that induce sensory afferents of one modality to project to central targets of a different modality (reviewed by Sur et al, 1990). In spite of such manipulations (including cortical lesions), there is very little change in the distribution of thalamocortical pathways, suggesting that these have already been specified at the time of these manipulations. After a lesion of the visual cortex, superior colliculus and LGN of a newborn ferret (corresponding in development to an E16 rat fetus), fibers of the optic tract terminate in the *'auditory'* thalamus and visual information is relayed to the region of temporal cortex (the original target of the 'auditory' thalamus) that would normally become primary auditory cortex.

Similarly, in the naturally blind mole rat, the auditory afferents project to the LGN and, through the otherwise normal geniculocortical projection, this auditory input is relayed to what would be the visual cortex (Bronchti et al, 1989, 1991; Heil et al, 1991). In each case, the matching of a given thalamic area to its original cortical region is unchanged, indicating that early specification of thalamocortical development is independent of the periphery, and is based on some kind of innate automatism.

Altered Thalamocortical Topography in Albinism and After Early Monocular Enucleation

Perhaps one of the most intriguing experiments of nature for studying the interplay between the autonomous and activity dependent processes is the albino mutant. Due to a gene defect, there are abnormally high levels of some intermediate metabolite of melanin synthesis, which might cause delay and disturbance in cell generation in the retina (Jeffery, 1997). Most probably due to this delay, many developing fibers from the temporal retina are inappropriately routed at the optic chiasm into the contralateral hemisphere (Lund, 1965; Guillery, 1986). Altering the ratio of the crossed and uncrossed retinal input into the thalamus will cause secondary changes in the thalamocortical topography. Interestingly, this alteration can manifest itself as cortical map reversal or compression (for review see Guillery, 1986). Early monocular enucleation can lead to similar misrouting of retinal fibers and to similar secondary changes in the thalamocortical topography (Trevelyan and Thompson, 1992) indicating that changes on the thalamocortical level are secondary and not directly linked to the mutation.

We still do not know why the same abnormal routing of retinal fibers in albinos can lead to different patterns of thalamocortical connections in some animals and not in others. Guillery (1986) suggested that the "relevant variable may well be environmental, involving diet, temperature" or the relative dominance of the retinal or the non-retinal process in producing Boston or Midwestern patterns in Siamese cats. It is indeed conceivable that the final pattern depends on the relative maturity of the thalamocortical connections at the time when the retinal influence is beginning to exert its influence and begins to signal the conflicting information from the sensory periphery. If this happens relatively early then there is scope for modification and will lead to the Boston pattern, where the geniculocortical projections are reversed, of the abnormal segment of layer A1. When it is too late to modify the thalamocortical topography, the abnormal geniculate representation will reach the cortex without correction and this leads to the intracortical supression of the entire layer A1 (see Fig. 10.1). It would be interesting to follow the initial phases of development in both animals. At present we know very little about the nature and mechanisms of these secondary changes. At what stage and how will the periphery influence thalamocortical topography? What is the anatomical substrate of the reversal in the Boston pattern and where do the thalamocortical fibers reverse?

I mentioned (in chapter 8) that in the adult macaque monkey the map between LGN and cortex is reversed mediolaterally but not anteroposteriorly (Conolly and Van Essen, 1984) and that the geniculocortical fibers in the white matter must reverse positions along one major axis but not the orthogonal one. Nelson and Le Vay (1985), using pairs of tracer injections to reveal fiber order in the adult cat, did indeed report crossing of thalamic axons in the mediolateral but not in the rostrocaudal axis of the optic radiation. This anisotropic decussation occurred, however, only a short distance (2-500 µm) below the visual cortex, at a depth that might well have been in the transient subplate during development (see Fig. 8.2). I argued in chapter 8 that the decussation pattern seen in the adult might represent a modification of the array of thalamocortical fibers that occurs during or at the end of the waiting period. The initial layout of thalamic axons might be very clear and regular in embryonic life, but might become substantially altered with further development *near the termination site*. However the fiber ordering at the *rest of the pathway* may preserve the simple juvenile topography.

This process might be general and employed in various different cortical regions. The body surface is represented as a sensory homunculus in the primary somatosensory cortex. In humans this map is fragmented in an interesting way, differently from the motor homonculus (see Szentágothai, 1977). It would be interesting to examine the fiber ordering all the way from the somatosensory thalamus, as Nelson and Le Vay (1985) did for the visual projection. Direct injection of tracers at the internal capsule would be necessary to resolve this question.

It would be equally interesting to examine whether these map reversals require activity or a specific receptor in normal and mutant animals. Further studies on these mutants could help determine the interplay between the autonomous and periphery-dependent processes of thalamocortical development and could even shed light on the pathology of strabismus.

Disturbance in the Thalamocortical Interactions and Cytoarchitectural Differentiation in *Barrelless* Mutant Mice

The earliest axonal inputs to the cortex could be capable of constituting an *extrinsic signal* that plays a part in triggering other specific regional differentiation, perhaps even leading to most of the repertoire of areal specializations (see chapter 1). In the extreme version of such a proposal, the cortical mantle might be generated as an

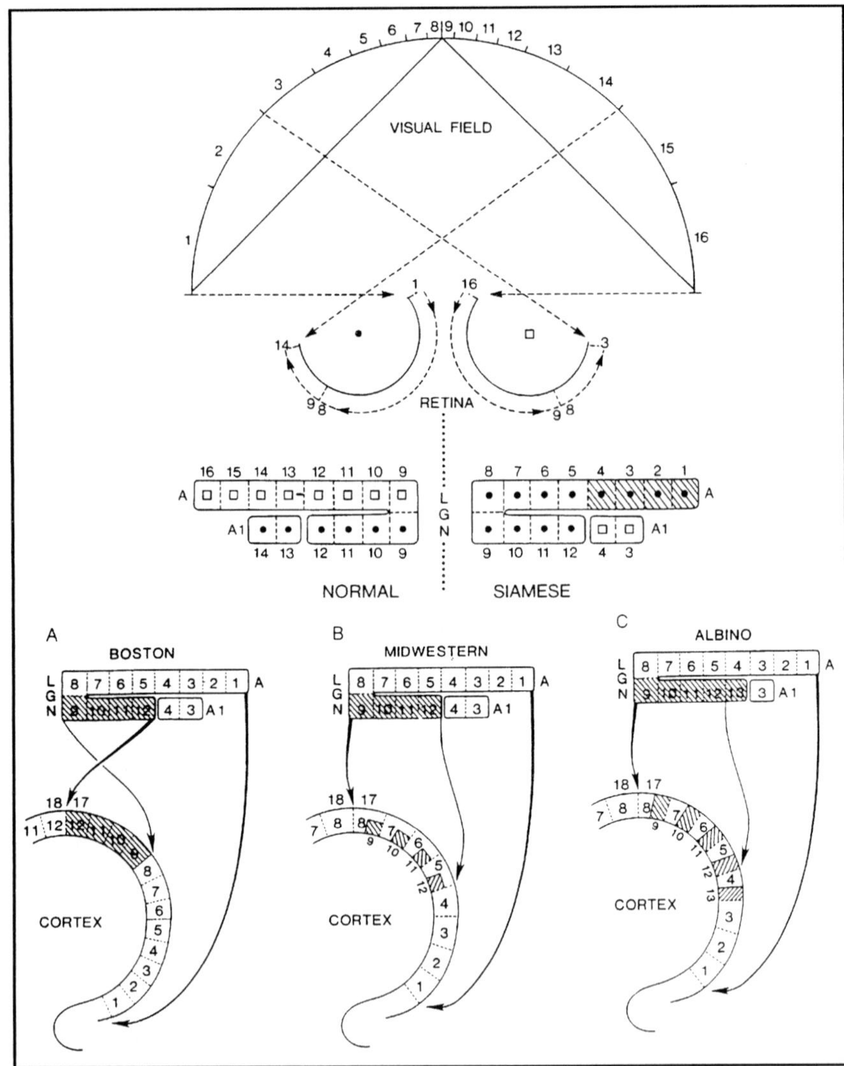

Fig. 10.1. Secondary changes in thalamocortical innervation patterns in albinos (Guillery, 1986). Upper diagram demonstrates the visual field and the normal (left) and abnormal (right, Siamese) representations in the lateral geniculate nucleus. Circles indicate sectors linked to left eye; squares indicate sectors linked to the right eye. In Siamese cats, due to abnormal routing of the retinogeniculate fibers, part of the ipsilateral visual field is abnormally represented in layer A1 (sectors 9-12) with reversed polarity. Lower diagrams demonstrate the Boston, Midwestern and albino patterns of thalamocortical projections (A,B and C respectively) altered secondarily. In the Boston pattern the thalamocortical projections reverse and change the pattern of visual representation in area 17, whereas in Midwestern or albino patterns the lateral geniculate nucleus innervates area 17 without correction. The Midwestern pattern is associated with intracortical supression of the input from the entire layer of A1 (indicated with

entirely uncommitted 'protocortex' (O'Leary, 1989) from which regional specialization is carved out by the influence of axons arriving at early stages. This hypothesis has aroused considerable interest in the relationship between the development of the first extrinsic connectivity of the cortex and its cytoarchitectonic differentiation.

In this context, the rodent somatosensory cortex has been extensively studied. The distinctive feature of this cortical area is the representation of individual whiskers as individual patternings of layer 4 cells and incoming fibers to form the whisker barrels. This has provided an excellent model system for numerous studies. These demonstrated that the postnatal maturation of the barrel field (van der Loos and Woolsey, 1973) requires an intact sensory periphery and intact afferentation during a critical period (Woolsey and Wann, 1976; Belford and Killackey, 1980; Jeanmonod et al, 1981). The normal barrel formation coincides with patterning of extracellular molecules like peanut agglutinin (Cooper and Steindler, 1986; McCandlish et al, 1989) and semaphorin (Skaliora et al, 1996), but this pattern appears subsequently to the patterning of the thalamic fibers. Application of more sensitive tracing techniques, including lipophilic carbocyanine dyes (Erzurumlu and Jhaveri, 1990) and acetylcholinesterase histochemistry (which provides an early transient marker of thalamocortical axons; see Robertson, 1987) has revealed the periphery-related patterning of thalamocortical afferent termination at least 1 day earlier than previous markers for barrels (reviewed by Schlaggar and O'Leary, 1993). Another group of experimental results in support of the external influence on cortical differentiation comes from transplantation studies. If an explant of E17 rat visual cortex is transplanted into the position of the putative barrel field in a newborn rat, then it is capable of developing at least some of the cytoarchitectonic features of barrels (patterned distribution of thalamocortical afferents, layer 4 cells and matrix

smaller hatched squares for the abnormal segment). In the albino pattern, where the chiasmatic misrouting is smaller and the reversed map involves only part of layer A1 in LGN, two distinct cortical maps will be established with reversed polarity. We still do not know why similar abnormal routing of retinal fibers in albinos can lead to different patterns of thalamocortical connections in some animals and not in others, but it might be related to the differences in the status and maturity of the geniculocortical projections at the onset of retinal dominance. Reproduced from Guillery (1986) with kind permission of Elsevier Science Ltd, Kidlington, England.

molecules) ordinarily unique to the primary somatosensory cortex (Schlaggar and O'Leary, 1991). A number of other cortical transplantation studies in rodents have demonstrated that heterotopic cortical afferent (Chang et al, 1986; O'Leary et al, 1992) and efferent (Stanfield and O'Leary, 1985; Barbe and Levitt, 1991) connections develop appropriately for the new location, rather than the original site of the cortical explant. Increasing evidence suggests that although a certain basic intrinsic cortical organization can develop without any afferent connectivity, some cytoarchitectural attributes of at least the major sensory cortical regions are somehow imposed by the influence of specific thalamocortical fibers (Woolsey et al, 1979; Killackey and Shinder, 1981; Welker and van der Loos, 1986; O'Leary, 1989). There are at least two types of mutants, where in spite of the presence of functioning thalamocortical connections the characteristic barrel pattern fails to appear (Cases et al, 1995; Welker et al, 1996). In one of them it is due to excess serotonin during the critical period (Cases et al, 1996). The other, although at first it seems similar, has a different, unknown origin. The locus was mapped to the proximal segment of chromosome 11 (Welker et al, 1996). Both strains of mutant mice could provide interesting model systems to study the interactions between the developing thalamic input and the forming cortical circuitry. They could provide further information on the role thalamic fibers play in triggering some aspects of regional cortical differentiation and defining boundaries of major cortical fields. I shall deal with the mutant mouse *barrelless* in more detail (Welker et al, 1996).

The Problem of *Barrelless*

It was demonstrated on Nissl-stained sections in some adult *barrelless* mice that barrels failed to appear. Labelling with dextran in primary somatosensory cortex (S1) of adult mutants revealed that thalamocortical connections terminate in layers IV and VI, but they are continuously distributed in layer IV, rather than being confined to barrel-like structures and leaving septa between barrels relatively free of label as in normal mice (Welker et al, 1996). Reconstructions of individual thalamocortical axons of the adult showed that, in normal animals, they terminate in a confined area of S1, whereas in *barrelless* they terminate in a large area (Welker et al, 1996; Gheorghita-Bächler et al, 1996).

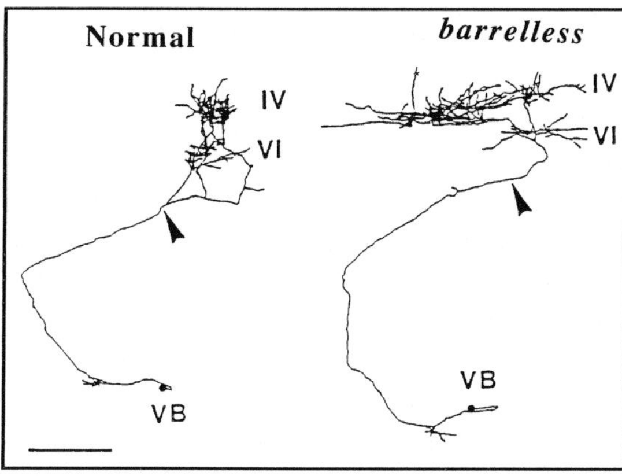

Fig. 10.2. Coronal view of individual thalamocortical axon reconstructions labelled with dextran in normal and *barelless* adult mice. The two examples demonstrate the abnormal thalamocortical fiber formation and the normal laminar targeting (layers IV and VI) of thalamocortical axons in *barrelless*. In normal, thalamocortical arbors terminate in a confined area of primary somatosensory area, whereas in *barrelless* they terminate in a large area. Arrowheads indicate the site of entering into the primary somatosensory cortex. Scale bar 500 µm. Reproduced from Welker et al (1996) with kind permission of the American Association for the Advancement of Science, Washington, D.C.

There are many questions that need to be answered. Are the earliest thalamocortical projections distributed in topographic order and without gross exuberance in *barrelless*? If so, when is that order established and why is it not maintained in the mutant? At what stage will the mutation affect thalamocortical development and why do the earliest input fibers fail to interact with the developing cortex to form the characteristic barrel pattern in the mutant?

We set out to examine the hypothesis that the cortical phenotype is due to a disorganized distribution of thalamocortical projection neurons in VB early in embryonic and postnatal development by tracing the thalamocortical connections with carbocyanine dyes anterogradely and retrogradely in embryonic, early postnatal and adult brains. We also examined the possible areal differences in the developing and adult thalamocortical topography of visual and somatosensory cortices in normal and *barrelless* animals (Molnár et al, 1996a).

Our study demonstrated that the initial laminar and tangential targeting of the thalamocortical connections is not affected by the mutation and is indistinguishable from the normal. Anterograde tracing from VB revealed an identical laminar pattern of thalamocortical fiber ingrowth at all stages examined. Thalamocortical fibers arrive at the cortex at E16, begin to invade the cortical plate at E18 and begin to arborize below the dense cortical plate at P0-P2 in both phenotypes. At early postnatal ages no complete barrel patterns were observed.

Lack of Formation and Stabilization of the Barrel Patterns in the Mutant

In Nissl-stained sections through layer IV of S1 in adult normal mice individual barrels can be observed, but they are missing in mutants. In certain species (e.g. rabbit and gray squirrel) barrels appear only transiently and the pattern disappears in adulthood (Woolsey et al, 1975; Rice et al, 1985; Rice, 1995). In *barrelless* no barrels appear at any time of development (Molnár et al, 1996a).

Agmon et al (1995) demonstrated that the barrel pattern is not the result of pruning. In normal mice the thalamic axons enter the cortex in an organized fashion and their ingrowth into the cortical plate is precise from the very beginning. There might be substantial rearrangement within the subplate and lower cortical plate by forming transient side branches, but this mechanism is somewhat earlier.

In spite of the substantial differences in the appearance of the adult thalamocortical arbors in normal and *barrelless* animals, there are parameters which remain constant for normal and *barrelless* thalamocortical axons. Individual thalamocortical axons of the normal adult mouse terminate in a confined area of S1, whereas in *barrelless* it terminates in a large area. In *barrelless* the total axon length was significanly longer and the tangential areal extent of arbors was approximately three times larger than those in normal animals. The number of boutons in *barrelless* is greater than in normal; however, the number of boutons per axonal length is 7-8 boutons/100 μm axon for both strains (Gheorghita-Bächler et al, 1996). This indicates that the *barrelless* mutation only affects certain aspects of thalamocortical arbor formation, while some remain unaffected. It seems that the extent of axon growth and bouton formation are regulated separately.

What Mechanisms Fail to Constrain the Growth of Thalamocotical Arbors Within the Barrels in Barrelless *?*

Tangential sections through layer IV of S1 of normal mice reveal labelling (dextran) that is confined to the inside of the barrels, leaving septa between barrels relatively free of label. In *barrelless* mice, labelling forms a continuous zone. What molecular clues limit the arbor formation? Are they activity related? Recent in vitro work of Bolz and colleagues demonstrated that thalamic fibers exhibit different growth patterns on membrane preparations from normal and *barrelless* animals (E. Welker, personal communication).

Is the Abnormality in Thalamocortical Topography Restricted to the Primary Somatosensory Cortex in barrelless*? Areal Differences in the Thalamocortical Innervation*

Putative somatosensory and visual cortices of fixed brains of normal (NOR) and *barrelless* mice showed no area-specific abnormality of thalamocortical projections. Pairs of single crystals of carbocyanine dyes (DiI and DiAsp) placed at 250 or 500 μm distance from each other into the putative visual cortex revealed very similar patterns of backlabelling in adult and normal early postnatal animals.

At early postnatal ages, quantitative analysis demonstrated a similar low percentage of double-labelled cells in VB and LGN of both groups of mice, both in somatosensory and visual thalamus (VB: NOR = 0.6%, brl = 0.7%; LGN: NOR = 0.5%, brl = 0.4%). However, in LGN, the number of labelled cells were 3-5 times smaller than in VB in b*oth* strains of mice (Molnár et al, 1996a). This suggests that the abnormality in *barrelless* is not due to spatial organization of the cell bodies of the projection neurons in VB or LGN, but merely due to a cortical defect of pattern formation after the initial correct ingrowth of thalamocortical axons. It is yet to be determined whether this error in pattern formation among thalamocortical axons is limited to somatosensory areas. Perhaps homologous errors occur in other cortical areas in adult *barrelless*.

The Role of Activity in Barrel Formation

Perhaps the best studied model system for the question of activity dependent modification of cortical cytoarchitecture is the barrel cortex of the rodent (van der Loos and Woolsey, 1973). Numerous studies demonstrated that blocking the flow of sensory information

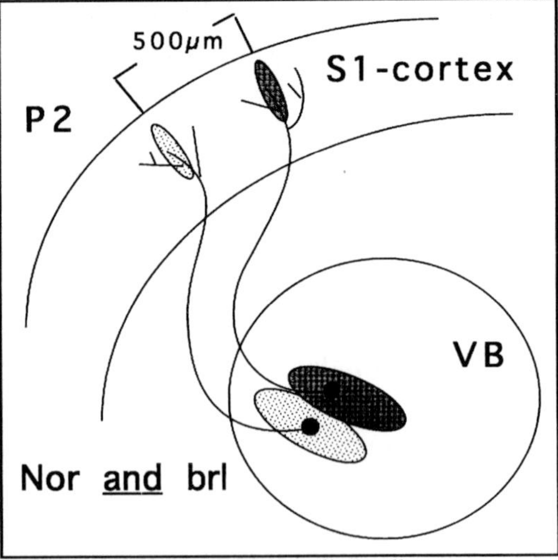

Fig. 10.3. Normal spatial organization of the cell bodies of thalamic projection neurons in VB in *barrelless* and normal mice revealed by retrograde labelling. At the second postnatal day, pairs of single crystals of carbocyanine dyes (DiI and DiAsp) placed at 500 μm distance from each other into the putative somatosensory cortex revealed very similar patterns of backlabelling in normal early postnatal animals. Quantitative analysis demonstrated a similar low percentage of double labelled cells in VB both groups of mice. Similar labelling from the visual cortex at postnatal day 2 showed no difference in the pattern of backlabelled LGN cells or in the number of double labelled cells. However, in LGN, the scatter was somewhat larger and the number of labelled cells were 3-5 times smaller than in VB in both strains of mice. These might suggest that the *barelless* abnormality is due to a cortical defect of pattern formation after the initial correct ingrowth of thalamocortical axons. Reproduced from Molnár et al (1996a) with kind permission of the Society for Neuroscience, Washington DC.

from the whiskers to the somatosensory cortex at various levels during the first postnatal week prevented the establishment of the characteristic patterning of layer 4 cells in S1 (see Schlaggar and O'Leary, 1992, 1993; Killackey et al, 1995). The cytoarchitectonic features of normal whisker barrels emerge in immature occipital cortex transplanted into the barrel area of newborn rats (Schlaggar and O'Leary, 1991), suggesting that barrel formation does not require pre-established intrinsic cues in the cortex itself. Interestingly the role of activity in the basic pattern formation has not been demonstrated convincingly. As I discussed in chapter 8, thalamic axons are capable of transmitting impulses to cortical cells during the embryonic and early prenatal period from E18/19 (Higashi et al, 1996) and this activity could begin to influence the target structure. However, blockade of postsynaptic activity from shortly after birth, by local application of an NMDA-receptor antagonist, does not prevent anatomical barrel formation in the normal barrel cortex of the rat (Fox et al, 1996), nor in the hamster, in which the cortex and thalamic innervation are less developed at birth (Chiaia et al, 1994). Fox et al (1996) does

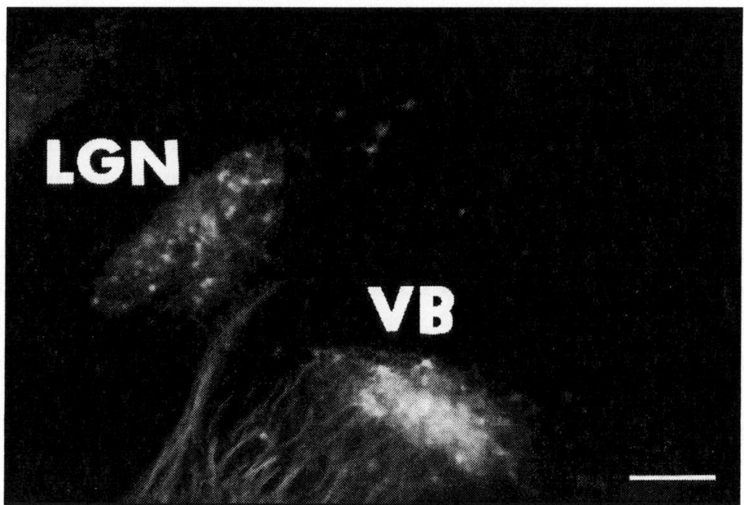

Fig. 10.4. Different spatial organization of the backlabelled cell bodies of thalamic projection neurons in VB and LGN in *barrelless* at postnatal day 2. Pairs of single crystals of carbocyanine dyes (DiI and DiAsp) placed at 500 µm distance from each other (medio-lateral) into the putative somatosensory and visual cortex revealed very similar patterns of backlabelling, but there was a considerable difference between LGN and VB in both genotypes. In both strains of mice, in LGN, the scatter of backlabelled thalamic projection cells was somewhat larger and the number of labelled cells were 3-5 times smaller than in VB. From Molnár et al, 1996a.

show disruption of thalamocortical and columnar organization at a functional level.

Nevertheless there is evidence from the NMDA receptor subunit knockout mice experiments (Li et al, 1994; Tonegawa et al, 1995; Kutsuwada et al, 1996) that the induction of barrelettes in the brainstem trigeminal nuclei requires NMDA-receptor activation and/or postsynaptic activity. None of these mice live long enough to examine development of patterened connectivity along the thalamocortical pathway. Recently, Iwasato et al rescued the NMDAR1 knockout mice by ectopic by ectopic expression of NMDAR1 transgene. These mice live for several weeks and whisker-specific patterning of pre and postsynaptic elements all along the trigeminal pathway, including the barrel cortex, is absent (Reha Erzurumlu, personal communication). In some forms of NMDA knockout mice Erzurumlu and colleagues recently showed that barreloids develop in the thalamus, but no barrels appear in the primary somatosensory cortex (R. Erzurumlu, personal communication).

The role of excess serotonin (5-HT) in barrel formation was elegantly demonstrated in a transgenic mouse line (Tg8) in which the integration of an interferon beta transgene has disrupted the gene encoding monoamine oxidase A (MAOA). As a result, 5HT concentrations in the brain increased up to 9-fold in pups while NA concentrations increased up to 2-fold (Cases et al, 1996). In this trangenic mouse, similarly to the *barrelless* mutant mouse (Welker et al, 1996), normal pattern formation exists in the thalamus and trigeminal nuclei, but the characteristic barrel pattern in layer 4 fails to appear in the primary somatosensory cortex. Thalamocortical projections labelled with biocytin arborized in a continuous band within layer 4 with gross topographic organization (Cases et al, 1996). Inhibition of the serotonin synthesis during the first postnatal week restored the barrel formation in the transgenic mouse line and transient MAOA inhibition, especially during thalamocortical fiber ingrowth (P0-P4) projections, phenocopying the barrelfield alterations of Tg8 mice. The role of serotonin in thalamocortical development is still not fully understood, but Lebrand et al (1996) recently demonstrated that developing thalamic neurons take up exogenous 5-HT through high affinity uptake sites located on thalamocortical axons and terminals and they might use it. Thalamic neurons do not synthetize 5-HT, but the internalized 5-HT might be used for signaling.

Conclusions

In this chapter I argued that studies on mutants could give insight into the mechanisms involved in the development of the thalamocortical pathway. It is very likely that further relevant mutants will be discovered. These mutations *directly* or *indirectly* will effect the normal development and will help us to understand further aspects of the development and plasticity of thalamocortical connections. In this chapter I discussed two major groups of mutations:

1) In the first group of mutants the abnormalities have *indirect* effects on the thalamocortical development. The initial thalamocortical development is not disturbed, but then the changed peripheral input will alter the thalamocortical projections in an early phase. The peripheral input might present the early autonomously developed thalamocortical connectivity with information which is in conflict with the one which developed autonomously. Depending on the relative dominance of the two

phases, the interaction can lead to the reduction of the primary cortical sensory area or to the altering of the cortical topography (reversal, map compression in albinism or monocular enucleation). These mutations could be extremely useful in understanding the interactions between the autonomous and the activity dependent phases of thalamocortical development.

2) In the second group I discussed various mutants where the thalamocortical projections fail to develop the characteristic arborization and in spite of the normal sensory periphery the characteristic cortical cytoarchitecture does not develop. In these mutants the mechanisms involved in the normal thalamocortical interaction are disturbed, primarily due to the lack of a receptor or ligand of unknown origin. These problems can lead to abnormal thalamocortical arbor formation and/or stabilization and a failure in cortical pattern formation. These mutants could reveal further important aspects of thalamocortical interactions.

There are numerous recently described spontaneous neurological mouse mutants where the gene defect has been identified (e.g. *tottering* and *opisthotonos*) and they might all be interesting in helping to understand other factors involved in cortical development (Flecher et al, 1996; Street et al, 1997). Perhaps genetic screening will be soon extended to mammals to identify genes responsible for formation of cortical connections and cytoarchitecture in the developing cortex as it was recently applied for the vertebrate retinotectal system by F. Bonhoeffer and his colleagues (Karlstrom et al, 1997). Obviously there is much to be learned about cortical development and adult plasticity from studying these mutants, including the multiple mechanisms involved in various phases of thalamocortical development.

Chapter 11

Mechanisms with Evolutionary Origin: Phylogeny Compared with Ontogeny

Last century a scientist called Farkas Kempelen claimed that he had constructed a machine which could play chess. He travelled around the World and the machine scored victory after victory. Everybody admired the clever machine. On one occasion, however, someone in the audience stood up, pointed to the machine and said: There must be a man in this machine. Everybody laughed at the suspicious person as he went to the machine and stabbed the model figure with his knife. The laughter became stronger, because the machine could still play. However, it turned out that in Farkas Kempelen's machine there was indeed a retired, legless Polish cavalry officer, a chess-playing expert, hidden under the chess board.

János Szentágothai, the celebrated neuroanatomist, once told this story in Budapest to illustrate our way of imagining the higher cognitive functions of the mind. The moral of this story is that we are all victims of the limitations of our ways of thinking, which is reflected in our behavioral repertoire, emotions, culture, history, art and even our science. Our instinctive ideas about the universe, evolution and even the structure of the brain are inevitably rather anthropomorphic. A similar story, entitled *Maelzel's Chess-Player*, was published in English by Edgar Allen Poe in his *Tales of Mystery and Imagination* (1845).

In this final chapter I shall be slightly more speculative and would like to expose some current ideas on thalamocortical development and on the possible relationship between ontogeny and phylogeny.

Development of Thalamocortical Connections,
by Zoltán Molnár. © 1998 Springer-Verlag and R.G. Landes Company.

Why Does Development Employ Transient Cell Populations?

During the early embryonic development of the mammalian pallium there are 'mysterious' transient cell populations in the cortical subplate and marginal zone and in the ganglionic eminence, the so-called perireticular cells (see chapter 6). These fascinating cells are believed to assist in pathfinding in the external cortical circuitry by providing temporary targets for corticopetal and corticofugal projections. They are believed to escort the developing corticopetal and possibly corticofugal projections by forming an early scaffold or acting as guidepost cells within the internal capsule (McConnell et al, 1989; Blakemore and Molnár, 1990; Ghosh et al, 1990; Mitrofanis and Guillery, 1993; Métin and Godement, 1996). Besides their long range connections, these cells seem to integrate into the early cortical circuitry, by establishing numerous local connections (Friauf et al, 1990; Higashi et al, 1996). Studies in this field have increased dramatically in the last ten years and we are beginning to understand the strategic position these cells have during cortical circuit formation. After apparently performing numerous tasks they largely disappear some time after birth (Luskin and Shatz, 1995a,b). We now believe that these cells are responsible for an increasing number of steps during the assembly of cortical circuitry in a variety of species.

To understand the mechanisms of each developmental step during the formation of thalamocortical connectivity, though itself a considerable challenge, is far simpler than to explain why Nature uses each particular solution and each particular mechanism, at each particular stage, in each particular species. Comparative developmental studies can bring us closer to understanding the various strategies discovered by different species to solve the same problems. Each solution to a new problem and each new solution to an existing problem are likely to have exploited and built on already existing mechanisms. Obviously, different processes could have evolved to solve the same kind of problem in different species, and some solutions may even employ redundant parallel mechanisms to give the system more stability and resistance to disturbance. This makes it difficult to envisage universal hypotheses of thalamocortical and cortical development. We have to be careful even with minor generalizations between different systems and between different species. Even the examination of the development of one particular structure in one chosen species can create difficult problems. Often, different temporally coincident but possibly independent mechanisms are responsible for the establishment of a certain structure. Then is it reasonable to ask which is the most important? Even if we could perturb the system in a clear-cut way (and often this is not the case) it might not be easy to choose from the many interpretations of the results. This makes studying the comparative aspects of brain development a frustrating task. Nevertheless it is one of the greatest challenges in developmental neurobiology to understand the evolution of brain development itself and the general building principles and homologies between the different subdivisions of the amphibian, reptile, bird and mammalian forebrains and the possible origin of the cerebral cortex.

Origin of the Mammalian Cerebral Cortex

The out-group and the recapitulation hypothesis are two alternative hypotheses which try to explain the possible way that the cerebral cortex originated from the roof of the cerebral hemispheric pallium (see Northcutt and Kaas, 1995).

According to the so-called 'out-group hypothesis' the cerebral cortex arose by an independent enlargement of the ancestral lateral pallium, and the cerebral hemispheres in living amphibians are comparable to the cerebral hemispheres of putative ancestral vertebrates. This hypothesis assumes that the dorsal ventricular ridge (DVR) arose independently in reptiles and birds from parts of the pallium. According to the 'recapitulation hypothesis' the dorsalventricular ridge represents a primitive ancestral condition, and this structure has been retained in living reptiles and birds; mammalian isocortex arose by the dual differentiation of the dorsal cortex and migration of the cells of the dorsal ventricular ridge. In this case, the dorsal ventricular ridge and dorsal cortex of reptiles are considered homologous to mammalian isocortex. This hypothesis also assumes that during mammalian development some of the cortical cells are generated in the lateral ganglionic eminence and migrate into the cortex. In the literature there is support for both theories. Cells generated in the lateral ganglionic eminence indeed migrate into the cerebral cortex (De Carlos et al, 1996; Kuan et al, 1997; Karten, 1997). However, we lack sufficient experimental evidence to arrive at a consensus regarding the homologies between the different subdivisions in the pallium and the origin of the mammalian neocortex.

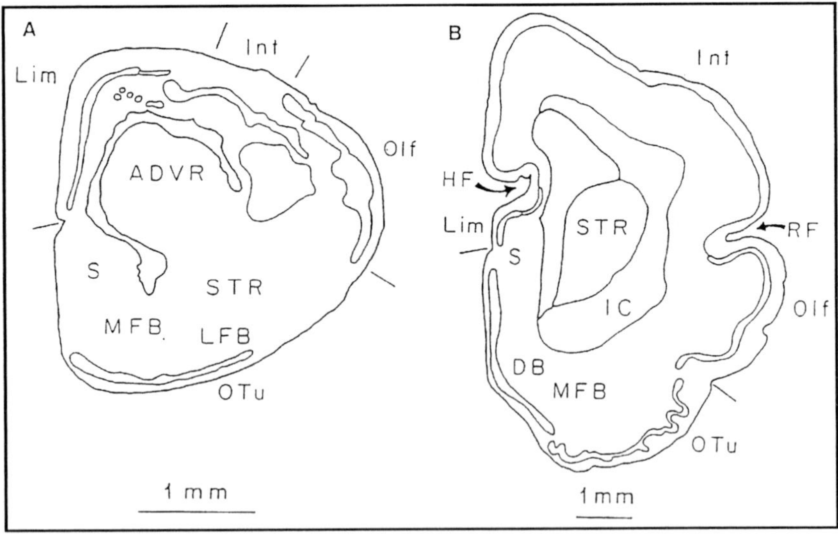

Fig. 11.1. Comparison of pallial organization in reptiles and mammals (Ulinski, 1990). This figure shows transverse sections through the telencephalon of (A) a tegu lizard, *Tupinambis*, and (B) an opossum, *nnn* . Unlike placental mammals, marsupials (and monotremes) lack a corpus callosum, thereby simplifying comparisons with the lizard brain. Both telencephalons have a lateral ventricle and are divided into three major parts. The septum (S) makes up the ventromedial face of the telencephalon. Olfactory structures such as the diagonal band (DB) or olfactory tubercle (OTu) are present on the ventral surface. The forebrain bundles run rostrally into the telencephalon. They consist of the medial forebrain bundle (MFB) and the lateral forebrain bundle (LFB) or internal capsule (IC), passing below the corpus striatum (STR). The pallium or roof of the telencephalon has two components in reptiles: a cortex and an anterior dorsal ventricular ridge (ADVR). The pallium in mammals has only a single component: the cortex. In both groups, the pallium can be divided into three functional segments. The olfactory cortex (Olf) is situated laterally. The intermediate segment (Int) is present dorsally. It includes part of dorsal cortex and ADVR in reptiles and the cortex between the rhinal fissure (RF) and the hippocampal formation (HF) in mammals. The limbic cortex (Lim) occupies the medial wall of the pallium in both groups. It is involuted by the hippocampal fissure in mammals. Note that the two sections are drawn to different scales. Reproduced from Ulinski (1990) with kind permission of Plenum Press, New York.

The Problem of Homology

In comparative neuroanatomy, the term homology can be used in a phylogenetic or a structural sense (Campbell and Hodos, 1970). The *structuralist* interpretation of the term homology refers to similarity in the morphology, topological relationships and connections of a given structure. In contrast, the *phylogenetic* meaning is based on the assumed evolutionary origin. Which groups of cells in modern species are derived from a single group of cells in a common ancestor? To avoid confusion I shall use the term homology in this phylogenetic sense: "*Structures and other entities are homologous when they could, in principle, be traced back through a genealogical series to a stipulated common ancestral precursor irrespective of morphological similarity*" (from Campbell and Hodos, 1970; based on Ghiselin, 1966, 1969; Bock, 1969).

Homologies can exist at any level of organization, between genes, cell types, tissues, organs and organ systems. If one keeps the above definition in mind we are less likely to get misled by superficial similarities. Nevertheless, similarity is a very important basis for the recognition of homology: the more similarities we find, from various different aspects, the greater the probability of homology. It was Marin-Padilla (1971, 1978) who pointed out (based on rapid Golgi preparations) that the mammalian neocortex develops on a reptilian framework. Today we have much more information about the ontogenetic steps during the development of the mammalian cortex and comparative studies on non-mammals have also progressed enormously (Ulinski, 1983; Jones and Peters, 1990; Northcutt and Kaas, 1995; Karten, 1997).

In this last chapter, I wish to examine the possibility that some of the first generated cells of the mammalian pallium (preplate, cells of the perireticular and thalamic reticular nuclei) are homologous to some neurons of the reptilian pallium. I shall argue that some of the early steps involved in the development of thalamocortical connections in mammals can be understood in terms of their evolutionary origins. Comparative developmental studies may reveal the workings of evolution, and through them a common developmental plan for such radically different brains.

Recent Revival of Developmental Evolutionary Biology

The ultimate ideal in phylogenetics would be to reconstruct all the branches of an evolutionary process in detail, as it actually happened. Now, when the ancestors have perished, we can only speculate about what might have happened: fossils give very limited information about the connectivity of brains!

The examination of existing species can, of course, provide some interesting clues. But we have to avoid simply looking at the end branches of the evolutionary tree, setting up a morphological sequence and fooling ourselves into believing that this was the historical

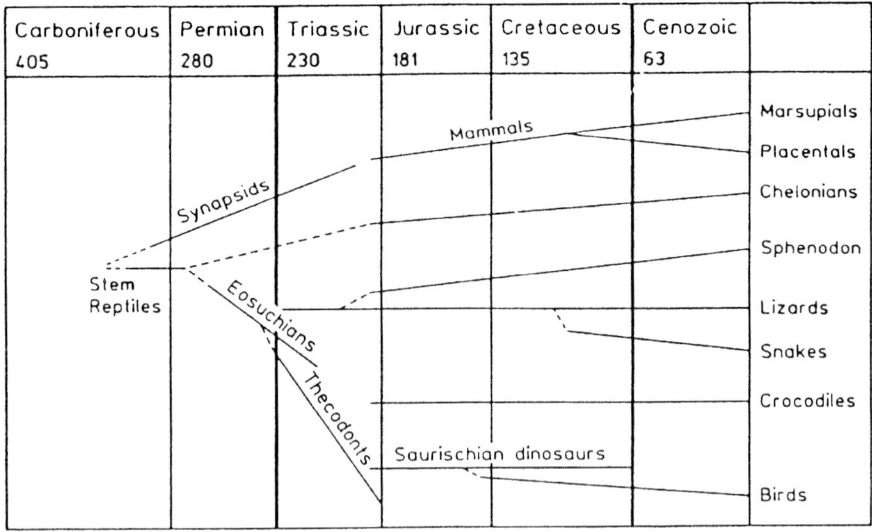

Fig. 11.2. Evolutionary filiations of upper vertebrates. Time in millions of years since the beginning of the geological period. Reproduced from Goffinet (1990) with kind permission of Springer-Verlag, Heidelberg.

sequence. If we are constantly aware of this danger, the comparative approach can provide important information. Such a comparison shows that many strategies in development are well conserved (or evolved in parallel) across numerous different phylogenetic levels. Sometimes the repetition of certain stereotyped developmental steps might not seem to be the most economical way of making of a nervous system, but we have to keep in mind that these steps evolved over millions of years and that each new development had to be based, to a certain extent, on existing structures. Thus it is even more important for us to understand the logic of development. Since it is not possible to examine this particular question from fossil records, I shall concentrate on developmental and structural similarities. The idea of relating development with evolution is not new. Before the turn of this century evolutionary biology and developmental biology were much more closely related, and studying homologies between parts of animals was considered the crucial way to understand anatomy (for a brilliant review, see Raff, 1996).

Haeckel (1866) stated that 'ontogenesis is the recapitulation of phylogenesis'. In this crude form, his statement is surely incorrect, but it is indeed very striking that most early developmental steps are highly conserved in all vertebrates (Romanes, 1901). There is no

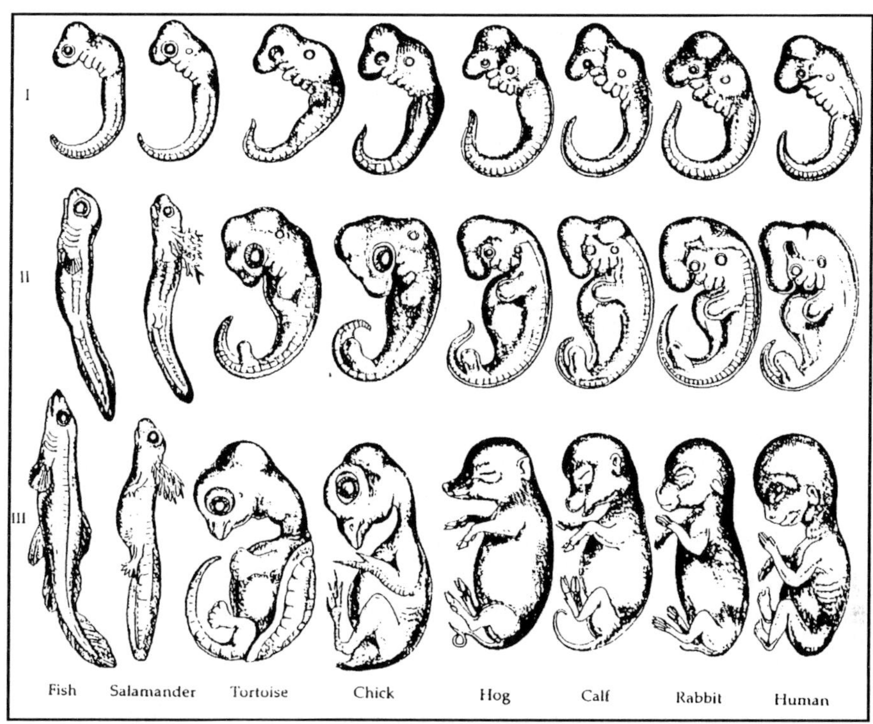

Fig. 11.3. The appearance of vertebrate embryos at various stages of development (from Romanes, 1901). The striking similarity of the different embryos during early development led Haeckel (1866) to suggest that the key to the "ontogeny lies in phylogeny". Although he put far too much emphasis on phylogeny as an *explanation* of ontogeny, it is indeed rather suggestive that these different animals share both a common ancestor and the same basic mechanisms of development. Reproduced from Romanes (1991).

reason to imagine that neural development is fundamentally different from the generation of other organ systems.

In 1828 Karl Ernst von Baer began to contradict the dominating concept of the Great Chain of Being by Aristotle (circa. 350 BC) and "*showed that there is no strict recapitulation of the development of a more primitive form by the embryo of a more advanced one. Instead, organisms diverge from one another in development. The embryos of higher forms do not duplicate the adults of more primitive relatives, but in their stages of development they do resemble the stages of development of the related lower forms.*" (quoted from Raff, 1996). These ideas suddenly became irrelevant when developmental biologists began to concentrate on understanding developmental mechanisms and the work of Roux (1894), Spemann (1938) and Hamburger (1988) revolutionized embryology. Nevertheless, the ideas of von Baer and Haeckel remain relevant to the study of homologies. Although ontogenetic algorithms can evolve, some stages of development employ highly conserved mechanisms. Since most if not all of these developmental patterns are inherited, they strongly support the notion of homology. I therefore made the assumption that comparisons could be made easier by searching for homologies in *developing* brains. Another reason why these early generated cells should be examined during development is that most of them disappear by adulthood (see chapters 3 and 8). To establish homology in developing brains, the same principles and approaches apply: position within a complex of features, similarities in structure and composition and historical transition (Raff, 1996). We were particularly interested in the earliest generated cells of the mammalian forebrain and felt that detailed specific connectional analysis combined with immunohistochemistry of specific markers (GABA, neuropeptide-Y, calretinin and parvalbumin) could help us to follow these cells even if they drastically change their position within different vertebrates. We made the assumption that, at the early stages of development, cells in homologous subdivisions of the pallium might establish connections with homologous cell groups. We concentrated on the partly transient populations of early-generated neurons in the cortical subplate, marginal zone, perireticular and thalamic reticular nuclei of mammals, which contain distinctive molecular markers (e.g. GABA, calretinin and parvalbumin) and form early axonal connections (Molnár, 1994; Cordery et al, 1996, 1997). I shall review some of the results of our comparative studies on early thalamocortical and corticofugal connectivity in developing reptiles in the light of the recent developmental studies on mammals (Mitrofanis and Guillery, 1991, 1993; Métin and Godement, 1996; see chapter 6). My major goal is to systematically analyze the hodological relationships in the developing reptilian pallium, similarly to chapter 6 for mammals (see Fig. 6.1) and to examine the possibility that the transient cells in the developing mammalian ganglionic eminence might be related to the DVR and transient preplate might be related to dorsal cortex in reptiles. On other equally exciting aspects of comparative forebrain development I would like to refer to recent reviews by Northcutt and Kaas (1995), and Karten (1997).

Early Thalamic Connections in Mammals

The advance of the embryonic tracing techniques (Godement et al, 1987) makes it relatively easy to examine the early connectivity patterns in the developing pallium in many different species and facilitates the understanding of the general mechanisms. In mammals there are numerous recent studies on the emerging early connectivity pattern in developing forebrain (Marin Padilla, 1971; McConnell et al, 1989; Blakemore and Molnár, 1990; Mitrofanis and Guillery, 1993; Métin and Godement, 1996). These studies revealed a very early network, including cells of the thalamic reticular nucleus, the medial and lateral ganglionic eminence and preplate (see chapter 6). These cells have numerous similarities: they are among the earliest generated cells in the developing pallium, express common neurotransmitters and form early functional connections before they presumably undergo preferential cell death (Shatz et al, 1990; Mitrofanis and Guillery, 1993). Métin and Godement (1996) described the earliest connectivities of the diencephalon with ganglionic eminence cells and proposed that they might form temporary targets for thalamocortical and corticofugal connections in the hamster. They also suggested that only medial ganglionic eminence cells project to the thalamus whereas lateral ganglionic eminence cells exclusively develop their connections with the developing cerebral cortex. We could not confirm this observation in other species. At corresponding ages in the developing rat, mouse and marsupial brains (*Monodelphis domestica, Dysaurus hallucatus*) we found very similar pattern of backlabelling, but *both* medial and lateral ganglionic eminence cells project to both dorsal and ventral thalamus, and there is no sharp boundary between medial and lateral ganglionic eminence in establishing connections with thalamus or cortex (see Fig. 6.1; Molnár et al, 1995b; Knott et al, 1996; Molnár et al, 199c). Chapter 6 describes some of the basic patterns of early forebrain connectivity in rat. The early generated cells in the preplate, ganglionic eminence and thalamic reticular nucleus have numerous similarities, but their connectivities respect certain boundaries. The axons of perireticular cells reach both dorsal and ventral thalamus before birth, but no cells reach the dorsal thalamus from the cortex. Nevertheless, the first cortical axons to pass into the diencephalon arise from preplate neurons, even before the appearance of the cortical plate, as suggested previously (McConnell et al, 1989). We suggest that their growth is delayed not in the white matter (Clasca et al,

1995), but closer to the thalamus, perhaps in the thalamic reticular nucleus (Molnár, 1994). The corticofugal projections are only delayed temporarily in the ganglionic eminence from E14-15. Interestingly the early generated cells of preplate and ganglionic eminence show striking resemblance to reptilian structures (Cajal, 1909-1911; Marin-Padilla, 1971, 1978, 1988; Reiner, 1990, 1991).

Early Thalamic Connections in Reptiles

In the light of the recent hypotheses about the possible origin of the cerebral cortex, further comparative studies on these issues in mammals and reptiles could have major importance. Chelonians are believed to derive from ancestors closest to the stem reptiles, which also give rise to Synapsids and Eosuchians. The living representatives of the Chelonians are thought to be the turtles (see Fig. 11.1). They therefore could provide an interesting system to study since they might show more similarities to the presumed ancestors from which the major phylogenetic categories diverged (Goffinet, 1990). Unfortunately there are no studies on developing reptile pallium along these lines. We therefore embarked on studying red-eared slider turtle embryos (*Pseudemys scripta elegans*) at stages 17, 20, 21, 22, 23, 24 and 25. This staging was also used for other species (*Chelydra serpentina* and *Emys orbicularis*) according to Yntema (1968). We were very interested in the possibility of finding a homologous cell group to the cells of the perireticular nucleus and the cortical preplate.

In rat, from E14, carbocyanine dye crystal placements in the thalamus backlabelled cells of the perireticular nucleus (PRN) within the internal capsule, but never beyond the border between striatum and the cortical intermediate zone (Fig. 6.1). In turtle, thalamic DiI crystal placements backlabelled cells in the striatum, of similar morphology to those in rat PRN, but very occasionally or never in the dorsal ventricular ridge (DVR).

In rat, cells of the PRN project to perirhinal and ventral cortices, but not to dorsal cortex until E17-18. In turtle, crystal placements in the dorsal cortex labelled cells in the DVR, while placements into ventral cortex labelled many cells in striatum and a few in the DVR. In E16 rat, calretinin immunohistochemistry revealed cells in the thalamic reticular nucleus and PRN, perirhinal cortex, marginal zone and subplate, but few in cortical plate. In turtle, at stages 22 and 24, calretinin-positive cells appeared in progression from the internal

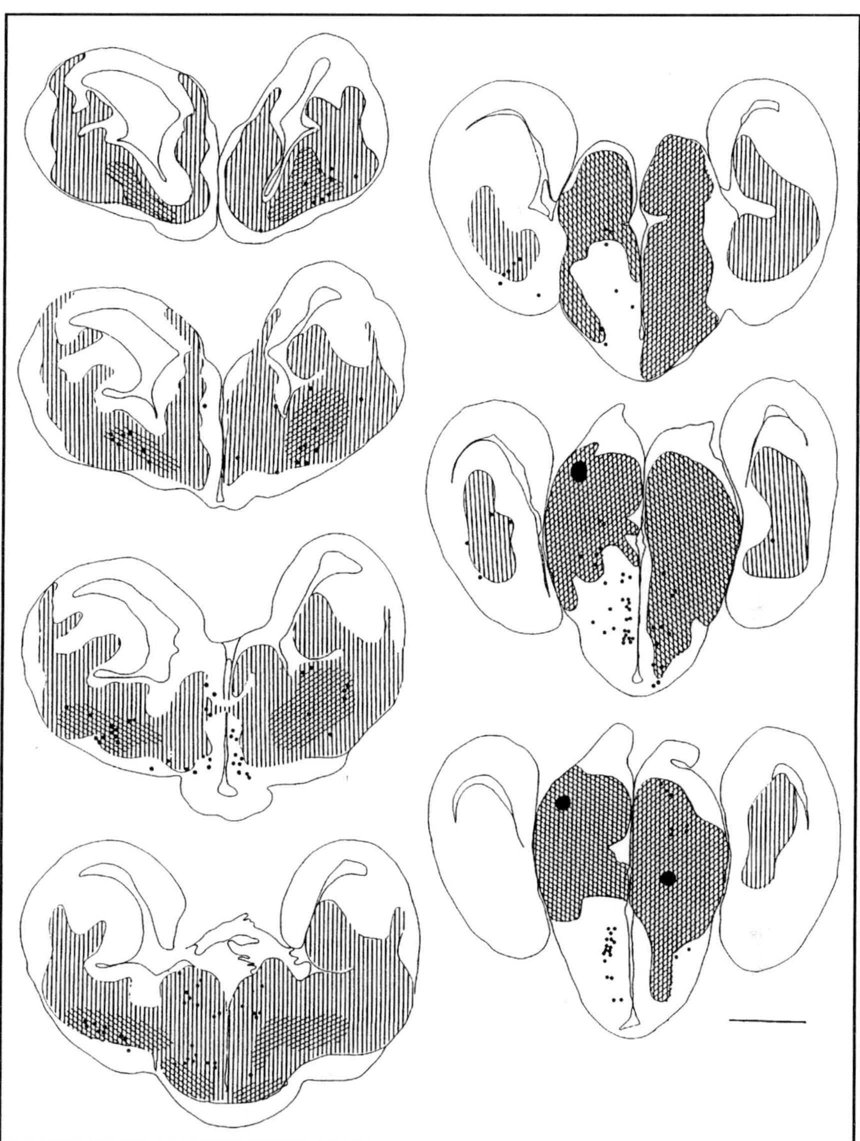

Fig. 11.4. In a stage 25 turtle embryo, our DiI labelling from dorsal and ventral thalamus (crystal placement sites indicated with large black dots) revealed backlabelled cells in striatum (indicated with small dots). These cells had similar morphology to backlabelled PRN cells in rat (Molnár, 1994; Cordery et al, 1996), but only very occasionally or never in dorso-ventricular ridge (DVR). Crystal placements into the dorsal and lateral cortex both labelled cells in the DVR (not shown). Every third 100 μm thick section is presented on an antero-posterior sequence. Shading indicates the distribution of labelled thalamic fibers in the internal capsule, striatum, DVR and dorsal cortex. Scale bar: 1 mm. Based on the results of Cordery et al (1997).

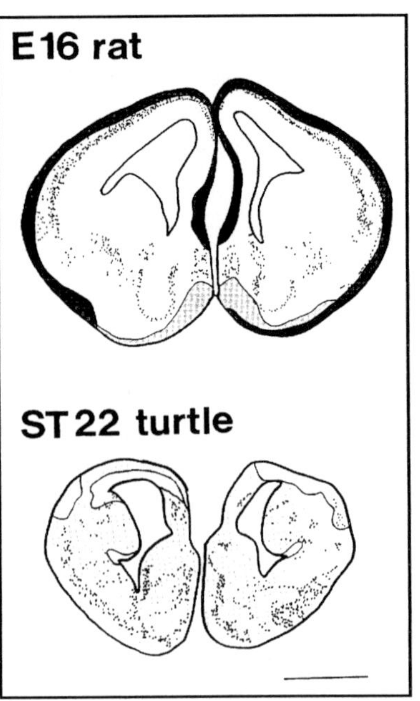

Fig. 11.5. In E16 rat brain calretitin immunohistochemistry revealed cells in the thalamic reticular and perireticular nuclei, perirhinal cortex, marginal zone and subplate but not in cortical plate. In an embryonic turtle, at stages 22 and 24, calretinin-positive cells appeared in a progression from the nucleus sphericus and the striatum towards the lateral cortex, but not within the dorsal cortex itself. In turtle, lateral and dorsal cortex contained no neuropil staining in contrast to all other surrounding regions. Few scattered calretinin-positive cells were observed in the DVR. Based on the results of Cordery et al (1997).

capsule and striatum towards lateral cortex, but never within the dorsal cortex. A few scattered calretinin-positive cells were observed in the DVR.

Our study demonstrates similar developmental patterns in mammals and reptiles and suggests that the mammalian PRN is homologous to a cell group in the reptilian striatum, but not to the DVR. In developing turtle brains NPY immunostaining revealed cells scattered in all regions of ventral and dorsal pallium, including dorsal cortex, and the neuropil staining did not spare the dorsal and lateral cortex, but appeared more homogeneous.

Our carbocyanine dye tracing study demonstrated that axons of similar cells are among the first to reach the dorsal and ventral thalamus in mammals and reptiles and our connectional analysis strongly suggests that the transient structure of the perireticular nucleus in mammals is homologous to a cell group within the reptilian striatum, but not with the cells of the DVR. The notion that the transient cells in the developing mammalian ganglionic eminence might be related to the DVR does not seem to get support from our hodological analysis of developing turtle forebrain. It would be of major interest to continue these studies in developing amphibians, where there is no DVR in the adult. It would be very important to examine whether amphibians have DVR at any point of their development. There are modern techniques for the study of cell lineage (chapter 2) which have not been employed for reptilian development. I expect a revolution in this field over the next few years.

Is Preplate Homologous to Reptilian Cortex? Comparison of Pallial Organization in Reptiles and Mammals

According to Herrick (1933), the reptilian cortex is characterized by superficial laminae of grey substance in the pallium of the cerebral hemispheres, divided into 3 longitudinal stripes: a mediodorsal cortex, a dorsal cortex and a lateral cortex. By tracing their connections, hodological arguments suggest that the lateral cortex is homologous to the pyriform cortex and the mediodorsal to the hippocampus (reviewed by Ulinski, 1983). The mammalian cortex is most probably related to the medial part of the reptilian pallium, the dorsal cortex (Goffinet, 1983, 1984). The DVR (dorsal ventricular ridge) is included among cortical areas because it contains several sensory fields and is presumably phylogenetically related to pallial structures (Ulinski, 1983).

Evidence for the Homology

Ontogenetic Similarity Between Reptilian and Mammalian Cortices

In the medial, dorsal and lateral supraventricular cortical zones of reptiles, neurogenesis follows a lateral to medial, and an anterior to posterior gradient (Goffinet, 1990). These tangential gradients correlate with the pattern of development seen in mammalian embryos (reviewed by Bayer and Altman, 1991). Nonmammalian cortices develop in an outside-first, inside-last fashion (Goffinet, 1983, 1990). Only in mammals is the cortical plate assembled as a 'sandwich', with the cortical plate itself splitting the primordial plexiform zone into two and developing in an inside-first, outside-last pattern (Marin-Padilla, 1971; Luskin and Shatz, 1985a,b; see chapter 3). It is interesting to point out here that the primordial plexiform zone itself (before it is split into subplate and marginal zone by the cortical plate) is generated in the 'reptilian' outside-first, inside-last fashion (Bayer and Altman, 1991). In reptiles an outside-first, inside-last morphogenetic gradient has been demonstrated at the ultrastructural level (Goffinet, 1990): synapses appear first in the molecular layer, then in the subcortex, and finally within the cortical ribbon.

Minuteness and Multiplicity of Similarities Between Reptilian Cortex and Primordial Plexiform Zone Cells (Similarities in Morphology and Histochemical Properties)

Cajal (1909-1911), already noticed the similarities between the polymorph cells of layer 1, and cells of the lizard cortex. The work of Marin-Padilla demonstrated that these marginal zone cells are indeed descendants of the primordial plexiform zone. He described the splitting of the first generated, polymorphous cell layer (the primordial plexiform zone or preplate) into the marginal zone and the subplate (Marin-Padilla, 1971, 1972, 1978, 1988) and he called these premature zones 'the reptilian framework' of the developing mammalian brain (Fig. 3.2).

The subplate cells have surprisingly similar transmitter systems to those of reptilian cortical neurons: Neuropeptide Y (NPY), cholecystokinin (CCK), somatostatin (SOM), glutamic acid decarboxilase (GAD) and GABA (Somogyi et al, 1984; Reiner et al, 1984; Chun et al, 1987; Chun and Shatz, 1989; Antonini and Shatz, 1990; Reiner, 1990) have all been demonstrated in mammalian preplate/subplate cells. Dávila et al (1991) demonstrated that cells of the cortical regions of turtles, lizards and mammals universally contain both NPY, SOM and GABA. Surprisingly, these neurons are only found at telencephalic levels, not in any other cerebral regions. Dávila et al (1991, 1997) took this as evidence for the conservative features of evolution, and believed that these similarities support the view that these cortical neurons were present in stem reptiles. All these similarities strongly suggest homology of the mammalian preplate. A. Goffinet recently observed reelin positive cells in the dorsal cortex of the developing turtle (personal communication).

Principle of Radial Organization

If the preplate is the equivalent of the reptilian cortex, where do the extra cells of the cortical plate come from? Shatz et al (1990) proposed that a mutation in the cell cycle produces extra cells and that this is responsible for the production of the cortical plate. The idea that preplate and plate cells come from the same ventricular and subventricular stem-cells (same cell lineage) can now be tested directly with retrovirus gene transfer technique (Price et al, 1987; Cepko et al, 1990). Do preplate cells and the cortical plate cells belong to common or separate clones?

My own collaborative autoradiographic studies suggest that radioactive thymidine administered at E12 (hence producing strong labelling of preplate cells) also produces dilute label in the later generated cells of the cortical plate (Molnár et al, 1991), although the thymidine disappears from the circulatory system within 2 hours (Luskin and Shatz, 1985a,b). This could indicate that the same progenitors generate firstly the subplate neurons, which become postmitotic at E12-14 (thus they have high concentration of [^{3}H]thymydine), and then cortical cells proper, through several mitotic divisions (therefore with lower [^{3}H]thymidine content). However it is equally possible that the progenitors of subplate neurons and cortical cells represent two distinct populations, one of which (the progenitors of the cortical neurons) generate only other progenitor cells and not postmitotic cells at E12. The diluted [^{3}H] thymidine label then appears in the cortical cell progeny of the late-generated progenitor cells. The investigation of the precise lineage relationships with recombinant retrovirus will be important in deciding this issue and might also cast light on the control of subplate death.

To accommodate the extra cells of the mammalian cortex, there must have been some mechanism of vertical organization. This raises another mystery, namely the transition from the outside-first, inside-last reptilian pattern of cell accumulation (still seen for the mammalian preplate) into the inside-first outside-last sequence of the mammalian plate. Teleologically, simple continuation of the outside-in arrangement (as is found in *reeler*, see chapter 7) would seem to be a more straightforward way to have accommodated the extra cortical plate. Why is the primordial plexiform zone split into two and the setting pattern reversed? Is this the result of another mutation?

Goffinet (1990) argues that the radial sequence in which the cortical plate is laid down is not an all-or-none property. There are various degrees of organization in the different segments of the embryonic pallium in different reptiles. On the basis of examination of the various descendants of the stem reptiles he suggested that radial cytoarchitectonics must be regarded as a "property which has been acquired gradually, independently and to variable extents after phylogenetic divergence" (Goffinet, 1990). He also noticed a rough negative correlation between laminar organization and the presence of an outside to inside gradient in reptiles. The inside-first, outside-last pattern, together with the prelaying of 'highways' of subplate efferents, may provide certain advantages. Perhaps cortico-cortical connectivity would be physically constrained in certain ways if outside-in accumulation of neurons were continued. Maybe *reeler* is a step-back mutation in the sequence of change: *reeler* is certainly disadvantaged in that its survival is poor (though this may be due to disorganization elsewhere in the brain than in the cerebral cortex). Also, in some of the different *reeler* strains there is no corpus callosum (Goffinet and Caviness, personal communications).

Highly advantageous mechanisms are usually discovered early and well conserved throughout the different branches of the evolutionary tree (or they develop separately and convergently). In all mammals, for instance, there is a remarkable uniformity in the number of cortical neurons in a given columnar volume (with the major exception of the primate striate cortex; see Rockel et al, 1980). The radial organizing principles are probably rather universal among mammals.

Perhaps the emergence of the various forms of interaction with preplate neurons was the prerequisite that made it possible for a bigger and more powerful cerebral cortex to emerge. It is also possible that "some events observed during development are not relevant to the final function of the organism, and that some of these events represent ways of moving around obstacles, which are laid down early in evolution and may constrain or channel future development, rather than being necessary steps that ultimately generate a particular function" (Krubitzer, 1995a). Nevertheless, during development the old and new organization co-exist and morphogenesis relies on the old system. Thalamic fibers might not be able to reach the appropriate cortical areas without the prelaid trajectories of preplate axons and without the subplate compartment to wait on. This stage of thalamic axon guidance all happens when the embryonic cerebrum is still lisencephalic (in all mammalian species), and the morphological context is therefore simple and the distances are small. The waiting period (Rakic, 1977; Lund and Mustari, 1977) may be essential in the development of bigger brains, since, if the thalamic fibers did not reach the appropriate cortical area early enough, the pathways might become too tortuous for them ever to arrive in accurate order.

Emergence of Afferentation of the Cortex

Finally, I would like to examine the development of the different sensory inputs to the cortex in various vertebrates. At which phylogenetic stage did the cortex begin to receive sensory afferentation? How was the cortical representation of the external world achieved? Why did evolution produce primary and secondary cortical areas? What do we know about the nature of this representation in the brain?

The cerebral cortices of fish, birds, reptiles and amphibians all receive inputs from sensory systems. The vast majority of the cortical input in these Classes is from the olfactory system. Other modalities (vision, etc.) are also represented, but most of those fibers terminate in the anterior dorsal ventricular ridge (Ulinski, 1983). Most of the information other than olfactory is believed to be processed subcortically in the colliculi or the striatum. In these species the intermediate cortex (dorsal pallium) is just a layer of a few pyramidal cells organized into 1 or 2 plexiform zones depending on the species. However, even this intermediate cortex receives input from the perirotundal nuclei of the dorsal thalamus (Heller and Ulinski, 1987; Cordery et al, 1997) and some visual responses were recently described within dorsal cortex of the iguana, together with at least two topographically organized representations in DVR (Manger et al, 1997).

When and how did the first afferents, carrying sensory information other than olfactory, reach the cortex? Did corticothalamic projections precede thalamocortical projections in phylogeny? How did the cortex process the first thalamic input? Did cortical differentiation progress as a result of this new afferentation, or did the two processes proceed in parallel?

Thalamus as a Developmental Device for the Expanding Cortex

The answers to these questions have yet to emerge from comparative studies, but I suspect that the expansion of cortical areas required an increase in the matching thalamic regions, and an increase in their interconnection. Finlay and Darlington (1995) found no correlation between the increase of size of the cerebral cortex and the thalamus, suggesting that the two processes are not linked and cortical growth surpasses that of the thalamus. It would be very interesting to perform further analysis on the possible correlation between cortical expansion and the increase of non-specific thalamic systems. The two probably appeared in parallel on the evolutionary time scale. It was recently proposed that the 'higher order' nuclei play a key role in corticocortical communication (Guillery, 1996).

Among living mammalian species there seems to be a close correlation between the cortical surface area and the volume of the non-specific thalamic nuclei. When the so-called association cortices increased in size, so too did the 'non-primary' and 'non-specific' thalamic regions. This might have functional relevance in thalamocortical and cortico-cortical integration (Guillery, 1995; Sherman and Guillery, 1996).

During the ontogenesis of the visual pathways, the different levels of the pathways interact horizontally and vertically. Dynamic interactions can result from changes at different levels. For instance, Guillery (1972) demonstrated that cells of the dLGN are protected from the normal failure of growth caused by monocular deprivation in the region corresponding to an 'artificial monocular segment' made by destroying a small area of temporal retina in the non-deprived eye shortly after birth. Because 'competitive' interactions between dLGN cells of different laminae almost certainly occur at their endings in the visual cortex, the shrinkage of dLGN cell bodies during monocular deprivation is probably a retrograde effect, originating in competitive interactions in the striate cortex (Guillery, 1972). Maybe similar dynamic interactions between cortex and thalamus occurred during phylogenesis, with thalamus and cortex developing together.

Cortico-Cortical Interactions Via Thalamus

The neuronal and synaptic morphology of the pulvinar and the mediodorsal nucleus closely resembles that of other 'specific' sensory and motor relay nuclei (Mathers, 1972; Schwartz et al, 1991). The most distinct difference, however, is that the synaptic sites in the glomeruli, generally reserved for ascending sensory afferents in major relay nuclei, are occupied by corticothalamic terminals in the non-primary nuclei (Mathers, 1972; Schwartz et al, 1991). Different cortical regions in the peristriate, posterior temporal or posterior parietal cortex converge on the same regions in the pulvinar and the outputs of different corticofugal layers (5, 6) might terminate on different regions of the dendritic tree of cells in the pulvinar and mediodorsal nucleus (Mathers, 1972; Schwartz et al, 1991; Kuroda and Price, 1991). It is an interesting possibility that different cortical regions interact through these nuclei (Chalupa et al, 1983; Robinson and Petersen, 1992; Mitrofanis and Guillery, 1993; Guillery, 1995; Sherman and Guillery, 1996).

The thalamus, structurally and functionally, is more closely associated with the phylogenetic expansion of the cerebral cortex than was previously thought. All this suggests that "the thalamus is not the intellectual desert of neurobiology" (Guillery, personal communication).

Multiple Constraints During Map Formation in the 'Protocortex'

Thalamic connections seem to be laid out in an orderly fashion on the basis of simple spatial and partly chronological algorithms (see chapter 6; Blakemore and Molnár, 1990; Shatz, 1992a,b). The topography of the initial intercortical connectivity is also rather simple and may depend on very similar mechanisms (Hankin and Silver, 1986). Fiber ordering preserves the chronological order of its establishment. Since the final representation is dependent on and complicated by many other factors, these initial clear-cut basic principles have gone unnoticed.

If we could examine the fiber ordering in the adult internal capsule or in the adult corpus callosum (away from the target area) the fibers might still show their basic topographic arrangement. Interestingly, there has been such a study on the adult callosum in the cat (but not yet on the internal capsule)(Molnár, 1995b). Nakamura and Tanakesi (1989) made small, discrete injections of HRP into different parts of the callosum of adult cats. Although there was no correlation between functional areas, there was a strict sectorial topographical correspondence between the corpus callosum and the cortical areas from which the cells were sending their projections (Fig. 8.4). Small injections into the adult cat presplenium labelled a zone of cortex very similar to that resulting from a neonatal injection into areas 17 and 18. Kind and Innocenti (1990) note that the topography of the juvenile callosal projection neurons may be related to that of their axons in the corpus callosum. I would rather say that the initial fiber ordering of the intercortical connections is *preserved* in the callosum, but substantially modified at the target fields.

Such experiments, examining fiber arrangements in tracts distant from origins and targets, would be extremely interesting in developing and adult albinos or after early enucleation. I suspect that the initial connectivity between the different cortical areas might also be explained with a few very simple principles. Interestingly, the corpus callosum is absent in few placental mammals: marsupials and monotremes use only the anterior commissure for inter-hemispheric connection, even though the general construction of the neocortex is very similar (see Mark and Marotte, 1992).

Killackey (1990) argues that there was evolutionary pressure for the 'uncoupling' of a portion of the vertebrate telencephalon, the neocortex, from strict genetic determination in favor of environmental determination. The very high degree of plasticity that seems to be such an important feature of this uncoupling of the maturing cortex may tend to blur the degree of simplicity and order present at earlier stages, especially for intrinsic and long-range cortico-cortical connections.

Young and Scannell (1993) demonstrated with quantitative topological analysis that the vast majority of corticocortical connections within the same hemisphere in adult mammals are formed between neighboring areas, and follow an anteroposterior sequence. Excluding the cingulate and the entorhinal cortex from this analysis, the correlation is even more apparent. The developing forward and backward cortico-cortical projections may be laid down through simple algorithms similar to those postulated for thalamocortical (chapter 6; Blakemore and Molnár, 1990) and intercortical connectivity (Hankin and Silver, 1986). However the final pattern that is carved out from a simpler initial organization is complicated by morphological change, selective elimination and activity-dependent plasticity. Fortunately, the degree of fiber order still present in the major pathways of the adult can stand as a witness to its earlier simplicity (Nakamura and Kanaseki, 1989).

Scannell et al (1997) recently used non-metric multidimensional scaling, non-parametric cluster analysis, and optimal set analysis to visualize and quantify the connectional architecture and organization of the thalamocortical system of the cat. He and his colleagues examined numerous anatomical studies and found 892 cortico-cortical, 379 thalamo-cortical, and 295 cortico-thalamic connections between 55 cortical areas and 41 thalamic nuclei within each hemisphere. Their analysis suggests that a small number of developmental factors (e.g. chemical or temporal gradients) could account for much of the variability in extrinsic connectivity within the entire thalamo-cortical network in the cat (Scannell et al, 1997). They proposed that "these factors could correlate with the spatial location of the thalamic nuclei whose location is an excellent predictor of their cortical connectivity; variability in the position of thalamic nuclei accounts for 62% of the variability in their pattern of cortical connectivity ($p < 0.001$). Furthermore, similarity in thalamic input to cortical areas predicts 66% of the similarity of their cortico-cortical connectivity ($p < 0.001$). These results suggest a developmental model in which thalamo-cortical projections, depending on the 3-D location of thalamic nuclei, shape the cortico-cortical network via correlated input" (Scannell et al, 1997).

Anatomical and Functional Units in the Cortex

There is now a good deal of evidence, at least for lower mammals, that the various areas of the cerebral cortex are not entirely pre-determined for any particular input and connectivity, cytoarchitectonic appearance or function (reviewed by O'Leary, 1989). Neuronal self-organization and the influence of the environment play an important part in carving out the connectivity and internal organization of areas, and hence their functions. Functional specialization becomes expressed in the emerging cytoarchitecture of the cortex (Brodmann, 1909), although the correlation between anatomical and functional subdivisions may be less precise than is sometimes imagined. For instance, area 18 of the primate extrastriate field is now known to contain a substantial number of separate representations of the visual field, at least some of which are highly specialized for particular functions (see Zeki, 1990). In the opossum, the primary somatosensory cortex and motor cortex were thought to be non-separate entities: instead they were believed to form a 'somatosensory-motor' amalgam. Lende (1963) described that the two areas entirely overlap, and they perform the same functions as the two separate regions in the rat. This overlap of the sensory and motor systems in the opossum is restricted to the cortical level. VP and VL receive separate, non-overlapping inputs, but both project to the same cortical region. Recent mapping studies of Beck et al (1996) with more sensitive techniques confirmed this overlap. In some marsupials partially separated motor area has been reported (Rowe, 1990). Krubitzer et al (1995) described separated sensory and motor representation in monotremes, where, similarly to marsupials, there is no callosal interhemispheric connections. It is still not clear whether the somatosensory amalgam is the ancestral condition (Beck et al, 1996).

Why does Nature sometimes employ different and separate neuronal representations and sometimes not? In the primate striate cortex the representation of the two eyes, different orientations, color, motion, etc. use the same basic mapping in a single cortical field. Mitchison (1992) argued that this results in a reduction in the length of wiring of the brain. However, beyond a certain limit of overlapping or closely interdigitated function, the brain appears to add extra functionally and morphologically discrete areas to carry out particular functions within the framework of a large functional system. Keeping several functions in a single area reduces cortico-cortical connectivity; dividing them between neighboring areas reduces local connectivity within each. Perhaps new areas appear when it becomes more economical to process newly-discovered properties of the sensory input in a different network (perhaps because that processing requires the proximity of cells wThis is remembering, of course, that evolution is driven by random mutations, selected (in the case of the brain) by the variable behavioral phenotypes/capacities they produce. Usually this 'analytical delegation' does not spread far away in the hemisphere. It stays in the closest available target space, often with reversals of mapping at borders, and mirror-image topography in neighboring areas. Van Essen (1997) proposed that many gross structural features, like the appearance of sulci and gyri in a characteristic species-specific pattern can be explained with the forming cortical connectivity and with the associated morphogenetic mechanisms. He advocates the role of the tension along axons, dendrites and glial processes during cortical development in the formation of the folds. This could produce a more compact cortico-cortical and extracortical wiring.

The streams within the cortical representation of vision can be largely explained by neighborhood relations; these streams are highly interconnected and might even reconverge (Young, 1992). However, one must also keep in mind the rich back-projections between visual areas, informing V1 of the primate cortex of the 'higher' functions analyzed in the extrastriate belt. Perhaps V1 (and the other primary sensory areas) should not be regarded as the most 'simple' and peripheral sensory areas just because they receive the primary thalamic input. Perhaps we should turn around the concept of higher and lower levels of visual processing and consider the primary areas to be both the major input to and the major output of each sensory field. Perhaps the 'primary' cortical areas evolved after the 'secondary' areas. Indeed we recently found functional subdivisions in the iguana dorsal ventricular ridge (Manger et al, 1997). Using microelectrode mapping techniques we could demonstrate that there are at least two physiologically defined visual subdivisions in the iguana DVR. Similarly to mammalian cerebral cortex, these subdivisions contain cells with different physiological properties; the representations are topographically organized and show reversals at their boundaries. This suggests that some of the organizing principles of representation appear similar in mammals and reptiles (Manger et al, 1997).

Conclusions

This book has discussed some of the mechanisms involved in the earliest phases of thalamocortical development. But thalamocortical connectivity itself provides only an initial framework for further organization of the neocortex. Although the thalamic fibers are undoubtedly the key factors in the delivery of information to the cortex, the final cortical representation of the external world is dependent on many other additional factors. The initial topography of the projection to the cortex may be explained by a cascade of relatively few, individually simple mechanisms, but the number of additional constraints rapidly increases during subsequent maturation. Our present knowledge of the interactions between these numerous constraints is far from complete; undoubtedly there are many more "men in the machine" waiting to be discovered.

References

Ackers, R.M., and Killackey, H.P. (1978) Organization of corticocortical connections in the parietal cortex of the rat. *J. Comp. Neurol.*, 181:513-538.

Acklin, S.E., and van der Kooy, D. (1993) Clonal heterogeneity in the germinal zone of the developing rat telencephalon. *Development*, 118:175-192.

Adams, N.C., and Baker, G.E. (1995) Cells of the perireticular nucleus project to the developing neocortex of the rat. *J. Comp. Neurol.*, 359:613-626.

Adams, N.C., Baker, G.E., and Guillery, R.W. (1993) Perireticular cell bodies are labelled following DiI or HRP injections into the cortex of perinatal rats. *Soc. Neurosci. Abstr.*, 19:45.

Adams, N.C., Lozsádi, D.A., and Guillery, R.W. (1997) Complexities in the thalamocortical and corticothalamic pathways. *Eur. J. of Neurosci.*, 9:204-209.

Adams, R., Molnár, Z., Nesbit, M., and Blakemore, C. (1993) Vital assay for neuronal cell death in organotypic cultures. *Soc. Neurosci. Abstr.*, 19:1506.

Aglioti, S., Cortese, F., Franchini, C. (1994) Rapid sensory remapping in the adult human brain as inferred from phantom breast perception. *Neuroreport*, 5:473-476.

Agmon, A. (1995) Time-lapse confocal microscopy of developing thalamocortical axons in mouse barrel cortex. *Soc. Neurosci. Abstr.*, 21:798.

Agmon, A., and Connors, B.W. (1991) Thalamocortical responses of mouse somatosensory (barrel) cortex in vitro. *Neuroscience*, 41:365-379.

Agmon, A., Yang, L.T., Jones, E.G., and O'Dowd, D.K. (1995) Topological precision in the thalamic projection to neonatal mouse barrel cortex. *J. Neurosci.*, 15:549-561.

Agmon, A., Yang, L.T., O'Dowd, D.K., and Jones, E.G. (1993) Organized growth of thalamocortical axons from the deep tier of terminations into layer IV of mouse barrel cortex. *J. Neurosci.*, 13:5365-5382.

Al-Ghoul, W.M., and Miller, M.W. (1989) Transient expression of Alz-50 immunoreactivity in the developing rat neocortex: A marker for naturally occurring neuronal death? *Brain Research*, 481:361-367.

Allendoerfer, K.L., and Shatz, C.J. (1994) The subplate, a transient neocortical structure: Its role in the development of connections between thalamus and cortex. *Ann. Rev. Neurosci.*, 17:185-218.

Allendoerfer, K.L., Shelton, D.L., Shooter, E.M., and Shatz, C.J. (1990) Nerve growth factor receptor immunoreactivity is transiently associated with the subplate neurons of the mammalian cerebral cortex. *Proc. Natl. Acad. Sci. USA*, 87:187-190.

Allman, J.M., and Kaas, J.M. (1971) Representation of the visual field in striate and adjoining cortex of the owl monkey *(Aotus trivirgatus). Brain Res.*, 35:89-106.

Altman, J., and Bayer, S.A. (1979) Development of the diencephalon in the rat. V. Thymidine radiographic observations on internuclear and intranuclear gradients in the thalamus. *J. Comp. Neurol.*, 188:473-500.

Anderson, J.C., Martin, K.A.C., and Picanco-Diniz, C.W. (1992) The neurons in layer 1 of cat visual cortex. *Proc. R. Soc. Lond. B*, 248:27-33.

Angevine, J.B., Jr., and Sidman, R.L. (1961) Autoradiographic study of cell migration during histogenesis of cerebral cortex in the mouse. *Nature*, 192:766-768.

Angevine, J.B., Jr., Bodian, D., Coulombre, A.J., Edds, M.V., Hamburger, V., Jacobson, M., Lyser, K.M., Prestige, M.C., Sidman, R.L., Varon, S., and Weiss, P. (Boulder Committee) (1970) Embryonic vertebrate central nervous system: Revised terminology. *Anat. Rec.*, 166:257-262.

Annis, C.M., Kageyama, G.H., Robertson, R.T., and Yu, J. (1990) Subplate cortical neurons of the rat: studies in vivo and in vitro. *Soc. Neurosci. Abstr.*, 16:1151.

Antonini, A., and Shatz, C.J. (1990) Relation between the putative transmitter phenotypes and connectivity of subplate neurons during cerebral cortical development. *Eur. J. Neurosci.*, 2:744-761.
Antonini, A., and Stryker, M.P. (1993) Rapid remodeling of axonal arbors in the visual cortex. *Science*, 260:1819-1821.
Ariano, M.A., Lewicki, J.A., Brandwein, H.J., and Murad, F. (1982) Immunohistochemical localization of guanyl cyclase within neurons of rat brain. *Proc. Natl. Acad. Sci. USA*, 79:1316-1320.
Armstrong-James, M. (1975) The functional status and columnar organization of single cells responding to cutaneous stimulation in neocortical rat somatosensory cortex S1. *J. Physiol.*, 246:501-538.
Armstrong-James, M., and Johnson, R., (1970) Qualitative studies of postnatal changes in synapses in rat superficial motor cerebral cortex. *Z. Zellforsch.*, 110:559-568.
Arnod, W.P., Mittal, C.K., Katsuki, S., and Murad, F. (1977) Nitric oxide activates guanylate cyclase and increases guanosine 3':5'-cyclic monophosphate levels in various tissue preparations. *Proc. Natl. Acad. Sci. USA*, 74:3203-3207.
Astrom, K.E. (1967) On the early development of the isocortex in fetal sheep. *Prog. Brain Res.*, 26:1-59.
Bagnard, D., Mann, F., Henke-Fahle, S., and Bolz, J. (1995) Developmental mechanisms underlying the segregation of afferent and efferent cortical projections. *Soc. Neurosci. Abstr.*, 21:1285.
Baimbridge, K.G., Celio, M.R., and Rogers, J.H. (1992) Calcium-binding proteins in the nervous system. *Trends in Neurosci.*, 15:303-308.
Baird, D.H., Baptista, C.A., Wang, L.C., and Mason, C.A. (1992) Specificity of a target cell-derived stop signal for afferent axonal growth. *J. Neurobiol.*, 234:579-591.
Baird, D.H., Hatten, M.E., and Mason, C.A. (1992) Cerebellar target neurons provide a stop signal for afferent neurite extension in vitro. *J. Neurosci.*, 12:619-634.
Baker, R.E., Bingmann, D., and Ruiter, J.M. (1989) Electrophysiological properties of neurons in neonatal rat occipital cortex slices grown in a serum-free medium. *Neurosci. Lett.*, 97:310-315.
Bar, I., Lambert de Rouvroit, C., Royaux, I., Kritzman, D.B., Dernoncourt, C., Ruelle, D., Beckers, M.C., and Goffinet, A.M. (1995) A YAC contig containing the *reeler* locus with preliminary characterization of candidate gene fragments. *Genomics*, 26:543-549.
Barbe, M.F., and Levitt, P. (1991) The early commitment of fetal neurons to limbic cortex. *J. Neurosci.*, 11:519-533.
Barbe, M.F., and Levitt, P. (1995) Age-dependent specification of the corticocortical connections of cerebral grafts. *J. Neurosci.*, 15:1819-1834.
Bastiani, M.J., du Lac, S., and Goodman, C.S. (1986) Guidance of neuronal growth cones in the grasshopper embryo. I. Recognition of a specific axonal pathway by the pCC neuron. *J. Neurosci.*, 6:3518-3531.
Bastmayer, M., and O'Leary, D.D.M. (1995) Dynamics of target recognition by interstitial axon branching along developing cortical axons. *J. Neurosci.*, 16:1450-1459.
Bastmayer, M., and O'Leary, D.D.M. (1996) Dynamics of target recognition by interstitial axon branching along developing cortical axons. *J. Neurosci.*, 16:1450-1459.
Bate, C.M. (1976) Pioneer neurons in the insect embryo. *Nature*, 260:54-56.
Bayer, S.A. (1980) The development of the hippocampal region in the rat. I. Neurogenesis examined with [^{3}H]thymidine autoradiography. *J. Comp. Neurol.*, 190:87-114.

Bayer, S.A. (1984) Neurogenesis in the rat neostriatum. *Int. J. dev. Neurosci.*, 2:163-175.

Bayer, S.A. (1985) Neurogenesis in the olfactory tubercle and islands of Calleja in the rat. *Int. J. dev. Neurosci.*, 3:229-243.

Bayer, S.A. (1986) Neurogenesis of the magnocellular basal telencephalic nuclei in the rat. *Int. J. dev. Neurosci.*, 4:251-271.

Bayer, S.A., and Altman, J. (1987) Directions in neurogenetic gradients and patterns of anatomical connections in the telencephalon. *Progress in Neurobiology*, 29:57-106.

Bayer, S.A., and Altman, J. (1990) Development of layer I and the subplate in the rat neocortex. *Exp. Neurol.*, 107:48-62.

Bayer, S.A., and Altman, J. (1991) *Neocortical Development*, Raven Press, New York.

Bear, M.F., Kleinschmidt, A., Gu, Q.A., and Singer, W. (1990) Disruption of experience-dependent synaptic modifications in striate cortex by infusion of an NMDA receptor antagonist. *J. Neurosci.*, 10:909-925.

Beck, P.D., Pospichal, M.W., and Kaas, J.H. (1996) Topography, architecture, and connections of somatosensory cortex in opossums: Evidence for five somatosensory areas. *J. Comp. Neurol.*, 366:109-133.

Behan, M., Kroker, A., and Bolz, J. (1991) Cortical barrelfields in organotypic slice cultures from rat somatosensory cortex. *Neurosci. Lett.*, 133:191-194.

Belford, G.R., and Killackey, H.P. (1980) The sensitive period in the development of the trigeminal system of the neonatal rat. *J. Comp. Neurol.*, 193:335-350.

Bentley, D., and Keshishian, H. (1982) Pioneer neurons and pathways in insect appendages. *Trends in Neurosci.*, 5:354-358.

Berman, N.E.J., and Hogan, D. (1991) Development of calbindin-d and parvalbumin immunoreactivity in interneurons of kitten visual cortex. *Soc. Neurosci. Abstr.*, 17:367.

Bernardo, K.L., and Woolsey, T.A. (1987) Axonal trajectories between mouse somatosensory thalamus and cortex. *J. Comp. Neurol.*, 258:542-564.

Berry, M., and Eayrs, J.T. (1963). Histogenesis of the cerebral cortex. *Nature*, 197:884-885.

Berry, M., and Rogers, A.W. (1965) The migration of neuroblasts in the developing cerebral cortex. *J. Anat.*, 99:691-709.

Berry, M., Rogers, A.W., and Eayrs, J.T. (1964) Pattern of cell migration during cortical histogenesis. *Nature*, 203:591-593.

Bicknese, A.R, and Pearlman, A.L. (1992) Growing corticothalamic and thalamocortical axons interdigitate in a restricted portion of the forming internal capsule. *Soc. Neurosci. Abstr.*, 18:778.

Bicknese, A.R., Sheppard, A.M., O'Leary, D.D.M., and Pearlman, A. L. (1991) Thalamocortical axons preferentially extend along a chondroitin sulfate proteoglycan enriched pathway coincident with the neocortical subplate and distinct from the efferent path. *Soc. Neurosci. Abstr.*, 17:764.

Bicknese, A.R., Sheppard, A.M., O'Leary, D.D.M., and Pearlman, A. L. (1994a) Thalamocortical axons extend along a chondroitin sulfate proteoglycan-enriched pathway coincident with the neocortical subplate and distinct from the efferent path. *J. Neurosci.*, 14:3500-3510.

Bicknese, A.R., Wang, W., and Sharma, A. (1994b) Multiple guidance cues are involved in the segregation of pioneering pathways in the cortex and internal capsule: evidence from *reeler*. *Soc. Neurosci. Abstr.*, 20:1683.

Blakemore, C. (1974) Effects of visual experience on the developing brain. *Mod. Probl. Paediat.*, 13:229-233.

Blakemore, C. (1978) Maturation and modification in the developing visual system. In *Perception*, eds. S.M. Anstis, R. Held, H.W. Leibowitz and H.-L. Teuber, Springer-Verlag, Berlin (Handbook of Sensory Physiology, vol. 8), 377-436.

Blakemore, C., and Cooper, G.F. (1970) Development of the brain depends on the visual environment. *Nature*, 228:477-478.

Blakemore, C., and Greenfield, S. (1987) *Mindwaves. Thoughts on Intelligence, Identity and Consciousness*, Blackwell, Oxford.

Blakemore, C., and Molnár, Z. (1990) Factors involved in the establishment of specific interconnections between thalamus and cerebral cortex. *Cold Spring Harbor Symposia on Quantitative Biology*, 55:491-504.

Blakemore, C., and Van Sluyters, R.C. (1975) Innate and environmental factors in the development of the kitten's visual cortex. *J. Physiol.*, 248:663-716.

Blakemore, C., Molnár, Z., Knott, G.W., and Saunders, N.R. (1997) Development of thalamocortical projections in the South American grey short-tailed opossum (Monodelphis domestica). *Soc. Neurosci. Abstr.*, 23: (in press).

Blakemore, C., Vital-Durand, F., and Garey, L.J. (1981) Recovery from monocular deprivation in the monkey. I. Reversal of physiological effects in the visual cortex. *Proc. R. Soc. Lond. B Biol. Sci.*, 213:399-423.

Blue, M.E., and Parnavelas, J. G. (1983) The formation and maturation of synapses in the visual cortex of the rat. II. Quantitative analysis. *J. Neurocytology*, 12:697-712.

Bock, W.J. (1969) Discussion: The concept of homology. In Petras and Noback, *Comparative and Evolutionary Aspects of the Vertebrate Central Nervous System*, Ann. N.Y. Acad. Sci., 167:71-73.

Bode-Greuel, K.M., Singer, W., and Aldenhoff, J.B. (1987) A current source density analysis of field potentials evoked in slices of visual cortex. *Exp. Brain Res.*, 69:213-219.

Bolz, J. (1995) Discussion on Subplate neurons and thalamocortical connections. In *Development of the Cerebral Cortex*, eds. G. Bock and G. Cardew, Wiley, Chichester (Ciba Foundation Symposium, 193), 168.

Bolz, J., and Götz, M. (1992) Mechanisms to establish specific thalamocortical connections in the developing brain. In *Development of the Central Nervous System in Vertebrates*, eds. S.C. Sharma and A.M. Goffinet, Plenum Press, New York-London, 179-192.

Bolz, J., Castellani, V., Mann, F., and Henke-Fahle, S. (1996) Specification of layer-specific connections in the developing cortex. *Prog. Brain Res.*, 108:41-54.

Bolz, J., Götz, M., Hübener, M., and Novak, N. (1993) Reconstructing cortical connections in a dish. *Trends in Neurosci.*, 16:310-316.

Bolz, J., Novak, N., and Staiger, V. (1992) Formation of specific afferent connections in organotypic slice cultures from rat visual cortex cocultured with lateral geniculate nucleus. *J. Neurosci.*, 12:3054-3070.

Bolz, J., Novak, N., Götz, M., and Bonhoeffer, T. (1990) Formation of target-specific neuronal projections in organotypic slice cultures from rat visual cortex. *Nature*, 346:359-362.

Bornstein, M.B. (1958) Reconstituted rat tail collagen as a substrate for tissue cultures on conventional coverslips in Maximov slides and roller tubes. *Lab. Invest.*, 7:134-140.

Bornstein, M.B. (1964) Morphological development of neonatal mouse cerebral cortex in tissue culture. In *Neurological and Electroencephalographic Correlative Studies in Infancy*, Grune and Stratton, New York, 1-11.*

Bornstein, M.B. (1973) The immunopathology of demyelinative disorders examined in organotypic cultures of mammalian central nerve tissues. In *Progress in Neuropathology*, vol. 2, ed. H. Zimmermann, Grune and Stratton, New York, 69-90.

Bottenstein, J.E., and Sato, G.H. (1979) Growth of a rat neuroblastoma cell line in serum-free supplemented medium. *Proc. Natl. Acad. Sci. USA*, 76:514-517.

Boulder Committee, see Angevine et al. (1970).

Braisted, J.E., and O'Leary, D.D.M. (1995) Axons from the ventrobasal thalamic nucleus pioneer the thalamocortical pathway to rat neocortex. *Soc. Neurosci. Abstr.*, 21:798.

Bredt, D.S., Hwang, P.M., and Snyder, S.H. (1990) Localization of nitric oxide synthase indicating a neural role for nitric oxide. *Nature*, 347:768-770.

Breer, H., Klemm, T., and Boekhoff, I. (1992) Nitric oxide mediated formation of cyclic GMP in the olfactory system. *Neuroreport*, 3:1030-1032.

Brodal, P. (1992) *The Central Nervous System: Structure and Function*, Oxford University Press, New York-Oxford.

Brodmann, K. (1909) *Vergleichende Localisationslehre der Grosshirnride in ihren Prinzipien dargestellt auf Grund des Zellenbaues.* J.A. Barth, Leipzig.

Bronchti, G., and Welker, E. (1995) Barrel cortex projects to subcortical stations of the 'visual' pathway in blind mice: A tracing study. *Eur. J. Neurosci., Suppl.*, 8:32.

Bronchti, G., Heil, P., Scheich, H., and Wollberg, Z. (1989) Auditory pathways and auditory activation of primary visual targets in the blind mole rat (*Spalax ehrenbergi*). I. 2-deoxyglucose study of subcortical centres. *J. Comp. Neurol.*, 284:253-274.

Bronchti, G., Rado, R., Terkel, J., and Wollberg, Z. (1991) Retinal projections in the blind mole rat: a WGA-HRP tracing study of a natural degeneration. *Dev. Brain Res.*, 58:159-170.

Bronchti, G., Schonenberger, N., Welker, E., and van der Loos, H. (1992) Barrelfield expansion after neonatal eye removal in mice. *Neuroreport*, 3: 489-492.

Brown, M.C. (1984) Sprouting of motor nerves in adult muscles: A recapitulation of ontogeny. *Trends in Neurosci.*, 7:10-14.

Brückner, G., Mares, V., and Biesold, D. (1976) Neurogenesis in the visual system of the rat: An autoradiographic investigation. *J. Comp. Neurol.*, 166:245-256.

Brune, B., and Lapetina, E.G. (1989) Activation of a cytosolic ADP-ribosyltransferase by nitric oxide-generating agents. *J. Biol. Chem.*, 264:8455-8458.

Brunner, G., Lang, K., Wolfe, R.A., McClure, D.B., and Sato, G.H. (1982) Selective cell culture of brain cells by serum-free, hormone-supplemented media: A comparative morphological study. *Dev. Brain Res.*, 2:563-575.

Brustle, O., Maskos, U., and McKay, R.D.G. (1995) Host guided migration allows targeted introduction of neurons into the embryonic brain. *Neuron*, 15:1275-1285.

Burger, M.M. (1974) Role of the cell surface in growth and transformation. In *Macromolecules Regulating Growth and Development*, eds. E.D. Hay, T.J. King and J. Papaconstantinou, Academic Press, New York, 3-24.

Butler, A.B., and Hodos, W. (1996) *Comparative Vertebrate Neuroanatomy. Evolution and Adaptation*, Wiley-Liss, New York-Chichester.

Caeser, M., Bonhoeffer, T., and Bolz, J. (1989) Cellular organization and development of slice cultures from rat visual cortex. *Exp. Brain. Res.*, 77:234-244.

Cajal, S.R. (1893) La rétine des vertébrés. *La Cellule*, 9:119-258.

Cajal, S.R. (1909-1911) *Histologie du Système Nerveux de l'Homme et des Vertébrés*, 2 vols. (trans. L. Azoulay), repr. Instituto Ramón y Cajal del C.S.I.C., Madrid, 1952-1955.

Cajal, S.R. (1928) *Degeneration and Regeneration of the Nervous System*, 2 vols., trans. R.M. May, repr. Hafner, New York, 1959.

Calarco, C.A., and Robertson, R.T. (1995) Development of basal forebrain projections to visual cortex: DiI studies in rat. *J. Comp. Neurol.*, 354:608-626.

Callaway, E.M., and Katz, L.C. (1990) Emergence and refinement of clustered horizontal connections in cat striate cortex. *J. Neurosci.*, 10:1134-1153.

Callaway, E.M., and Katz, L.C. (1992) Development of axonal arbors of layer 4 spiny neurons in cat striate cortex. *J. Neurosci.*, 12:570-582.

Campbell, C.B.G., and Hodos, W. (1970) The concept of homology and the evolution of the nervous system. *Brain Behav. Evol.*, 3:353-367.

Campbell, C.B.G., and Hodos, W. (1993) The *scala naturae* revisited: Evolutionary scales and anagenesis in comparative psychology. *J. Comp. Psychology.*

Campbell, K., Olsson, M., and Bjorklund, A. (1995) Regional incorporation and site-specific differentiation of striatal precursors transplanted into the embryonic forebrain ventricle. *Neuron*, 15:1259-1273.

Caric, D., Rennie, S., and Price, D.J. (1991) *In vivo* and *in vitro* studies of the development of cortical connections. *Soc. Neurosci. Abstr.*, 17:767.

Caroni, P., and Schwab, M.E. (1988a) Two membrane protein fractions from rat central myelin with inhibitory properties for neurite growth and fibroblast spreading. *J. Cell Biol.*, 106:1281-1288.

Caroni, P., and Schwab, M.E. (1988b) Antibody against myelin-associated inhibitor of neurite growth neutralizes non-permissive substrate properties of CNS white matter. *Neuron*, 1:85-96.

Carpenter, R.H.S.(1984) *Neurophysiology*, Edward Arnold, London.

Casagrande, V.A. (1994) A third parallel visual pathway to primate area V1. *Trends in Neurosci.*, 17:305-310.

Cases, O., Vitalis, T., Seif, I., De Maeyer, R., Sotelo, C., and Gaspar, P. (1996) Lack of barrels in the somatosensory cortex of monoamine oxidase A deficient mice: Role of serotonin excess during the critical period. *Neuron*, 16:297-307.

Catalano, S., Robertson, R.T., and Killackey, H.P. (1991) Early ingrowth of thalamocortical afferents to the neocortex of the prenatal rat. *Proc. Natl. Acad. Sci. USA*, 88:2999-3003.

Catalano, S.M., Robertson, R.T., Killackey, H.P. (1996) Individual axon morphology and thalamocortical topography in developing rat somatosensory cortex. *J. Comp. Neurol.*, 367:36-53.

Catania, K.C., and Kaas, J.H. (1995) Organization of the somatosensory cortex of the star-nosed mole. *J. Comp. Neurol.*, 351:549-567.

Catsicas, S., and Clarke, P.H. (1987) Spatiotemporal gradients of cainate-sensitivity in the developing chicken retina. *J. Comp. Neurol.*, 262:512-522.

Caviness, V.S., Jr. (1976) Patterns of cell and fiber distribution in the neocortex of the *reeler* mutant mouse. *J. Comp. Neurol.*, 170:435-448.

Caviness, V.S., Jr. (1982) Neocortical histogenesis in normal and *reeler* mice: A developmental study based upon [^{3}H] thymidine autoradiography. *Dev. Brain Res.*, 4:293-302.

Caviness, V.S., Jr. (1988) Architecture and development of the thalamocortical projection in the mouse. In *Cellular Thalamic Mechanisms*, eds. M. Bentivoglio and R. Spreafico, Excertpa Medica, Amsterdam-New York, 489-499.

Caviness, V.S., Jr., and Frost, D.O. (1980) Tangential organization of thalamic projections of the neocortex in the mouse. *J. Comp. Neurol.*, 194:355-367.

Caviness, V.S., Jr., and Korde, M.G. (1981) Monoaminergic afferents to the neocortex: A developmental histofluorescence study in normal and *reeler* mouse embryos. *Brain Res.*, 209:1-9.

Caviness, V.S., Jr., and Rakic, P. (1978) Mechanisms of cortical development: A view from mutations in mice. *Ann. Rev. Neurosci.*, 1:297-326.

Caviness, V.S., Jr., and Sidman, R.L. (1973) Time of origin of corresponding cell classes in the cerebral cortex of normal and *reeler* mutant mice: An autoradiographic analysis. *J. Comp. Neurol.*, 148:141-151.

Caviness, V.S., Jr., and Yorke, C.H. (1976) Interhemispheric neocortical connections of the corpus callosum in the *reeler* mutant mouse: a study based on anterograde and retrograde methods. *J. Comp. Neurol.*, 170: 449-460.

Caviness, V.S., Jr., Crandall, J.E., and Edwards, M.A. (1988) The *reeler* malformation, implications for neocortical histogenesis. In *Cerebral Cortex*, vol. 7: *Development and Maturation of Cerebral Cortex*, eds. E.G. Jones and A. Peters, Plenum Press, New York-London, 59-89.

Caviness, V.S., Jr., Pinto-Lord, M.C., and Evrard, P. (1981) The development of laminated pattern in mammalian neocortex. In *Morphogenesis and Pattern Formation*, eds. L.L. Brinkley, B.M. Carlson and T.G. Connelly, Raven Press, New York, 103-126.

Caviness, V.S., Jr., So, K.-F., Sidman, R.L. (1972) The hybrid *reeler* mouse. *J. Hered.*, 63:241-246.

Celio, M.R. (1986) Parvalbumin in most gamma amino butyric acid containing neurons in the rat cerebral cortex. *Science*, 23:995-997.

Celio, M.R. (1990) Calbindin D-28k and parvalbumin in the rat nervous system. *Neuroscience*, 35:375-475.

Cellerino, A., Siciliano, R., Domenici, L., and Maffei, L. (1992) Parvalbumin immunoreactivity: A reliable marker for the effects of monocular deprivation in the rat visual cortex. *Neuroscience*, 51:749-753.

Cepko, C.L., Austin, C.P., Walsh, C., Ryder, E.F., Halliday, A., and Fields-Berry, S. (1990) Studies of cortical development using retrovirus vectors. *Cold Spring Harbor Symposia on Quantitative Biology*, 55:265-278.

Chalupa, L.M., Coyle, R.S., and Lindsey, D.B. (1976) Effects of pulvinar lesions on visual pattern discrimination in monkeys. *J. Neurophysiol.*, 39:354-369.

Chang, F.F., Steedman, J.G., and Lund, R.D. (1986) The lamination and connectivity of embryonic cerebral cortex transplanted into newborn rat cortex. *J. Comp. Neurol.*, 244:401-411.

Changeux, J., and Danchin, A. (1976) Selective stabilization of developing synapses as a mechanism for the specification of neuronal networks. *Nature*, 264:705-711.

Chapman, P.F., Atkins, C.M., Allen, M.T., Haley, J.E., and Steinmetz, J.E. (1992) Inhibition of nitric oxide synthesis impairs two different forms of learning. *Neuroreport*, 3:567-570.

Chen, S., and Hillman, D.E. (1986) Selective ablation of neurons by methylazoxymethanol during pre- and postnatal brain development. *Exp. Neurol.*, 94:103-119.

Chiaia, N.L., Fish, S.E., Bauer, W.R., Figley, B.A., Eck, M., Bennett-Clarke, C.A., and Rhodes, R.W. (1994) Effects of postnatal blockade of cortical activity with tetrodotoxin upon the development and plasticity of vibrissa-related patterns in the somatosensory cortex of hamsters. *Somatosens. Mot. Res.*, 11:219-228.

Choi, D.W. (1991) Excitotoxicity on cultured cortical neurons in glutamate. In *Cell Death and Memory*, eds. P. Ascher, D.W. Choi and Y. Christen, Springer-Verlag, Berlin-Heidelberg-New York, 125-136.

Choi, D.W. (1992) Excitotoxic cell death. *J. Neurobiol.*, 23:1261-1276.

Chun, J.J.M., and Shatz, C.J. (1988a) A fibronectin-like molecule is present within the developing cat cerebral cortex and is correlated with subplate neurons. *J. Cell Biol.*, 106:857-872.

Chun, J.J.M., and Shatz, C.J. (1988b) Distribution of synaptic vesicle antigens is correlated with the disappearance of a transient synaptic zone in the developing cerebral cortex. *Neuron*, 1:297-310.

Chun, J.J.M., and Shatz, C.J. (1989) Interstitial cells of the adult neocortical white matter are the remnant of the early generated subplate neuron population. *J. Comp. Neurol.*, 282:555-569.

Chun, J.J.M., Nakamura, M.J., and Shatz, C.J. (1987) Transient cells of the developing mammalian telencephalon are peptide immunoreactive neurons. *Nature*, 325:617-620.

Chung, W.W., Lagenaur, C.F., and Lund, J.S. (1990) Distribution of adhesion and cell surface molecules in developing mouse neocortex. *Soc. Neurosci. Abstr.*, 16:178.

Chung, W.W., Lagenaur, C.F., Yan, Y.M., and Lund, J.S. (1991) Developmental expression of neuronal cell adhesion molecules in the mouse neocortex and olfactory bulb. *J. Comp. Neurol.*, 314:290-305.

Clasca, F., Angelucci, A., and Sur, M. (1994) Layer 5 neurons establish the first cortical projection to the dorsal thalamus in ferrets. *Soc. Neurosci. Abstr.*, 20:98.

Clasca, F., Angelucci, A., and Sur, M. (1995) Layer-specific programs of development in neocortical projection neurons. *Proc. Natl. Acad. Sci. USA*, 92:11145-11149.

Clemence, A.E., and Mitrofanis, J. (1992) Cytoarchitectonic heterogeneities in the thalamic reticular nucleus of cats and ferrets. *J. Comp. Neurol.*, 322:167-180.

Cobcroft, M., Vaccaro, T, and Mitrofanis, J. (1989) Distinct patterns of distribution among NADPH-diaphorase neurones of the guinea pig retina, *Neurosci. Lett.*, 103:1-7.

Cogeshall, R.E. (1964) A study of diencephalic development in the albino rat. *J. Comp. Neurol.*, 122:241-269.

Cohen-Tannoudji, M., Babinet, C., and Wassef, M. (1994) Early determination of a mouse somatosensory cortex marker. *Nature*, 368:460-463.

Cole, G.J., and McCabe, C.F. (1991) Identification of a developmentally regulated keratan sulfate proteoglycan that inhibits cell adhesion and neurite outgrowth. *Neuron*, 7:1007-1018.

Colello, R.J., and Guillery, R.W (1987) The distribution of axonal profiles within the optic tract of the embryonic mouse. *Neuroscience*, 22 (Suppl.): 807.

Colello, R.J., and Guillery, R.W. (1990) The early development of retinal ganglion cells with uncrossed axons in the mouse: Retinal position and axonal course. *Development*, 108:514-523.

Colello, R.J., and Guillery, R.W. (1992) Observations on the early development of the optic nerve and tract of the mouse. *J. Comp. Neurol.*, 317:357-378.

Conolly, M., and Van Essen, D. (1984) The representation of the visual field in parvicellular and magnocellular layers of the lateral geniculate nucleus in the macaque monkey. *J. Comp. Neurol.*, 226:544-564.

Cooper, N.G., and Steindler, D.A. (1986) Lectins demarcate the barrel subfield in the somatosensory cortex of the early postnatal mouse. *J. Comp. Neurol.*, 249:157-169.

Cordery, P., Blakemore, C., and Molnár, Z. (1997) Comparison of mammalian and reptilian pallium during development. *Soc. Neurosci. Abstr.*, 23:320:19.

Cordery, P., Molnár, Z., and Blakemore, C. (1996) Origin of the first projections reaching the internal capsule and thalamus during embryonic development. *Brain Res. Assoc. Abstr.*, 13:52.

Crabtree, J.W. (1992a) The somatotopic organization within the cat's thalamic reticular nucleus. *Eur. J. Neurosci.*, 4:1352-1361.

Crabtree, J.W. (1992b) The somatotopic organization within the rabbit's thalamic reticular nucleus. *Eur. J. Neurosci.*, 4:1343-1351.

Cragg, B.G. (1975) The development of synapses in kitten visual cortex during visual deprivation. *Exp. Neurol.*, 46:445-451.

Crain, S.M. (1972) Tissue culture models of epileptiform activity. In *Experimental Models of Epilepsy*, eds. D.P. Purpura, J.K. Penry, D.B. Tower, D.M. Woodbury and R.D. Walter, Raven, New York, 291-316.

Crain, S.M. (1974) Tissue culture models of developing brain functions. In *Studies on the Development of Behavior and the Nervous System*, vol. 2, *Aspects of Neurogenesis*, ed. G. Gottlieb, Academic Press, New York, 69-114.

Crain, S.M. (1976) *Neurophysiologic Studies in Tissue Culture*, Raven, New York.

Crair, M.C., and Malenka, R.C. (1995) A critical period for long-term potentiation at thalamocortical synapses. *Nature*, 375:277-278.

Crair, M.C., Molnár, Z., Higashi, S., Kurotani, T., and Toyama, K. (1993) The development of thalamocortical and intracortical connectivity in rat somatosensory "barrel" cortex imaged by optical recording. *Soc. Neurosci. Abstr.*, 19:1705.

Crair, M.C., Molnár, Z., Higashi, S., Kurotani, T., and Toyama, K. (1997) The development of thalamocortical and intracortical connectivity in rat somatosensory "barrel" cortex imaged by optical recording. Manuscript submitted to *Current Biology*.

Crandall, J. E., and Caviness, V.S., Jr. (1984b) Thalamocortical connections in newborn mice. *J. Comp. Neurol.*, 228:542-556.

Crandall, J.E., and Caviness, V.S., Jr. (1984a) Axon strata of the cerebral wall in the embryonic mice. *Dev. Brain Res.*, 14:185-195.

Crandall, J.E., Hassinger, L.C., and Bonacorso, J. (1992) Afferents to preplate neurons in embryonic mouse cortex. *Soc. Neurosci. Abstr.*, 18:778.

Cullen, M.J., and Kaiserman-Abramoff, I.R. (1986) Cytological organization of the dorsal lateral geniculate nuclei in mutant anophtalmic and postnatally enucleated mice. *J. Neurocytol.*, 5:407-424.

Cunningam, T.J., Haum, F., and Chantler, P.D. (1987) Diffusible proteins prolong survival of dorsal lateral geniculate neurons following occipital cortex lesions in newborn rats. *Devel. Brain Res.*, 37:133-141.

D'Arcangelo, G., Miao, G.G., Chen, S.C., Soares, H.D., Morgan, J.I., and Curran, T. (1995) A protein related to extracellular matrix proteins deleted in the mouse mutant *reeler*. *Nature*, 374:719-723.

D'Arcangelo, G., Nakajima, K., Miyata, T., Ogawa, M., Mikoshiba, K., and Curran, T. (1997) Reelin is a secreted glycoprotein recognized by the CR-50 monoclonal antibody. *J. Neurosci.*, 17:23-31.

Dávila, J.C., de la Calle, A., Gutierrez, A., Megias, M., Andreu, M.J., and Guirado, S. (1991) Distribution of neuropeptide Y (NPY) in the cerebral cortex of the lizards *psammodromus algirus* and *podarcis hispanica*: Co-localisation of NPY, somatostatin, and GABA. *J. Comp. Neurol.*, 308: 397-408.

Dávila, J.C., Padial, J., Andreu, M.J., Ángeles, Real, and Guirado, S. (1997) Calretinin immunoreactivity in the cerebral cortex of the lisard *Psammodromus algirus*: a light and electron microscopic study. *J. Comp. Neurol.*, 382:382-393.

Dawson, D.W., and Killackey, H.P. (1987) The organization and mutability of the forepaw and hindpaw representation in the somatosensory cortex of the neonatal rat. *J. Comp. Neurol.*, 256:246-256.

Dawson, V.L., Dawson, T.M., London, E.D., and Bredt, D.S. (1991) Nitric oxide mediates glutamate neurotoxicity in primary cortical cultures. *Proc. Natl. Acad. Sci. USA*, 88:6368-6371.

De Carlos, J.A., and O'Leary, D.D.M. (1990) Subplate neurons "pioneer" the output pathway of rat cortex but not pathways to brainstem or spinal targets. *Soc. Neurosci. Abstr.*, 16:311.

De Carlos, J.A., and O'Leary, D.D.M. (1992) Growth and targeting of subplate axons and establishment of major cortical pathways. *J. Neurosci.*, 12:1194-1211.

De Carlos, J.A., Lopez-Mascaraque, L., and Valverde, F. (1996) Dynamics of cell migration from the lateral ganglionic eminence in the rat. *J. Neurosci.*, 16:6146-6156.

De Carlos, J.A., Schlaggar, B.L., and O'Leary, D.D.M. (1995) Development of acetylcholinesterase-positive thalamic and basal forebrain afferents to embryonic rat neocortex. *Exp. Brain Res.*, 104:385-401.

De Jong, B.M., Ruiter, J.M., and Romijn, H.J. (1988) Cytoarchitecture in cultured rat neocortex explants. *Int. J. Dev. Neurosci.*, 6:327-339.

DeBruyn, E.J., and Casagrande, V.A. (1981) Demonstration of ocular dominance columns in a New World primate by means of monocular deprivation. *Brain Res.*, 207:453-458.

DeFelipe, J., and Jones, E.G. (eds.) (1988) *Cajal on the Cerebral Cortex: An Annotated Translation of the Complete Writings*, Oxford University Press, New York-Oxford, 188.

Dehay, C., Giroud, P., Berland, M., Killackey, H., and Kennedy, H. (1996) Contribution of thalamic input to the specification of cytoarchitectonic cortical fields in the primate: Effects of bilateral enucleation in the fetal monkey on boundaries, dimensions, and gyrification of striate and extrastriate cortex. *J. Comp. Neurol.*, 367:70-89.

Dehay, C., Giroud, P., Berland, M., Smart, I., Kennedy, H. (1993) Modulation of the cell cycle contributes to the primate visual cortex. *Nature*, 366:464-466.

Dehay, C., Horsburgh, G., Berland, M., Killackey, H., and Kennedy, H. (1989) Maturation and connectivity of the visual cortex in monkey is altered by prenatal removal of retinal input. *Nature*, 337:265-267.

Del Río, J.A., Heimrich, B., Borrell, V., Fröster, E., Drakew, A., Alcántara, S., Nakajima, K., Miyata, T., Ogawa, M., Mikoshiba, K., Derer, P., Frotscher, M., and Soriano, E. (1997) The role for Cajal-Retzius cells and *reelin* in the development of hippocampal connections. *Nature*, 385:70-74.

Del Río, J.A., Heimrich, B., Super, H., Borrell, V., Frotscher, M., and Soriano, E. (1996) Differential survival of Cajal-Retzius cells in organotypic cultures of hippocampus and neocortex. *J. Neurosci.*, 16:6896-6907.

Del Río, J.A., Martinez, A., Fonseca, M., Auladell, C., and Soriano, E. (1995) Glutamate-like immunoreactivity and the fate of Cajal-Retzius cells in the murine cortex as identified with calretinin antibody. *Cerebral Cortex*, 5:13-21.

Derer, P., and Derer, M. (1990) Cajal-Retzius cell ontogenesis and death in mouse brain visualised with horseradish peroxidase and electron microscopy. *Neuroscience*, 36:839-856.

Derer, P., and Derer, M. (1992) Development and fate of Cajal-Retzius cells *in vivo* and *in vitro*. In *Development of the Central Nervous System in Vertebrates*, eds. S.C. Sharma and A.M. Goffinet, Plenum, New York-London.

Derer, P., and Nakanishi, S. (1983) Extracellular matrix distribution during neocortical wall ontogenesis in normal and *reeler* mice. *J. Hirnforsch.*, 24:209-224.

Dermietzel, R., and Spray, D.C., (1993) Gap junctions in the brain: where, what type, how many and why? *Trends Neurosci.*, 16:186-192.

Dermon, C.R., and Barbas, H. (1994) Contralateral thalamic projections predominantly reach transitional cortices in the rhesus monkey. *J. Comp. Neurol.*, 344:508-531.

Dixon, G. (1990) Prenatal development of the geniculo-cortical pathway in the rat. *Proc. Aust. Neurosci. Soc.*, 1:73.

Dräger, U.C. (1976) *Reeler* mutant mice: Physiology in primary visual cortex. *Exp. Brain Res., Suppl.*, 1:274-276.

Driesch, H. (1914) *The History and Theory of Vitalism*, Macmillan, London.

Driscoll, M., and Chalfie, M. (1992) Developmental and abnormal cell death in C. elegans. *Trends in Neurosci.*, 15:15-19.

Droogleever-Fortuyn, A.B. (1912) Die Ontogenie der Kerne des Zwischenhirns beim Kaninchen. *Arch. Anat. Physiol. (Anat. Abt.)*, 1912:303-352.

Druga, R., and Syka, J. (1993) NADPH-diaphorase activity in the central auditory structures of the rat. *NeuroReport*, 4:999-1002.

Dubinsky, J.M., and Rothman, S.M. (1991) Intracellular Calcium concentration during 'chemical hypoxia' and excitotoxic neuronal injury. *J. Neurosci.*, 11:2545-2551.

Dunn, G.A. (1971) Mutual contact inhibition of extention of chick sensory nerve fibers *in vitro. J. Comp. Neurol.*, 143:491-508.

Eagle, H. (1959) Amino acid metabolism in mammalian cell cultures. *Science*, 130:432-437.

Easter, S.S., Jr., Rusoff, A.C., and Kish, E.P. (1981) The growth and organization of the optic nerve and tract in juvenile and adult goldfish. *J. Neurosci.*, 1:793-811.

Easter, S.S., Ross, L.S., and Frankfurter, A. (1993) Initial tract formation in the mouse-brain. *J. Neurosci.*, 13:285-299.

Eccles, J.C. (1983) Calcium in long term potentiation as a model for memory. *Neuroscience*, 10:1071-1081.

Edinger, L. (1885) *Vorlesungen über den Bau der nervösen Zentralorgane des Menschen und der Tier*, Vogel, Leipzig.

Elbert, T., Pantev, C., Wienbruch, C, Rockstroh, B., and Taub, E. (1995) Increased cortical representation of the fingers of the left hand in string players. *Science*, 270:305-307.

Emerling, D.E., and Lander, A.D. (1994) Laminar specific attachment and neurite outgrowth of thalamic neurons on cultured slices of developing cerebral neocortex. *Development* , 120:2811-2822.

Emerling, D.E., and Lander, A.D. (1996) Inhibitors and promoters of thalamic neuron adhesion and outgrowth in embryonic neocortex: Functional association with chondroitin sulfate. *Neuron*, 17:1089-1100.

Erzurumlu, R.S., and Jhaveri, S. (1990) Thalamic axons confer a blueprint of the sensory periphery onto the developing rat somatosensory cortex. *Dev. Brain Res.*, 56:229-234.

Erzurumlu, R.S., and Jhaveri, S. (1992) Emergence of connectivity in the embryonic rat parietal cortex. *Cerebral Cortex*, 2:336-352.

Evans, V.J., Bryant, J.C., Fioramonti, M.C., McQuilkin, W.T., Stanford, K.K., and Earle W.R. (1956) Studies of nutrient media for tissue cells in vitro. I. A protein free chemically defined medium for cultivation of strain L cell. *Cancer Res.*, 16:77-86.

Fazeli, M.S. (1992) Synaptic plasticity: on the trail of the retrograde messenger. *Trends in Neurosci.*, 15:115-117.

Feng, L., and Sretavan, D.W. (1991) A midline neuronal population is present at the mammalian optic chiasm during early retinal ganglion cell axon ingrowth. *Soc. Neurosci. Abstr.*, 17:182.

Ferrer, I., Soriano, E., Del Río, J.A., Alcántara, S., and Auladell, C. (1992) Cell death and removal in the cerebral cortex during development. *Progress in Neurobiology*, 39:1-43.

Finlay, B.L., and Darlington, R.B. (1995) Linked regularities in the development and evolution of mammalian brains. *Science*, 268:1578-1584.

Fishell, G. (1995) Striatal precursors adopt cortical identities in response to local cues. *Development* , 121:803-812.

Fishell, G., Mason, C.A., and Hatten, M.E. (1993) Dispersion of neural progenitors within the germinal zones of the forebrain. *Nature*, 362:636-638.

Flecher, C.F., Lutz, C.M., O'Sullivan, T.N., Saughnessy, F.D., Hawkes, R., Frankel, W.N., Copeland, N.G., and Jenkins, N.A. (1996) Absence epilepsy in *tottering* mutant mice is associated with calcium channel defects. *Cell*, 87:607-617.

Floeter, M.K., and Jones, E.G. (1985) Transplantation of fetal postmitotic neurons to rat cortex: survival, early pathway choices and long-term projections of outgrowing axons. *Brain Res.*, 353:19-38.

Florence, S.L., Conley, M., and Casagrande, V.A. (1986) Ocular dominance columns and retinal projections in New World spider monkeys *(Ateles ater)*. *J. Comp. Neurol.*, 243:234-248.

Fox, K. (1995) The critical period for long-term potentiation in primary sensory cortex. *Neuron*, 15:485-488.

Fox, K., Schlaggar, B.L., Glazewitski, S., and O'Leary, D.D.M. (1996) Glutamate receptor blockade at cortical synapses disrupts development of thalamocortical and columnar organization in somatosensory cortex. *Proc. Natl Acad Sci. USA*, 93:5584-5589.

Franklin, J.L., and Johnson, E.M., Jr. (1992) Suppression of programmed neuronal death by sustained elevation of cytoplasmic calcium. *Trends in Neurosci.*, 15:501-508.

Freshney, R.I. (ed) (1986) *Animal Cell Culture. A Practical Approach*, IRL Press, Oxford-Washington DC.

Friauf, E., and Shatz, C.J. (1991) Changing patterns of synaptic input to subplate and cortical plate. *J. Neurophysiol.*, 66:2059-2071.

Friauf, E., McConnell, S.K., and Shatz, C.J. (1990) Functional synaptic cirquits in the subplate during fetal and early postnatal development of cat visual cortex. *J. Neurosci.*, 10:2601-2613.

Friedlander, M.J., Martin, K.A.C., and Wassenhowe-McCarthy, D. (1991) Effects of monocular visual deprivation on geniculocortical innervation of area 18 in cat. *J. Neurosci.*, 11:3268-3288.

Frost, D.O., and Caviness, V.S., Jr. (1980) Radial organization of thalamic projections to the neocortex in the mouse. *J. Comp. Neurol.*, 194:369-393.

Frost, D.O., and Métin, C. (1985) Induction of functional retinal projections to the somatosensory system. *Nature*, 317:162-164.

Frotcher, M., and Gähwiler, B.H. (1988) Synaptic organization of intracellularly stained CA3 pyramidal neurons in slice cultures of rat hippocampus. *Neuroscience*, 24:541-551.

Fukuda, T., Kawano, H., Ohyama, K., Li, H.P., Takeda, Y., Oohira, A., and Kawamura, K. (1997) Immunohistochemical localisation of neurocan and L1 in the formation of thalamocortical pathway of developing rats. *J. Comp. Neurol.*, 382:141-152.

Gähwiler, B.H. (1988) Organotypic cultures of neural tissue. *Trends in Neurosci.*, 11:484-489.

Gähwiler, B.H., Capogna, M., Debanne, R.A., McKinney, R.A., and Thompson, S.M. (1997) Organotypic cultures: A technique has come of age. *Trends in Neurosci.* (in press).

Galli, L., and Maffei, L. (1988) Spontaneous impulse activity of rat retinal ganglion cells in prenatal life. *Science*, 242:90-91.

Gally, J.A., Montague, P.R., Reeke, G.N., and Edelman, G.M. (1990) The NO hypothesis: Possible effects of a short-lived, rapidly diffusible signal in the development and function of the nervous system. *Proc. Natl. Acad. Sci. USA*, 87:3547-3551.

Garey, L.J. (1994) *Brodmann's 'Localisation in the cerebral cortex'*, Smith-Gordon, London.

Garraghty, P.E., and Kaas, J.H. (1991) Functional reorganization in adult monkey thalamus after peripheral nerve injury. *NeuroReport*, 2:747-750.

Garraghty, P.E., and Muja, N. (1996) NMDA receptors and plasticity in adult primate somatosensory cortex. *J. comp. Neurol.*, 367:319-326.

Garraghty, P.E., Sur, M., Weller, R.E., and Sherman, S.M. (1986) Morphology of retinogeniculate X and Y axon arbors in monocularly enucleated cats. *J. Comp. Neurol.*, 251:198-215.

Garthwaite, J., Garthwaite, G., and Williams, G.D. (1991) Excitatory amino acid neurotoxicity in rat brain slices in glutamate. In *Cell Death and Memory*, eds P. Ascher, D.W. Choi and Y. Christen, Springer-Verlag, Berlin-Heidelberg-New York, 117-125.

Gaze, R.M., and Grant, P. (1978) The diencephalic course of regenerating retinotectal fibers in the Xenopus tadpoles. *J. Embryol. Exp. Morphol.*, 44:201-216.

Gaze, R.M., Keating, M.J., Ostberg, A., and Chung, S.H. (1979) The relationship between retinal and tectal growth in larval Xenopus: implications for the development of the retinotectal projection. *J. Embryol. Exp. Morphol.*, 53:103-142.

Gheorghita-Bächler, F., DuBois, R., and Welker, E. (1996) Morphology of thalamocortical axons in the somatosensory cortex (S1) of the mouse mutant *barrelless*. *Eur.J. Neurosci., Suppl.*, 9:93.

Ghiselin, M.T. (1966) An application of the theory of definitions to systematic principles. *Syst. Zool.*, 15:127-130.

Ghiselin, M.T. (1969) The principles and concepts of systematic biology. In *Systematic Biology*, Proc. of an Internat. Conf., National Academy of Sciences, Washington DC, 45-55.

Ghosh, A., and Shatz, C.J. (1992a) Pathfinding and target selection by developing geniculocortical axons. *J. Neurosci.*, 12:39-55.

Ghosh, A., and Shatz, C.J. (1992b) Involvement of subplate neurons in the formation of ocular dominance columns. *Science*, 255:1441-1443.

Ghosh, A., and Shatz, C.J. (1993) A role for subplate neurons in the patterning of connections from thalamus to neocortex. *Development*, 117:1031-1047.

Ghosh, A., and Shatz, C.J. (1994) Segragation of geniculocortical afferents during the critical period: A role for subplate neurons. *J. Neurosci.*, 14:3862-3880.

Ghosh, A., Antonini, A., McConnell, S.K., and Shatz, C.J. (1990) Requirement for subplate neurons in the formation of thalamocortical connections. *Nature*, 347:179-181.

Gil, Z., and Amitai, Y. (1995) NMDA and nonNMDA receptors contribution to thalamocortical and intracortical transmission. *Soc. Neurosci. Abstr.*, 21:116.

Gil, Z., and Amitai, Y. (1996) Adult thalamocortical transmission involves both NMDA and non-NMDA receptors. *J. Neurophysiol.*, 76:2547-2554.

Gilbert, C.D. (1983) Microcircuitry of the visual cortex. *Ann. Rev. Neurosci.*, 6:217-247.

Gilbert, C.D., and Wiesel, T.N. (1992) Receptive field dynamics in adult primary visual cortex. *Nature*, 356:150-152.

Gilles, K., and Price, D. (1993) Cell migration and subplate loss in explant cultures of murine cerebral cortex. *Neuroreport*, 4:911-914.

Gillespie, D.C., Ruthazer, E.S., Dawson, T.M., Snyder, S.H., and Stryker, M.P. (1993) Inhibition of nitric oxide synthase does not prevent ocular dominance plasticity in cat visual cortex. *Soc. Neurosci. Abstr.*, 19:893.

Gilmore, E.D., and Herrup, K. (1997) Cortical development: Layers of complexity. *Current Biology*, 7:231-234.

Godement, P., Saillour, P., and Imbert, M. (1979) Thalamic afferents to the visual cortex in congenitally anophtalmic mice. *Neurosci. Lett.*, 13:271-278.

Godement, P., Vanselow, J., Thanos, S., and Bonhoeffer, F. (1987) A study in developing visual systems with a new method of staining neurones and their processes in fixed tissue. *Development*, 101:697-713.

Godfraind, C., Schachner, M., and Goffinet, A.M. (1988) Immunohistochemical localisation of cell adhesion molecules L1, J1, N-CAM and their common carbohydrate L2 in the embryonic cortex of normal and *reeler* mice. *Dev. Brain Res.*, 42:99-111.

Goffinet, A.M. (1979) An early developmental defect in the cerebral cortex of the *reeler* mouse. *Anat. Embryol.*, 157:205-216.

Goffinet, A.M. (1980) The cerebral cortex of the *reeler* mouse embryo: An electron microscopic analysis. *Anat. Embryol.*, 159:199-210.

Goffinet, A.M. (1983) The embryonic development of the cortical plate in reptiles: A comparative analysis. *J. Comp. Neurol.*, 219:10-24.

Goffinet, A.M. (1984) Events governing organization of postmigratory neurons: Studies on brain development in normal and *reeler* mice. *Brain Res. Rev.*, 7:261-296.

Goffinet, A.M. (1990) Cortical architectonic development: A comparative study in reptiles. *Experimental Brain Research Series*, 19:135-144.

Goffinet, A.M. (1992) The *reeler* gene: A clue to brain development and evolution. *Int. J. Dev. Biol.*, 36:101-107.

Goffinet, A.M. (1995) A real gene for *reeler*. *Nature*, 374:675-676.

Goffinet, A.M., and Lyon, G. (1979) Early histogenesis in the mouse cerebral cortex: A Golgi study. *Neurosci. Lett.*, 14:61-66.

Goffinet, A.M., Daumerie, C.H., Langerwerf, B., and Pieau, C. (1986) Neurogenesis in reptilian cortical structures: ^{3}H-thymidine autoradiographic analysis. *J. Comp. Neurol.*, 243:106-116.

Goldman-Rakic, P.S. (1982) Neuronal development and plasticity of association cortex in primates. *Neurosci. Res. Prog. Bull.*, 20:520-532.

Goldowitz, D., Laywell, E., Steindler, D., D'Arcangelo, G., Sheldon, M., Curran, T., Davisson, M., Johnson, K., and Sweet, H. (1996) *Scrambler* and *reeler*: Morphomolecular analysis of mutant cerebella indicate one phenotype from two different genes. *Soc. Neurosci. Abstr.*, 22:961.

Gonzalez, J.L., Russon, C., Sweet, H., Davisson, M., Goldowitz, D., and Walsh, C. (1996) Neocortical birthdating analysis on *scrambler*: A novel mouse mutant reveals a *reeler*-like phenotype. *Soc. Neurosci. Abstr.*, 22:987.

Goodman, C.S., and Shatz, C.J. (1993) Developmental mechanisms that generate precise patterns of neuronal connectivity. *Cell*, 72: *Neuron, Suppl.*, 10:77-98.

Goodman, C.S., Bastiani, M.J., Doe, C.Q., du Lac, S., Helfand, S.L., Kuwada, J.Y., and Thomas, J.B. (1984) Cell recognition during neuronal development. *Science*, 225:1271-1279.

Gordon, B., Kinch, G., Kato, N., Keele, C., Lissman, T., and Fu, L.N. (1997) Development of MK-801, kainate, AMPA, and muscimol binding sites and the effect of dark rearing in rat visual cortex. *J. Comp. Neurol.*, 383:73-81.

Götz, M., Novak, N., and Bolz, J. (1991) Development of afferent projections in rat visual cortex in vivo and in vitro. *Soc. Neurosci. Abstr.*, 17:898.

Götz, M., Novak, N., Bastmayer, M., and Bolz, J. (1992) Membrane-bound molecules in rat cerebral cortex regulate thalamic innervation. *Development*, 116:507-519.

Götz, M., Williams, B.P., Bolz, J., and Price, J. (1995) The specification of neural fate: A common precursor for neurotransmitter subtypes in the rat cerebral cortex *in vitro*. *Eur. J. Neurosci.*, 7:889-898.

Gray, E.G. (1959) Axo-somatic and axo-dendritic synapses of the cerebral cortex: An electron microscope study. *J. Anat.*, 93:420-433.

Green, M.C. (1981) Catalogue of mutant genes and polymorphic loci. In *Genetic Variants and Strains of the Laboratory Mouse*, ed. M.C. Green, Fisher Verlag, New York, 8-27.

Grinvald, A. (1984) From growth cones to the intact brain: Real-time optical imaging of neuronal activity. *Trends in Neurosci.*, 16:143-150.

Guillery, R.W. (1972) Binocular competition in the control of geniculate cell growth. *J. Comp. Neurol.*, 144:117-130.

Guillery, R.W. (1986) Neural abnormalities of albinos. *Trends in Neurosci.*, 9:364-367.

Guillery, R.W. (1995) Anatomical evidence concerning the role of the thalamus in corticocortical communication: A brief review. *J. Anat.*, 187:583-592.

Guillery, R.W., and Walsh, C. (1987) Changing glial organization relates to changing fiber order in the developing optic nerve of ferrets. *J. Comp. Neurol.*, 265:203-217.

Guillery, R.W., Ombrellaro, M., and LaMantia, A.L. (1985) The organization of the lateral geniculate nucleus and the geniculocortical pathway that develops without retinal afferents. *Dev. Brain Res.*, 20:221-233.

Gutnick, M.J., Wolfson, B., and Baldino, F., Jr. (1989) Synchronised neuronal activities in neocortical explant cultures. *Exp. Brain Res.*, 76:131-140.

Haeckel, E. (1866) *Generelle Morphologie der Organismen*, G. Reimer, Berlin.

Haley, J.E., Wilkox, G.L., and Chapman, P.F. (1992) The role of nitric oxide in hippocampal long-term potentiation. *Neuron*, 8:211-216.

Halloran, M.C., and Kalil, K. (1994) Dynamic behaviors of growth cones extending in the corpus callosum of living cortical brain slices observed with video microscopy. *J. Neurosci.*, 14:2161-2177.

Ham, R.G. (1965) Clonal growth of mammalian cells in a chemically defined, synthetic medium. *Microbiology*, 53:288-293.

Hamasaki, T., Komatsu, Y., Yamamoto, N., Nakajima, S., Hirakawa, K., and Toyama, K. (1987) Electrophysiological study of synaptic connections between a transplanted lateral geniculate nucleus and the visual cortex of the host rat. *Brain Res.*, 422:172-177.

Hamburger, V. (1988) *The Heritage of Experimental Embryology: Hans Spemann and the Organizer*, Oxford University Press, Oxford.

Hankin, M.H., and Lund, R.D. (1991) How do retinal axons find their targets in the developing brain? *Trends in Neurosci.*, 14:224-228.

Hankin, M.H., and Silver, J. (1986) Mechanisms of axonal guidance. The problem of intersecting fiber system. In *Developmental Biology*, ed. L.W. Browder, vol. 2, Plenum Publishing Corporation.

Harrison, R.G. (1907) Observations on the living developing nerve fiber. *Anat. Rec.*, 1:116-118.

Hata, Y., and Stryker, M.P. (1994) Control of thalamocortical afferent rearrangement by postsynaptic activity in developing visual cortex. *Science*, 265:1732-1735.

Heil, P., Bronchti, G., Wollberg, Z., and Scheich, H. (1991) Invasion of visual cortex by the auditory system in the naturally blind mole rat. *NeuroReport*, 2:735-738.

Heller, S.B., and Ulinski, P.S. (1987) Morphology of geniculocortical axons in turtles of the genera *Pseudemys* and *Chrysemys*. *Anat. Embryol.*, 175:505-515.

Henderson, Z., and Blakemore, C. (1986) Organization of the visual pathways in the newborn kitten. *Neurosci. Res.*, 3:628-659.

Hendrickson, A.E. (1983) Pyramidal and stellate neurons in monkey visual cortex label for different peptides. *Soc. Neurosci.Abstr.*, 9:821.

Hendrickson, A.E., Hunt, S.P., and Wu, J.Y. (1981) Immunocytochemical localization of glutamic acid decarboxylase in monkey striate cortex. *Nature*, 292:605-607.

Hendry, S.H.C., Jones, E.G., DeFilipe, J., Schmechel, D., Brandon, C. and Emson, P.C. (1984a) Neuropeptide containing neurons of the cerebral cortex are also GABAergic. *Proc. Natl. Acad. Sci. USA*, 81:6526-6530.

Hendry, S.H.C., Jones, E.G., Emson, P.C. (1984b) Morphology, distribution, and synaptic relations of somatostatin- and neuropeptide Y- immunoreactive neurons in rat and monkey neocortex. *J. Neurosci.*, 4:2497-2517.

Hendry, S.H.C., Jones, E.G., Emson, P.C., Lawson D.E.M., Heizmann, C.W., and Streit P. (1989) Two classes of cortical GABA neurons defined by differential calcium binding protein immunoreactivities. *Exp. Brain Res.*, 76:467-472.

Henke-Fahle, S., Mann, F., Götz, M., Wild, K., and Bolz, J. (1996) Dual action of a carbohydrate epitope on afferent and efferent axons in cortical development. *J. Neurosci.*, 16:4195-4206.

Herrick, C.J. (1917) The internal structure of the midbrain and thalamus of *Necturus. J. Comp. Neurol.*, 28:215-348.

Herrick, C.J. (1918) *An Introduction to Neurology*, second edition, Saunders, Phyladelphia.

Herrick, C.J. (1926) *Brains of Rats and Men*, University of Chicago Press, Chicago.

Herrick, C.J. (1933) The amphibian forebrain. VI. *Necturus. J. Comp. Neurol.*, 58:1-288.

Herrick, C.J. (1941) Development of the optic nerves of Amblystoma. *J. Comp. Neurol.*, 74:473-534.

Herrick, C.J. (1942) Optic and postoptic systems in the brain of Amblystoma trinum. *J. Comp. Neurol.*, 77:191-353.

Herrmann, K., and Shatz, C.J. (1992) Glutamate-induced calcium responses of developing subplate cells. *Soc. Neurosci. Abstr.*, 18:924.

Herrmann, K., and Shatz, C.J. (1995) Blockade of action potential activity alters initial arborization of thalamic axons within cortical layer 4. *Proc. Natl. Acad. Sci. USA*, 92:11244-11248.

Herrmann, K., Antonini, A., and Shatz, C.J. (1991) Thalamic axons make synaptic contacts with subplate neurons in cortical development. *Soc. Neurosci Abstr.*, 17:899.

Herrmann, K., Antonini, A., and Shatz, C.J. (1994) Ultrastructural evidence for synaptic interactions between thalamocortical axons and subplate neurones. *Eur. J. Neurosci.*, 6:1729-1742.

Hickey, T.L. (1980) Development of the dorsal lateral geniculate nucleus in normal and visually deprived cats. *J. Comp. Neurol.*, 189:467-481.

Hickey, T.L., and Cox, N.R. (1979) Cell birth in the lateral geniculate nucleus of the cat: A ^{3}H-thymidine study. *Soc. Neurosci. Abstr.*, 5:788.

Hickey, T.L., and Hitchcock, P.F. (1984) Genesis of neurons in the dorsal lateral nucleus of the cat. *J. Comp. Neurol.*, 228:186-199.

Hicks, S.P., and D'Amato, C.J. (1968) Cell migrations to the isocortex in the rat. *Anat. Rec.*, 160:619-634.

Higashi, S., Crair, M.C., Kurotani, T., Molnár, Z, and Toyama, K. (1993) Imaging neural excitation propagating from thalamus to somatosensory "barrel" cortex of the rat. *Soc. Neurosci. Abstr.*, 19:108.

Higashi, S., Molnár, Z., Kurotani, T., Inokawa, H., and Toyama, K. (1996) Functional thalamocortical connections develop during embryonic period in the rat: An optical recording study. *Soc. Neurosci. Abstr.*, 22:1976.

Hirotsune, S., Takahara, T., Sasaji, N., Hirose, K., Yoshiki, A., Ohashi, T., Kusakabe, M., Murakami, Y., Muramatsu, M., Watanabe, S., et al. (1995) The *reeler* gene encodes a protein with an EGF-like motif expressed by pioneer neurons. *Nat. Genet.*, 10:77-83.

His, W. (1893) Über das frontale Ende des Gehirnrohres. *Z. Anat. Entwicklungsgesch.*, 157:171-194.

His, W. (1904) *Die Entwicklung des Menschlichen Gehirns waehrend der ersten Monate.* Leipzig, Hirzel, 28.

Hochstetter, F. (1909) *Beiträge zur Entwicklungsgeschite des menschlichen Gehirns*, vol. 1, F. Deuticke, Wien-Leipzig, 77.

Hodos, W., and Campbell, C.B.G. (1969) *Scala Naturae.* Why there is no theory in comparative psychology. *Psychological Review*, 76:337-350.

Hoffarth, R., Johnston, J.G., Krushel, L.A., and van der Kooy, D. (1995) The mouse mutation *reeler* causes increased adhesion within a subpopulation of early postmitotic cortical neurons. *J. Neurosci.*, 15:4838-4850.

Hogan, D., and Berman, N.E.J. (1991) Calbindin-d is transiently expressed in pyramidal cells of neonatal kittens in an area dependent pattern. *Soc. Neurosci. Abstr.*, 17:367.

Hohn, A., Allendoerfer, K.L., Toroian-Raymond, A., and Shatz, C.J. (1993) Survival of subplate neurons in cultures of developing neocortex. *Soc. Neurosci. Abstr.*, 19:1504.

Honig, L.S., Herrmann, K., and Shatz, C.J. (1996) Developmental changes revealed by immunohistochemical markers in human cerebral cortex. *Cereb. Cortex*, 6:794-806.

Honig, M.G., and Hume, R.I. (1986) Fluorescent carbocyanine dyes allow living neurons of identified origin to be studied in long-term cultures. *J. Cell Biol.*, 103:171-187.

Honig, M.G., and Hume, R.I. (1989) DiI and DiO: Versatile fluorescent dyes for neuronal labelling and pathway tracing. *Trends in Neurosci.*, 12:333-335.

Honnegger, P., Lenoir, D., and Favrod, P. (1979) Growth and differentiation of aggregating fetal brain cells in a serum-free, defined medium. *Nature*, 282:305-308.

Hope, B.T., Michael, G.J., Knigge, K.M., and Vincent, S.R. (1991) Neuronal NADPH diaphorase is a nitric oxide synthase. *Proc. Natl. Acad. Sci. USA*, 88:2811-2814.

Horder, T.J., and Martin, K.A.C. (1978) Morphogenetics as an alternative to chemospecificity in the formation of nerve connections. In *Cell-Cell Recognition*, ed. A.S.G. Curtis, Cambridge University Press, Cambridge, 275-358.

Horsburgh, G.M., and Sefton, A.J. (1986) The early development of the optic nerve and chiasm in embryonic rat. *J. comp. Neurol.*, 243:547-560.

Horton, J.C., Greenwood, M.M., and Hubel, D.H. (1980) Non-retinotopic arrangement of fibers in the cat optic nerve. *Nature*, 282:720-722.

Huang, K., and Guillery, R. W. (1985) A demonstration of two distinct geniculocortical projection patterns in albino ferrets. *Dev. Brain Res.*, 20:213-220.

Hubel, D.H., and Wiesel, T.N. (1970) The period of susceptibility to the physiological effects of unilateral eye closure in kittens. *J. Physiol.*, 206:419-436.

Hubel, D.H., and Wiesel, T.N., (1971) Aberrant visual projections in the Siamese cat, *J. Physiol.*, 218:33-62.

Hübener, M., Götz, M., Klostermann, S., and Bolz, J. (1992) Growth behavior of thalamic axons on cortical membranes studied with time-lapse video microscopy. *Soc. Neurosci. Abstr.*, 18:924.

Hübener, M., Götz, M., Klostermann, S., and Bolz, J., (1995) Guidance of thalamocortical axons by growth-promoting molecules in developing rat cerebral cortex. *Eur. J. Neurosci.*, 7:1963-1972.

Humphrey, A.L., Sur, M., Uhlrich, D.J., and Sherman, S.M. (1984) Projection patterns of individual X- and Y-cell axons from the lateral geniculate nucleus to cortical area 17 in the cat. *J. Comp. Neurol.*, 233:159-189.

Humphreys, T. (1967) The cell surface and specific cell aggregation. In *The Specificity of Cell Surfaces*, eds. B.D. Davis and L. Warren, Prentice-Hall, Englewood Cliffs, 195-210.

Hunter, K.E., and Hatten, M.E., (1995) Radial glia cell transformation to astrocytes is bidirectional: Regulation by a diffusible factor in embryonic forebrain. *Proc. Natl. Acad. Sci. USA*, 92:2061-2065.

Iadecola, C. (1993) Regulation of the cerebral microcirculation during neural activity: Is nitric oxide the missing link? *Trends in Neurosci.*, 16:206-214.

Ignarro, L.J. (1990) Haem-dependent activation of guanylate cyclase and cyclic GMP formation by endogenous nitric oxide: a unique transduction mechanism for transcellular signaling. *Pharmacol. Toxicol.*, 67:1-7.

Ignarro, L.J., Byrns, R.E., and Wood, K.S. (1988) Biochemical and pharmacological properties of endothelium-derived relaxing factor and its similarity to nitric oxide radical. In *Vasodilatation: Vascular Smooth Muscle, Peptides, Autonomic Nerves and Endothelium*, ed. P.M. Vanhoutte, Raven Press, New York, 427-436.

Innocenti, G.M., and Caminiti, R. (1980) Postnatal shaping of callosal connections from sensory areas. *Exp. Brain Res.*, 38:381-394.

Innocenti, G.M., and Clarke, S. (1984) The organization of immature callosal connections. *J. Comp. Neurol.*, 230:287-309.

Insausti, R., Blakemore, C., and Cowan, W.N. (1984) Ganglion cell death during development of ipsilateral retino-collicular projections in golden hamster. *Nature*, 308:362-365.

Isaac, J.T., Crair, M.C., Nicoll, R.A., and Malenka, R.C. (1997) Silent synapses during development of thalamocortical inputs. *Neuron*, 18:269-280.

Ito, M. (1995) Barrelfield of the prenatally X-irradiated rat somatosensory cortex: A histochemical and electrophysiological study. *J. Comp. Neurol.*, 352:248-262.

Ito, M., and Karachot, L. (1990) Receptor subtypes involved in, and time course of, the long-term desensitization of glutamate receptors in cerebellar purkinje cells. *Neurosci. Res.*, 8:303-307.

Izumi, Y., and Zorumski, C.F. (1993) Nitric oxide and long-term synaptic depression in the rat hippocampus. *NeuroReport*, 4:1131-1134.

Jablonska, B., Gierdalski, M., Siucinska, E., Skangiel-Kramska, J., and Kossut, M. (1995) Partial blocking of NMDA receptors restricts plastic changes in adult mouse barrel cortex. *Behav. Brain Res.*, 66:207-216.

Jacobs, W., Farivar, N., and Butcher, L.L. (1985) Alzheimer dementia and reduced nicotinamide adenine (NADH)-diaphorase activity in senile plaques and the basal forebrain. *Neurosci. Lett.*, 53:39-44.

Jacobson, M. (1991) *Developmental Neurobiology*, third edition, Plenum Press, New York-London.

Jain, N., Florence, S.L., and Kaas, J.H. (1995) Limits on platicity in somatosensory cortex of adult rats: Hindlimb cortex is not reactivated after dorsal column section. *J. Neurophisiol.*, 73:1537-1546.

Jeanmonod, D., Rice, F.L., and van der Loos, H. (1977) Mouse somatosensory cortex: Development of the alterations in the barrel field which are caused by injury to the vibrissal follicles. *Neurosci. Lett.*, 6:151-156.

Jeanmonod, D., Rice, F.L., and van der Loos, H. (1981) Mouse somatosensory cortex: Alterations in the barrelfield following receptor injury at different early postnatal ages. *Neuroscience*, 6:1503-1535.

Jeffery, G. (1997) The albino retina: An abnormality that provides insight into normal retinal development. *Trends in Neurosci.*, 20:165-169.

Jensen, K.F., and Killackey, H.P. (1987a) Terminal arbors of axons projecting to the somatosensory cortex of the adult rat. I. The normal morphology of specific thalamocortical afferents. *J. Neurosci.*, 7:3529-3543.

Jensen, K.F., and Killackey, H.P. (1987b) Terminal arbors of axons projecting to the somatosensory cortex of the adult rat. II. The altered morphology of thalamocortical afferents following neocortical infraorbital nerve cut. *J. Neurosci.*, 7:3544-3553.

Jhaveri, S., Erzurumlu R.S., and Crossin, K. (1991) Barrel construction in rodent neocortex: Role of thalamic afferents versus extracellular matrix molecules. *Proc. Natl. Acad. Sci. USA*, 88:4489-4493.

Johnson, G.D., Davidson, R.S., McNamee, K.C., Russell, G., Goodwin, D., and Holbarow, E.J. (1982) Fading of immunofluorescence during microscopy: A study of the phenomenon and its remedy. *J. Immunol. Methods*, 55:231-242.

Jones, E.G. (1981) The development of connectivity in the cerebral cortex. In *Studies in Developmental Neurobiology: Essays in Honor of Viktor Hamburger*, ed. W.M. Cowan, Oxford University Press, New Yord-Oxford, 354-394.

Jones, E.G. (1983) Summing up. In *Somatosensory Integration in the Thalamus*, eds. G. Macchi, A. Rustioni and R. Spreafico, Elsevier Scientific, Amsterdam, 385-392.

Jones, E.G. (1985) *The Thalamus*, Plenum Press, New York.

Jones, E.G., and Hendry, S.H.C. (1986) Co-localisation of GABA and neuropeptides in neocortical neurons. *Trends in Neurosci.*, 9:71-76.

Jones, E.G., and Hendry, S.H.C. (1989) Differential Calcium binding protein immunoreactivity distinguishes classes of relay neurons in monkey thalamic nuclei. *Eur. J. Neurosci.*, 1:222-246.

Jones, E.G., and Peters, A. (1990) *Cerebral Cortex*, vols. 8A-8B: *Comparative Structure and Evolution of Cerebral Cortex, Part I/Part II*, Plenum Press, New York and London.

Jones, E.G., Schreyer, D.J., and Wise, S.P. (1982) Growth and maturation of the rat corticospinal tract. *Prog. Brain Res.*, 57:361-379.

Juliano, S.L., Palmer, S.L., Sonty, R.V., Noctor, S., and Hill, G.F. (1996) Development of local connections in ferret somatosensory cortex. *J. Comp. Neurol.*, 374:259-277.

Kaas, J.M., and Krubitzer, L.A. (1992) Area 17 lesions deactivate area MT in owl monkeys. *Vis. Neurosci.*, 9:399-407.

Kageyama, G.H., and Robertson, R.T. (1993) Development of geniculocortical projections to visual cortex in rat: Evidence for early ingrowth and synaptogenesis. *J. Comp. Neurol.*, 335:123-148.

Kaiserman-Abramoff, I.R. (1983) Intrauterine enucleation of normal mice mimics a structural compensatory response in the geniculate of eyeless mutant mice. *Brain Res.*, 270:149-153.

Kaiserman-Abramoff, I.R., Graybiel, A.M., and Nauta, W.J.H. (1980) The thalamic projection of cortical area 17 in a congenitally anophtalamic mouse strain. *Neuroscience*, 5:41-52.

Kalil, R. (1978) Development of the dorsal lateral geniculate nucleus in the cat. *J. Comp. Neurol.*, 182:265-292.

Kandel, E.R. (1981) Calcium and the control of synaptic strength by learning. *Nature*, 293:697-700.

Karlsson, U. (1966) Observations on the postnatal development of neuronal structures in the lateral geniculate nucleus of the rat by electron microscopy. *J. Ultrastruct. Res.*, 17:158-175.

Karlstrom, R.O., Trowe, T., and Bonhoeffer, F. (1997) Genetic analysis of axon guidance and mapping in the zebrafish. *Trends in Neurosci.*, 20:3-8.

Karten, H. (1997) Evolutionary developmental biology meets the brain: The origins of mammalian cortex. *Proc. Natl. Acad. Sci. USA*, 94:2800-2804.

Kasper, E.M, Larkman, A.U., Lubke, J., and Blakemore, C. (1994) Pyramidal neurons in layer 5 of the rat visual cortex. III. Differential maturation of axon targeting, dendritic morphology and electrophysiological properties. *J. Comp. Neurol.*, 339:495-518.

Kato, N., Kawaguchi, S., and Hirofumi, M. (1984) Geniculo-cortical projection of layer I of area 17 in kittens: Orthograde and retrograde HRP studies. *J. Comp. Neurol.*, 225:441-447.

Katz, L.C. (1993) Coordinate activity in retinal and cortical development. *Current Opinion in Neurobiology*, 3:93-99.

Katz, L.C., and Callaway, E.M. (1992) Development of local circuits in mammalian visual cortex. *Ann. Rev. Neurosci.*, 15:31-56.

Katz, L.C., and Shatz, C.J. (1996) Synaptic activity and the construction of cortical circuits. *Science*, 274:1133-1138.

Killackey, H., and Shinder, A. (1981) Central correlates of peripheral pattern alterations in the trigeminal system of the rat. II. The effect of nerve section. *Dev. Brain Res.*, 1:121-126.

Killackey, H.P. (1990) Neocortical expansion: An attempt toward relating phylogeny and ontogeny. *J. Cognitive Neurosci.*, 2:1-17.

Killackey, H.P., Rhoads, R.W., and Bennett-Clarke, C.A. (1995) The formation of a cortical somatotopic map. *Trends in Neurosci.*, 18:402-407.

Kim, G.J., Shatz, C.J., and McConnell, S.K. (1991) Morphology of pioneer and follower growth cones in the developing cerebral cortex. *J. Neurobiol.*, 22:629-642.

Kind, P., and Innocenti, G.M. (1990) The development of cortical projections. In *Systems Approaches to Developmental Neurobiology*, eds. P.A. Raymond, S.S. Easter Jr. and G.M. Innocenti, Plenum , New York-London, 113-125.

Klauer, S. (1991) The corticotectal projection of the rat established in organotypic culture. *NeuroReport*, 2:569-572.

Knott, G.W., Molnár, Z., Saunders, N.R., and Blakemore, C. (1997) Development of thalamocortical projections in the South American grey short-tailed opossum *(Monodelphis Domestica). Proc. Australian Soc. Neurosci.*, 8:105.

Knyihár, E., Csillik, B., and Rakic, P. (1978) Transient synapses in the embryonic primate spinalcord. *Science*, 202:1206-1209.

Koester, S.F., and O'Leary, D.D.M. (1991) Subplate cells in medial cortex send the first axons across the corpus callosum. *Soc. Neurosci. Abstr.*, 17:41.

Koester, S.F., and O'Leary, D.D.M. (1992) Functional classes of cortical projection neurons develop dendritic distinctions by class-specific sculpting of an early common pattern. *J. Neurosci.*, 12:1382-1393.

Komatsu, Y., and Iwakiri, M. (1993) Long-term modification of inhibitory synaptic transmission in developing visual cortex. *Neuroreport*, 4:907-910.

Komuro, H., and Rakic, P. (1992) Selective role of N-type calcium channels in neuronal migration. *Science*, 257:806-809.

Komuro, H., and Rakic, P. (1993) Modulation of neuronal migration by NMDA receptors. *Science*, 260:95-97.

König, N., and Marty, R. (1981) Early neurogenesis and synaptogenesis in cerebral cortex. *Bibl. Anat.*, 19:152-160.

König, N., Valat, J., Fulcrand, J., and Marty, R. (1977) The time of origin of Cajal-Retzius cells in the rat temporal cortex: An autoradiographic study. *Neurosci. Lett.*, 4:21-26.

Kornack, D.R., and Rakic, P. (1995) Radial and horizontal deployment of clonally related cells in the primate neocortex: Relationship to distinct mitotic lineages. *Neuron*, 15:311-321.

Kossel, A., Lowel, S., and Bolz, J. (1995) Relationship between dendritic fields and functional architecture in striate cortex of normal and visually deprived cats. *J. Neurosci.*, 15:3913-3926.

Kostovic, I. (1986) Prenatal development of nucleus basalis complex and related fiber systems in man: A histochemical study. *Neuroscience*, 17:1047-1077.

Kostovic, I., and Molliver, M.E. (1974) A new interpretation of the laminar development of cerebral cortex: Synaptogenesis in different layers of neopallium in the human fetus. *Anat. Rec.*, 178:395.

Kostovic, I., and Rakic, P. (1980) Cytology and time of origin of interstitial neurons in the white matter in infant and adult human and monkey telencephalon. *J. Neurocytol.*, 9:219-242.

Kostovic, I., and Rakic, P. (1990) Developmental history of the transient subplate zone in the visual and somatosensory cortex of the macaque monkey and human brain. *J. Comp. Neurol.*, 297:441-470.

Kostovic, I., and Rakic, P. (1990) Developmental history of the transient subplate zone in the visual and somatosensory cortex of the macaque monkey and human brain. *J. Comp. Neurol.*, 297:441-470.

Kowall, N.W., Ferante, R.J., Bear, M.F., Richardson, E.P., Sofroniew, M.V., Cuello, A.C., and Martin, J.B. (1987) Neuropeptide Y, somatostatin and reduced nicotinamide adenine dinucleotide phosphate diaphorase in the human striatum: A combined immunocytochemical and enzyme histochemical study. *Neuroscience*, 20:817-828.

Kristt, D.A. (1978) Neuronal differentiation in somatosensory cortex of the rat. I. Relationship to synaptogenesis in the first postnatal week. *Brain Res.*, 150:467-486.

Krubitzer, L. (1995) General discussion II: Teleology for tangential migration. In *Development of the Cerebral Cortex*, eds. G. Bock and G. Cardew, Wiley, Chichester (Ciba Foundation Symposium, 193), 119.

Krubitzer, L. (1995) The organization of neocortex in mammals: Are species differences really so different? *Trends in Neurosci.*, 18:408-4017.

Krubitzer, L., Manger, P., Pettigrew, J., and Calford, M. (1995) Organization of somatosensory cortex in monotremes: In search of the prototypical plan. *J. comp Neurol.*, 351:261-306.

Kuan, C.Y., Elliott, E.A., Flavell, R.A., and Rakic, P. (1997) Restrictive clonal allocation in the chimeric mouse brain. *Proc. Natl. Acad. Sci. USA*, 94:3374-3379.

Kuroda, M., and Price, J.L. (1991) Synaptic organization of projections from basal forebrain structures to the mediodorsal thalamic nucleus of the rat. *J. Comp. Neurol.*, 303:513-533.

Kurotani, T., Crair, M.C., Higashi, S., Molnár, Z., and Toyama, K. (1996) The development of rat somatosensory (barrel) cortex visualized by optical recording. *Protein Nucleic Acid Enzyme*, 41: 758-765.

Kurotani, T., Yamamoto, N., and Toyama, K. (1993) Development of neural connections between visual cortex and transplanted lateral geniculate nucleus in rats. *Dev. Brain Res.*, 71:151-168.

Kurukulasuriya, N.C., and Salinger, W.L. (1996) Error correction in *reeler* brain development. *Soc. Neurosci. Abstr.*, 22:972.

Layer, P.G., Rommel, S., Bultfoff, H., and Hengstenberg, R. (1988) Independent spatial waves of biochemical differentiation along the surface of chicken brain as revealed by the sequential expression of acetylcholinesterase. *Cell Tissue Res.*, 251:587-595.

Le Gros Clark, W.E. (1932a) A morphological study of the lateral geniculate body. *Br. J. Ophtal.*, 16:264-284.

Le Gros Clark, W.E. (1932b) The structure and connections of the thalamus. *Brain* 55:406-470.

Lebrand, C., Cases, O., Adelbrecht, C., Doye, A., Alvarez, C., El-Mestikawy, S., Seif, I., and Gaspar, P. (1996) Transient uptake and storage of serotonin in developing thalamic neurons. *Neuron*, 17:823-835.

Lende, R.A. (1963) Cerebral cortex: A sensorimotor amalgam in the marsupialia. *Science*, 141:730-732.

Lerner-Natoli, M., Rondouin, G., De Bock, F., and Bockaert, J. (1992) Chronic NO synthase inhibition fails to protect hippocampal neurons against NMDA toxicity. *Neuroreport*, 3:1109-1112.

Letinic, K., and Kostovic, I., (1996) Transient neuronal population of the internal capsule in the developing human cerebrum. *Neuroreport*, 7:2159-2162.

LeVay, S., and Stryker, M.P. (1979) The development of ocular dominance columns in the cat. *Soc. Neurosci. Abstr.*, 4:83-98.

LeVay, S., Stryker, M.P., and Shatz, C.J. (1978) Ocular dominance columns and their development in layer IV of the cat's visual cortex: A quantitative study. *J. Comp. Neurol.*, 179:223-244.

LeVay, S., Wiesel, T.N., and Hubel, D.H. (1980) The development of ocular dominance columns in normal and visually deprived monkeys. *J. Comp. Neurol.*, 191:1-51.

Levi-Montalcini, R. (1952) Effects of mouse tumor transplantation on the nervous system. *Ann. N.Y. Acad. Sci.*, 55:330-343.

Levi-Montalcini, R. (1982) Developmental neurobiology and natural history of nerve growth factor. *Ann. Rev. Neurosci.*, 5:341-362.

Levitt, P., Barbe, M.F., and Eagleson, K.L. (1997) Patterning and specification of the cerebral cortex. *Ann. Rev. Neurosci.*, 18:419-439.

Li, Y., Erzurumlu, R.S., Chen, C., Jhaveri, S., and Tonegawa, S. (1994) Whisker-related neuronal patterns fail to develop in the trigeminal brainstem nuclei of NMDAR1 knockout mice. *Cell*, 76:427-437.

Lidov, H.G.W., M.E. Molliver, and Zecevic, N.R. (1978) Characterization of the monoaminergic innervation of immature neocortex: A histofluorescence analysis. *J. Comp. Neurol.*, 181:663-679.

Lindsay, K.W., Bone, I., and Callander, R. (1986) *Neurology and Neurosurgery Illustrated*, Churchill Livingstone, Edinburgh-London-Melbourne-New York.

Ling, J.J., Yu, J., and Robertson, R.T. (1995) Sustained inhibition of acetylcholinesterase activity does not disrupt early geniculocortical ingrowth to developing rat visual cortex. *Dev. Brain Res.*, 86:354-358.

Liu, L., and Layer, P.G. (1984) Spatio-temporal patterns of differentiation of whole heads of the embryonic chick as revealed by binding of a FITC-coupled peanut-agglutinin (FITC-PNA). *Dev. Brain Res.*, 12:173-182.

Lotto, R.B., and Price, D.J. (1995) The stimulation of thalamic neurite outgrowth by cortex-derived growth factors in vitro: The influence of cortical age and activity. *Eur. J. Neurosci.*, 7:318-328.

Lotto, R.B., and Price, D.J. (1996) Effects of subcortical structures on the growth of cortical neurites in vitro. *Neuroreport*, 7:1185-1188.

LoTurco, J.J., and Krigstein, A.R. (1991) Clusters of coupled neuroblasts in embryonic neocortex. *Science*, 252:563-566.

LoTurco, J.J., Owens, D.F., Heath, M.J.S., Davis, M.B.E., and Kriegstein, A.R. (1995) GABA and glutamate depolarize cortical progenitor cells and inhibit DNA synthesis. *Neuron*, 15:1287-1298.

Lowenstein, D.H., Miles, M.F., Hatam, F., and McCabe, T. (1991) Upregulation of Calbindin-D28K mRNA in the rat hippocampus following focal stimulation of the perforant path. *Neuron*, 6:627-633.

Lowenstein, P.R., and Shering, A.F. (1991) Synaptic input to subplate neurons in the cat: Infragranular neurons provide synaptic input to underlying subplate cells during early post-natal neocortical development: A light and electron microscopical study. *J. Anat.*, 180:383.

Lózsádi, D.A., Gonzalez-Soriano, J., and Guillery, R.W. (1996) The course and termination of corticothalamic fibers arising in the visual cortex of the rat. *Eur. J. Neurosci.*, 8:2416-2427.

Ludena, M.A. (1973) The growth of spinal ganglion neurons in a serum-free medium. *Dev. Biol.*, 33:470-476.

Luhmann, H.J., Martinez Millan, L., and Singer, W. (1986) Development of horizontal intrinsic connections in cat striate cortex. *Exp. Brain Res.*, 63:443-448

Lumsden, A.G.S. (1991) Chemotaxis in the developing nervous system. In *The Nerve Growth Cone*, eds. P. Letourneau, S.B. Kater and E.R. Macagno, Raven, New York, 167-180.

Lumsden, A.G.S., and Cohen, J. (1991) Axon guidance in vertebrate central nervous system. *Current Opinion in Genetics and Development*, 1:230-235.

Lund, R.D. (1965) Uncrossed visual pathways in the hooded and albino rats. *Science*, 149:1506-1509.

Lund, R.D., and Bunt, A.H. (1976) Prenatal development of central optic pathways in albino rats. *J. Comp. Neurol.*, 165:247-264.

Lund, R.D., and Mustari, M.J. (1977) Development of the geniculocortical pathway in rats. *J. Comp. Neurol.*, 173:289-305.

Luskin, M.B., and Shatz, C.J. (1985a) Neurogenesis of the cat's primary visual cortex. *J. Comp. Neurol.*, 242:611-631.

Luskin, M.B., and Shatz, C.J. (1985b) Studies of the earliest-generated cells of the cat's visual cortex: Cogeneration of subplate and marginal zones. *J. Neurosci.*, 5:1062-1075.

Lynch, G., and Baudry, M. (1984) The biochemistry of memory: A new and specific hypothesis. *Science*, 224:1057-1063.

Macchi, G. (1983) Old and new anatomo-functional criteria in the subdivision of the thalamic nuclei. In *Somatosensory Integration in the Thalamus*, eds. G. Macchi, A. Rustioni and R. Spreafico, Elsevier Scientific, Amsterdam, 3-16.

Mackintosh, N. (1992) Who is the brightest of them all? *The Higher*, March 6, 15 and 18.

Manger, P., Molnár, Z., Slutsky, D., and Krubitzer, L. (1997) Subdivisions of visually responsive regions of the dorsal ventricular ridge of the iguana *(Iguana iguana)*. *Soc. Neurosci. Abstr.*, 23: (in press).

Manger, P.R., Woods, T.M., and Jones, E.G. (1996a) Plasticity of the somatosensory cortical map in macaque monkeys after chronic partial amputation of a digit. *Proc. R. Soc. Lond. B. Biol. Sci.*, 263:933-939.

Manger, P.R., Woods, T.M., Jones, E.G. (1996b) Representation of face and intraoral structures in area 3b of macaque monkey somatosensory cortex. *J. comp. Neurol.*, 371:513-521.

Maranto, A.R., (1982) Neuronal mapping: A photooxidation reaction makes Lucifer yellow useful for electron microscopy. *Science*, 217:953-955.

Marin-Padilla, M. (1971) Early prenatal ontogenesis of the cerebral cortex (neocortex) of the cat *(Felis domestica)*: a Golgi study. I. The primordial neocortical organization. *Z. Anat. Entwicklungsgesch.*, 134:117-145.

Marin-Padilla, M. (1972) Prenatal ontogenetic history of the principal neurons of the neocortex of the cat *(Felis domestica)*: a Golgi study. II. Developmental differences and their significances. *Z. Anat. Entwicklungsgesch.*, 136:125-142.

Marin-Padilla, M. (1978) Dual origin of the mammalian neocortex and evolution of the cortical plate. *Anat. Embryol.*, 152:109-126.

Marin-Padilla, M. (1988) Early ontogenesis of the human cerebral cortex. In *Cerebral Cortex*, vol. 7: *Development and Maturation of Cerebral Cortex*, eds. A. Peters and E.G. Jones, Plenum Press, New York-London, 1-34.

Marin-Padilla, M. (1996) Neuronal changes in gray matter overlying white matter lesions in developing human neocortex. *Soc. Neurosci. Abstr.*, 22:1207.

Mark, R.F., and Marotte, L.R. (1992) Australian marsupials as models for the developing mammalian visual system. *Trends in Neurosci.*, 15:51-57.

Martin, K.A.C., and Perry, V.H. (1983) The role of fiber ordering and axon collateralization in the formation of topographic projections. In *Molecular and Cellular Interactions Underlying Higher Brain Functions*, eds. J.-P. Changeux, J. Glowinski, M. Imbert and F.E. Bloom, Elsevier, Amsterdam-Oxford (Progress in Brain Research, 58), 321-337.

Masiakowski, P., and Shooter, E.M. (1988) Nerve growth factor induces the genes for two proteins related to a family of calcium-binding proteins in PC12 cells. *Proc. Natl. Acad. Sci. (USA)*, 85:1277-1281.

Mason, C.A., and Godement, P. (1992) Growth cone form reflects interactions in visual pathways and cerebral targets. In *The Nerve Growth Cone*, eds. P.C. Letourneau, S.B. Kater and E.R. Macagno, Raven, New York, 405-423.

Mathers, L.H. (1972) The synaptic organization of the cortical projection to the pulvinar of the squirrel monkey. *J. Comp. Neurol.*, 146:43-60.

Maximov, A. (1925) Tissue cultures of young mammalian embryos. *Contr. Embryol. Carnegie Inst.*, 16:47-114.

McAllister, J.P., II, and Das, G.D. (1977) Neurogenesis in the epithalamus, dorsal thalamus and ventral thalamus of the rat: An autoradiographic and cytological study. *J. Comp. Neurol.*, 172:647-686.

McCandlish, C., Waters, R.S., and Cooper, N.G.F. (1989) Early development of the representation of the body surface in SI cortex barrel subfield in neonatal rats as demonstrated with peanut agglutinin binding: Evidence for differential development within the rattunculus. *Exp. Brain Res.*, 77:425-431.

McCasland, J.S., and Woolsey, T.A. (1988) High-resolution 2-deoxyglucose mapping of functional cortical columns in mouse barrel cortex. *J. Comp. Neurol.*, 278:555-569.

McConnell, S.K. (1988) Development and decision-making in the mammalian cerebral cortex. *Brain Res. Rev.*, 13:1-23.

McConnell, S.K. (1989) The determination of neuronal fate in the cerebral cortex. *Trends in Neurosci.*, 12:342-349.

McConnell, S.K. (1991) Specification of cerebral cortex during development. *Soc. Neurosci. Abstr.*, 17:1279.

McConnell, S.K. (1995) Strategies for the generation of neuronal diversity in the developing nervous system. *J. Neurosci.*, 15:6987-6998.

McConnell, S.K., Ghosh, A., and Shatz, C.J. (1989) Subplate neurons pioneer the first axon pathway from the cerebral cortex. *Science*, 245:978-982.

McConnell, S.K., Ghosh, A., and Shatz, C.J. (1994) Subplate pioneers and the formation of descending connections from cerebral cortex. *J. Neurosci.*, 14:1892-1907.

McCormick, D.A., Trent, F., and Ramoa, A.S. (1995) Postnatal development of synchronized network oscillations in the ferret dorsal lateral geniculate and perigeniculate nuclei. *J. Neurosci.*, 15:5739-5752.

McDonald, J.K., Parnavelas, J.G., Karamanlidis, A.N., and Brescha, N. (1982a) The morphology and distribution of peptide-containing neurons in the adult and developing visual cortex. IV. Avian pancreatic polypeptide. *J. Neurocytol.*, 11:985-995.

McDonald, J.K., Parnavelas, J.G., Karamanlidis, A.N., Brecha, N., Koening, J.I. (1982b) The morphology and distribution of peptide containing neurons in the adult developing visual cortex. I. Somatostatin. *J. Neurocytol.*, 11:829-824.

Meakin, S.O., and Shooter, E.M. (1992) The nerve growth factor family of receptors. *Trends in Neurosci.*, 15:323-331.

Meinecke, D.L., and Peters, A. (1987) GABA immunoreactive neurons in rat visual cortex. *J. Comp. Neurol.*, 261:388-404.

Meister, M., Wong, R.O.L., Baylor, D.A., and Shatz, C.J. (1991) Synchronous bursts or action potentials in ganglion cells of the developing mammalian retina. *Science*, 252:939-943.

Merrill, E.G., and Wall, P.D. (1978) Plasticity of connections in the adult neuron system. In *Neuronal Plasticity*, ed. C. Cotman, Raven Press, New York, 97-111.

Merzenich, M.M., Nelson, R.J., Sur, M., Wall, J.T., Felleman, D.H., and Kaas, J.H. (1983) Progression of change following median nerve section in the cortical representation of the hand in areas 3b and in 1 in adult owl and squirrel monkeys. *Neuroscience*, 10:639-666.

Merzenich, M.M., Recanzone, G.H., Jenkins, W.M., and Grajski, K.A. (1990) Adaptive mechanisms in cortical networks underlying cortical contributions to learning and nondeclarative memory. *Cold Spring Harbor Symposia on Quantitative Biology*, 55:873-887.

Mesulam, M.M. (1976) The blue reaction product in horseradish peroxidase neurohistochemistry: Incubation parameters and visibility. *J. Histochem. Cytochem.*, 74:1273-1280.

Mesulam, M.M. (1978) Tetramethylbenzidine for horseradish peroxidase neurochemistry: A non-carcinogenic blue reaction product with superior sensitivity for visualizing neuronal afferents and efferents. *J. Histochem. Cytochem.*, 26:106-117.

Métin, C., and Godement, P. (1994) Early development of thalamo-cortical and ganglionic eminence projections in the hamster. *Soc. Neurosci. Abstr.*, 20:1683.

Métin, C., and Godement, P. (1996) The ganglionic eminence may be an intermediate target for corticofugal and thalamocortical axons. *J. Neurosci.*, 16:3219-3235.

Miki, N., Kawabe, Y., and Kuriyama, K. (1977) Activation of cerebral guanylate cyclase by nitric oxide. *Biochem. Biophys. Res. Commun.*, 75:851-856.

Miller, B., Chou, L., and Finlay, B.L. (1991) The early development of visual corticothalamic and thalamocortical projections in the golden hamster. *Soc. Neurosci. Abstr.*, 17:764.

Miller, B., Chou, L., and Finlay, B.L. (1993) The early development of thalamocortical and corticothalamic projections. *J. Comp. Neurol.*, 335:16-41.

Miller, B., Sheppard, A.M., and Pearlman, A.M. (1992) Expression of two chondroitin sulfate proteoglycan core proteins in the subplate pathway of early cortical afferents. *Soc. Neurosci. Abstr.*, 18:778.

Miller, B., Sheppard, A.M., Bicknese, A.R., and Pearlman, A.M. (1995) Chondroitin sulfate proteoglycans in the developing cerebral cortex: The distribution of neurocan distinguishes forming afferent and efferent axonal pathways. *J. comp Neurol.*, 355:615-628.

Miller, B., Windrem, M.S., and Finlay, B.L. (1991) Thalamic ablations and neocortical development: Alterations in thalamic and callosal connectivity. *Cerebral cortex*, 1:230-240.

Miller, M.W. (1988) Development of projection and local circuit neurons in neocortex. In *Cerebral Cortex*, vol. 7: *Development and Maturation of Cerebral Cortex*, eds. A. Peters and E.G. Jones, Plenum Press, New York-London, 133-175.

Mione, M.C., Danevic, C., Boardman, P., Harris, B., and Parnavelas, J.G. (1994) Lineage analysis reveals neurotransmitter (GABA or glutamate) but not calcium-binding protein homogeneity in clonally related cortical neurons. *J. Neurosci.*, 14:107-123.

Mione, M.C., Danevic, C., Cavanagh, M.E., Boardman, P., and Parnavelas, J.G. (1992) Neurotransmitter expression in clonally related neurons in the rat cerebral cortex. *Soc. Neurosci. Abstr.*, 18:1111.

Mitchison, G. (1992) Axonal trees and cortical architecture. *Trends in Neurosci.*, 15:122-126.

Mitrofanis, J. (1989) Development of NADPH-diaphorase cells in the rat's retina. *Neurosci. Lett.*, 102:165-172.
Mitrofanis, J. (1992a) Patterns of antigenic expression in the thalamic reticular nucleus of developing rats. *J. Comp. Neurol.*, 320:161-181.
Mitrofanis, J. (1992b) NADPH-diaphorase reactivity in the ventral and dorsal lateral geniculate nuclei of rats. *Visual Neurosci.*, 9:211-216.
Mitrofanis, J. (1994) Development of the pathway from the reticular and perireticular nuclei to the thalamus in ferrets: A DiI study. *Eur. J. Neurosci.*, 6:1865-1882.
Mitrofanis, J., and Baker, G.E. (1993) Development of the thalamic reticular and perireticular nuclei in rats and their relationship to the course of growing corticofugal and corticopetal axons. *J. Comp. Neurol.*, 338:575-587.
Mitrofanis, J., and Guillery, R.W. (1991) Patterns of antigen expression in the developing rat thalamic reticular nucleus. *Soc. Neurosci. Abstr.*, 17:38.
Mitrofanis, J., and Guillery, R.W. (1993) New views of the thalamic reticular nucleus in the adult and developing brain. *Trends in Neurosci.*, 16:240-245.
Mitrofanis, J., Earle, K.L., and Reese, B.E. (1997) Glial organization and chondroitin sulfate proteoglycan expression in the developing thalamus. *J. Neurocytology*, 26:83-100.
Mitzdorf, U., and Singer, W. (1978) Prominent excitatory pathways in the cat visual cortex (A17 and 18): A current source density analysis of electrically evoked potentials. *Exp. Brain Res.*, 33:371-394.
Miyachi, E., Murakami, M., and Nakaki, T. (1990) Arginine blocks gap junctions between retinal horizontal cells. *Neuroreport*, 1:107-110.
Mizukawa, K., Vincent, S.R., McGeer, P.L., and Mc Geer, E.G. (1988) Reduced nicotinamide adenine dinucleotide phosphate (NADPH)-diaphorase-positive neurons in cat cerebral white matter. *Brain Res.*, 461:274-281.
Molliver, M.E., and Kristt, D.A. (1975) The fine structural demonstration of monoaminergic synapses in immature rat neocortex. *Neurosci. Lett.*, 1:305-310.
Molliver, M.E., and van der Loos, H. (1970) The ontogenesis of cortical circuitry: The spatial distribution of synapses in somesthetic cortex of newborn dog. *Ergeb. Anat. Entwickl. Gesch.*, 42:1-43.
Molliver, M.E., Kostovic, I., and van der Loos, H. (1973) The development of synapses in cerebral cortex of the human fetus. *Brain Res.*, 50:403-407.
Molnár, Z. (1994) Multiple mechanisms in the establishment of thalamocortical innervation. D.Phil. thesis, University of Oxford, Oxford, UK.
Molnár, Z. (1995a) Development of thalamocortical projections. *Eur. J. Neurosci., Suppl.*, 8:105.
Molnár, Z. (1995b) General discussion III: The handshake hypothesis. In *Development of the Cerebral Cortex*, eds. G. Bock and G. Cardew, Wiley, Chichester (Ciba Foundation Symposium, 193), 197-198.
Molnár, Z. (1997) Development of the cerebral cortex (Bursary Report). *Ciba Foundation Bulletin*, 47: (in press).
Molnár, Z., Adams, R., and Blakemore, C. (1993a) 3-D confocal microscopic study on the early development of thalamocortical projections in the rat. *Soc. Neurosci. Abstr.*, 19:1088.
Molnár, Z., Adams, R., and Blakemore, C. (1994) Multiple mechanisms in the establishment of thalamocortical innervation. *Brain Pathology*, 4:296.
Molnár, Z., Adams, R., and Blakemore, C. (1997a) Mechanisms underlying the establishment of topographically ordered early thalamocortical connections in the rat. Manuscript submitted and revised to *J. Neurosci.*
Molnár, Z., Adams, R., Goffinet, A. M., and Blakemore, C. (1997b) The role of the first postmitotic cells in the development of thalamocortical fiber ordering in the *reeler* mouse. Manuscript submitted and revised to *J. Neurosci.*

Molnár, Z., and Blakemore, C. (1990a) Relationship of corticofugal and corticopetal projections in the prenatal establishment of projections from thalamic nuclei to the specific cortical areas of the rat. *J. Physiol.*, 430:104.

Molnár, Z., and Blakemore, C. (1990b) Development of thalamocortical connectivity *in vivo* and *in vitro*. *Soc. Neurosci. Abstr.*, 16:1007.

Molnár, Z., and Blakemore, C. (1990c) Specificity of thalamo-cortical connections studied with co-culture techniques. *Neurosci. Lett., Suppl.*, 38:92.

Molnár, Z., and Blakemore, C. (1991) Lack of regional specificity for connections formed between thalamus and cortex in co-culture. *Nature*, 351:475-477.

Molnár, Z., and Blakemore, C. (1992) How are thalamocortical axons guided in the *reeler* mouse? *Soc. Neurosci. Abstr.*, 18:778.

Molnár, Z., and Blakemore, C. (1995a) How do thalamic axons find their way to the cortex? *Trends in Neurosci.*, 18:389-397.

Molnár, Z., and Blakemore, C. (1995b) Guidance of thalamocortical innervation. In *Development of the Cerebral Cortex*, eds. G. Bock and G. Cardew, Wiley, Chichester (Ciba Foundation Symposium, 193), 127-149.

Molnár, Z., Blakemore, C., and Adams, R. (1995a) Development of thalamocortical and intracortical circuitry studied with Ca^{2+} imaging in rat thalamocortical slices. *Soc. Neurosci . Abstr.*, 21:570.

Molnár, Z., Bronchti, G., Blakemore, C., and Welker, E. (1996a) Initial topological order in thalamocortical projections in the *barrelless* mutant mouse. *Soc. Neurosci. Abstr.*, 22:1013.

Molnár, Z., Knott, G.W., Blakemore, C., and Saunders, N.R. (1997c) Development of thalamocortical projections in the south American grey short-tailed opossum *(Monodelphis Domestica)*. Manuscript submitted for publication to *J. Comp. Neurol.*

Molnár, Z., Krubitzer, L., Nelson, J.E., and Blakemore, C. (1995b) Development of thalamocortical innervation in the marsupial northern native cat *(Dasyurus Hallucatus)*. *Eur. J. Neurosci., Suppl.*, 8:78.

Molnár, Z., Kurotani, T., Higashi, S., Blakemore, C., and Toyama, K. (1996b) Development of functional thalamocortical synapses studied with current source density analysis in whole forebrain slices. *Brain Res. Assoc. Abstr.*, 13:53.

Molnár, Z., Mitrofanis, J., and Blakemore, C. (1993b) Neurons producing nitric oxide in the developing rat cortex. *J. Physiol.*, 59:124.

Molnár, Z., Nesbit, M., Skaliora, I., Smith, A., Blakemore, C., and Gähwiler, B. (1997d) Morphological differentiation of neurons in layer 5 depends on activation of NMDA receptors. *Soc. Neurosci. Abstr.*, 23:11.

Molnár, Z., Yee, K., Lund, R., and Blakemore, C. (1991) Development of rat thalamus and cerebral cortex after embryonic interruption of their connections. *Soc. Neurosci. Abstr.*, 17:764.

Moncada, S., Palmer, R.M.J., and Higgs, E.A. (1991) Nitric oxide: Physiology, pathophysiology, and pharmacology. *Pharmacological Reviews*, 43:109-142.

Mooney, R., Penn, A.A., and Shatz., C.J. (1995) Periodic synaptic currents in the neonatal LGN are generated by retinal activity. *Soc. Neurosci. Abstr.*, 21:1504.

Mooney, R., Penn, A.A., Gallego, R., and Shatz, C.J. (1996) Thalamic relay of spontaneous retinal activity prior to vision. *Neuron*, 17:863-874.

Movshon, J.A., and Van Sluyters, R.C. (1981) Visual neural development. *Ann. Rev. Psychol.*, 33:477-522.

Nadarajah, B., Jones, A.M., Evans, W.H., and Parnavelas, J.G. (1997) Differential expression of connexins during neocortical development and neuronal circuit formation. *J. Nerurosci.*, 17:3096-3111.

Naegele, J.R., Jhaveri, S., and Schneider, G.E. (1988) Sharpening of topographical projections and maturation of geniculocortical axon arbors in the hamster. *J. Comp. Neurol.*, 277:593-607.

Nagata, I., and Nakatsuji, N. (1990) Granule cell behaviour on laminin in cerebellar microexplant cultures. *Dev. Brain Res.*, 52:63-73.

Naito, J. (1986) Course of retinogeniculate projection fibers in the cat optic nerve. *J. Comp. Neurol.*, 251:376-387.

Naito, J. (1989) Retinogeniculate projection fibers in the monkey optic nerve: A demonstration of the fiber pathways by retrograde axonal transport of WGA-HRP. *J. Comp. Neurol.*, 284:174-186.

Nakamura, H., and Kanaseki, T. (1989) Topography of the corpus callosum in the cat. *Brain Res.*, 485:171-175.

Nakane, M., Ichikawa, M., and Deguchi, T. (1983) Light and electron microscopic demonstration of guanylate cyclase in the rat brain. *Brain Res.*, 455:144-147.

Nangla, J., and Adams, R.J. (1997) A study of cell divisions in the ventricular zone of rat neocortex. *Soc. Neurosci. Abstr.*, 23: (in press).

Nelson, S.B., and LeVay, S. (1985) Topographic organization of the optic radiation of the cat. *J. Comp. Neurol.*, 240:322-330.

Nicholson, C., and Freeman, J.A. (1975) Theory of current source density and determination of conductivity tensor for anuran cerebellum. *J. Neurophysiol.*, 38:356-368.

Nishi, R., and Berg, D.K. (1981) Effects of high K^+ concentrations on the growth and development of ciliary ganglion neurons in cell culture. *Dev. Biol.*, 87:301-307.

Northcutt, R.G., and Kaas, J.H., (1995) The emergence and evolution of mammalian neocortex. *Trends in Neurosci.*, 18:373-379.

O'Leary, D.D.M. (1989) Do cortical areas emerge from a protocortex? *Trends in Neurosci.*, 12:400-406.

O'Leary, D.D.M., and Koester, S.E. (1993) Development of projection neuron types, axon pathways, and patterned connections of the mammalian cortex. *Neuron*, 10:991-1006.

O'Leary, D.D.M., and Stanfield, B.B. (1986). A transient pyramidal tract projection from the visual cortex in the hamster and its removal by selective collateral elimination. *Dev. Brain Res.*, 27:87-99.

O'Leary, D.D.M., and Stanfield, B.B. (1989). Selective elimination of axons extended by developing cortical neurons is dependent on regional locale: Experiments utilizing fetal cortical transplants. *J. Neurosci.*, 9:2230-2246.

O'Leary, D.D.M., Bicknese, A.R., De Carlos, J.A., Heffner, C.D., Koester, S.E., Kutka, L.J., and Terashima, T. (1990) Target delection by cortical axons: Alternative mechanisms to establish axonal connections in the developing brain. *Cold Spring Harbor Symposia on Quantitative Biology*, 55:453-468.

O'Leary, D.D.M., Heffner, C.D., Kutka, L., Lopez-Mascaraque, L., Missias, A., and Reinoso, B.S. (1991) A target-derived chemoattractant controls the development of the corticopontine projection by a novel mechanism of axon targeting. *Development, Suppl.*, 2:123-130.

O'Leary, D.D.M., Schlaggar, B.L., and Stanfield, B.B. (1992) The specification of sensory cortex: Lessons from cortical transplantation. *Exp. Neurol.*, 115:121-126.

O'Rourke, N.A., Dailey, M.E., Smith, S.J., and McConnell, S.K. (1991) Time-lapse imaging of migrating cells in cortical slices. *Soc. Neurosci. Abstr.*, 17:533.

O'Rourke, N.A., Dailey, M.E., Smith, S.J., and McConnell, S.K. (1992) Diverse migratory pathways in the developing cerebral cortex. *Science*, 258:299-302.

O'Rourke, N.A., Sullivan, D.P., Kazanowski, C.E., Jacobs, A.A., and McConnell, S.K. (1995) Tangential migration of neurons in the developing cerebral cortex. *Development*, 121:2165-2176.

Ogawa, M., Miyata, T., Nakajima, K., Yagyu, K., Seike, M., Ikenaka, K., Yamamoto, H., and Mikoshiba, K. (1995) The *reeler* gene-associated antigen on Cajal-

Retzius neurons is a crucial molecule for laminar organization of cortical neurons. *Neuron*, 14:899-912.

Ojemann, J.G., and Silbergeld, D.L. (1995) Cortical stimulation mapping of phantom limb rolandic cortex: Case report. *J. Neurosurg.*, 82:641-644.

Omidi, K., Hendry, S.H.C., Jones, E.G., and Emson, P.C. (1988) Organization and plasticity of GABA neuronal subpopulations in monkey area 17, identified by coexistence of calcium binding proteins. *Soc. Neurosci. Abstr.*, 14:188.

Oohira, A., Matsui, F., and Katoh-Semba, R. (1991) Inhibitory effects of brain chondroitin sulfate proteoglycans on neurite outgrowth from PC12D cells. *J. Neurosci.*, 11:822-827.

Oppenheim, R.W. (1989) The neurotrophic theory and naturally occurring motoneuron death. *Trends in Neurosci.*, 12:252-255.

Osen-Sand, A., Catsicas, M., Staple, J.K., Jones, K.A., Ayala, G., Knowles, J., Grenningloh, G., and Catsicas, S. (1993) Inhibition of axonal growth by SNAP-25 antisense oligonucleotides *in vitro* and *in vivo*. *Nature*, 364: 445-448.

Pallas, S.L., and Sur, M. (1993) Visual projections induced into the auditory pathway of ferret. II. Corticocortical connections of primary auditory cortex. *J. Comp. Neurol.*, 337:317-333.

Parnavelas, J.G., Bradford, R., Mounty, E.J., and Lieberman, A.R. (1978) The development of non-pyramidal neurons in the visual cortex of the rat. *Anat. Embryol.*, 155:1-14.

Parnavelas, J.G., Papadopoulos, G.C., and Cavanagh, M.E. (1988) Changes in neurotransmitters during development. In *Cerebral Cortex*, vol. 7: *Development and Maturation of Cerebral Cortex*, eds. A. Peters and E.G. Jones, Plenum Press, New York-London, 177-209.

Pasqualotto, B.A., and Vincent, S.R. (1991) Galanin and NADPH-diaphorase coexistence in colinergic neurons of the rat basal forebrain. *Brain Res.*, 55:78-86.

Peinado, A., and Katz, L.C. (1990) Development of cortical spiny stellate cells: Retraction of a transient apical dendrite. *Soc. Neurosci. Abstr.*, 16:1127.

Peters, A, and Jones, E.G. (eds.) (1985) *Cerebral Cortex*, vol. 3: *Visual Cortex*; vol. 4: *Association and Auditory Cortices*, Plenum Press, New York-London.

Pini, A. (1993) Chemorepulsion of axons in the developing mammalian central-nervous-system. *Science*, 261:95-98.

Pinto-Lord, M.C., and Caviness, V.S., Jr. (1979) Determinants of cell shape and orientations: A comparative Golgi analysis of cell-axon interrelationships in the developing neocortex of normal and *reeler* mice. *J. Comp. Neurol.*, 187:49-70.

Pitkänen, A., and Amaral, D.G. (1991) Distribution of reduced nicotinamide adenine dinucleotide phosphate diaphorase (NADPH-d) cells and fibers in the monkey amygdaloid complex. *J. Comp. Neurol.*, 313:326-348.

Placzek, M., Tessier-Lavigne, M., Yamada, T., Dodd, J., and Jessell, T.M. (1990) Guidance of developing axons by diffusible chemoattractants. *Cold Spring Harbor Symposia on Quantitative Biology*, 55:279-289.

Poe, E.A. (1845) Maelzel's Chess-Player. In *The Complete Illustrated Stories and Poems of Edgar Allan Poe*, Chancellor Press, London 1988, 685-702.

Poyak, S. (1957) *The Vertebrate Visual System*, Chicago University Press, Chicago.

Price, D.J. (1991) The development of visual cortical afferents. In *Development and Plasticity of the Visual System*, ed. J. Cronly-Dillon, Macmillan, Basingstoke (Vision and Visual Dysfunction, 11), 337-352.

Price, D.J., and Lotto, R.B. (1996) Influences of the thalamus on the survival of subplate and cortical cortical plate cells in cultured embryonic mouse brain. *J. Neurosci.*, 16:3247-3255.

Price, D.J., Lotto, R.B., Warren, N., Magowan, G., and Clausen, J. (1995) The roles of growth factors and neural activity in the development of the neocortex. In *Development of the Cerebral Cortex*, eds. G. Bock and G. Cardew, Wiley, Chichester (Ciba Foundation Symposium, 193), 231-250.

Price, D.J., Aslam S., Tasker, L., and Gillies, K. (1997) Fates of the earliest generated cells in the developing murine neocortex. *J. Comp. Neurol.*, 377:414-422.

Price, J., Turner, D., and Cepco, C. (1987) Lineage analysis in the vertebrate nervous system by retrovirus-mediated gene transfer. *Proc. Natl. Acad. Sci. USA*, 84:156-160.

Price, R.H., Jr., Mayer, B., and Beitz, A.J. (1993) Nitric oxide synthase neurons in rat brain express more NMDA receptor mRNA than non-NOS neurons. *NeuroReport*, 4:807-810.

Provis, J.M., and Mitrofanis, J. (1990) NADPH-diaphorase neurons of human retina have a uniform topographical distribution. *Visual Neurosci.*, 4:619-623.

Purves, D. (1988) *Body and Brain. A Trophic Theory of Neural Connections*, Harvard University Press, Cambridge Mass., 433.

Purves, D., and Lichtman, J.W. (1985) *Principles of Neuronal Development*, Sinauer Association INC. Publishers, Sunderland Mass.

Raedler, E., and Raedler, A. (1978) Autoradiographic study of early neurogenesis in rat neocortex. *Anat. Embryol.*, 154:267-284.

Raff, M.C. (1992) Social controls on cell survival and cell death. *Nature*, 356:397-399.

Raff, R.A. (1996) *The Shape of Life*, University of Chicago Press, Chicago-London.

Rakic, P. (1971) Guidance of neurons migrating to the fetal monkey neocortex. *Brain Res.*, 33:471-476.

Rakic, P. (1972) Mode of cell migration to the superficial layers of monkey neocortex. *J. Comp. Neurol.*, 145:61-84.

Rakic, P. (1974) Neurons in rhesus monkey visual cortex: Systemic relation between time of origin and eventual disposition. *Science*, 183:425-427.

Rakic, P. (1976) Prenatal genesis of connections subserving ocular dominance in the rhesus monkey. *Nature*, 261:467-471.

Rakic, P. (1977) Prenatal development of the visual system in the rhesus monkey. *Philos. Trans. R. Soc. Lond. B Biol. Sci.*, 278:245-260.

Rakic, P. (1982) Early development events: Cell lineages, acquisition of neuronal positions, and areal and laminar development. *Neurosci. Res. Program Bull.*, 20:439-451.

Rakic, P. (1988) Specification of cerebral cortical areas. *Science*, 241:170-176.

Rakic, P. (1995) Radial versus tangential migration of neuronal clones in the developing cerebral cortex. *Proc. Natl. Acad. Sci. USA*, 92:11323-11327.

Rakic, P., and Caviness, V.S., Jr. (1995) Cortical development: View from neurological mutants two decades later. *Neuron*, 14:1101-1104.

Rakic, P., and Singer, W. (1988) *Neurobiology of Neocortex* (Report of the Dahlem Workshop, Berlin 1987, May 17-22), Wiley, Chichester-New York-Brisbane-Toronto-Singapore (Life Sciences Research Reports, ed. S. Bernhard).

Ramcharan, E.J., and Guillery, R.W. (1997) Membrane specializations in the developmentally transient perireticular nucleus of the rat. *J. Comp. Neurol.*, 380:435-448.

Rasmusson, D. (1996) Changes in the organization of the ventroposterior lateral thalamic nucleus after digit removal in adult raccoon. *J. comp. Neurol.*, 364:92-103.

Rauschecker, J.P., and Marler, P. (eds.) (1987) *Imprinting and cortical plasticity. Comparative aspects of sensitive periods*, Wiley, New York-Chichester-Brisbane-Toronto-Singapore (Wiley Series in Neuroscience, 1).

Recanzone, G.H., Merzenich, M.M., Jenkins, W.M., Grajski, K.A., and Dinse, H.R. (1992) Topographic reorganization of the hand representation in cortical area 3b owl monkeys trained in a frequency-discrimination task. *J. Neurophisiol.*, 67:1031-1056.

Reese, B.E., and Baker, G.E. (1992) Changes in fiber organization within the chiasmatic region of mammals. *Visual Neurosci.*, 9:527-533.

Reh, T.A., Pitts, E., and Constantine-Paton, M. (1983) The organization of fibers in the optic nerve of normal and tectum-less Rana pipiens. *J. Comp. Neurol.*, 218:282-296.

Reid, C.B., Liang, I., and Walsh, C. (1995) Systematic widespread clonal organization in cerebral cortex. *Neuron*, 15:299-310.

Reiner, A. (1990) Neurotransmitter organization and connections of turtle cortex: Implications for the evolution of mammalian isocortex. *Com. Biochem. Physiol.*, 104:735-748.

Reiner, A. (1991) A comparison of neurotransmitter-specific and neuropeptide-specific neuronal cell types present in the dorsal cortex in turtles with those present in the isocortex in mammals: Implications for the evolution of isocortex. *Brain Behav. Evol.*, 38:53-91.

Reiner, A., Krause, J.E., Keyser, K.T., Eildred, W.D., and McKelvy, J.F. (1984) The distribution of substance P in the turtle nervous system: A radioimmunoassay and immunohistochemical study. *J. Comp. Neurol.*, 226:50-75.

Reinoso, B.S., and O'Leary, D.D.M. (1990) Correlation of geniculocortical growth into the cortical plate with the migration of their layer 4 and 6 target cells. *Soc. Neurosci. Abstr.*, 16:493.

Rennie, S., Lotto, R.B., and Price, D.J. (1994) Growth-promoting interactions between the murine neocortex and thalamus in organotypic co-cultures. *Neuroscience*, 61:547-564.

Reznikov, K.Y., Fülöp, Z., and Hajós, F. (1984) Mosaicism of the ventricular layer as the developmental basis of neocortical columnar organization. *Anat. Embryol.*, 170:99-105.

Rhodes, R.W., and Fish, S.E. (1983) Bilateral enucleation alters visual callosal but not cortico-tectal or corticogeniculate projections in hamsters. *Exp. Brain Res.*, 51:451-462.

Rice, F.L. (1995) Comparative aspects of barrel structure and development. In *Cerebral Cortex*, vol. 11: *The Barrel Cortex of Rodents*, eds. by E.G. Jones and I.T. Diamond, Plenum Press, New York, 1-75.

Rice, F.L., Gómez, C., Barstow, C., Burnet, A., and Sands, P. (1985) A comparative analysis of the development of the primary somatosensory cortex: Interspecies similarities during barrel and laminar development. *J. Comp. Neurol.*, 236:477-495.

Richards, L.J., Koester, S.E., Tuttle, R., and O'Leary, D.D.M. (1997) Directed growth of early cortical axons is influenced by a chemoattractant released from an intermediate target. *J. Neurosci.*, 17:2445-2458.

Rickmann, M., Chronwall, M., and Wolff, J.R. (1977) On the development of non-pyramidal neurons and axons outside the cortical plate: The early marginal zone as a pallial anlage. *Anat. Embryol.*, 151:285-307.

Robertson, R.T. (1987) A morphogenetic role for transiently expressed acetylcholinesterase in developing thalamocortical systems? *Neurosci. Lett.*, 75:259-264.

Robinson, D.L., and Petersen, S.E. (1992) The pulvinar and visual salience. *Trends in Neurosci.*, 15:127-132.

Rockel, A.J., Hiorns, R.W., and Powell, T.P.S. (1980) The basic uniformity in structure of the neocortex. *Brain*, 103:221-244.

Romanes, G.J. (1901) Darwin and After Darwin. Open Court Publishing Co, London.

Romijn, H.J. (1988) Development and advantages of serum-free, chemically defined nutrient media for culturing of nerve tissue. *Biology of the Cell*,

63:263-268.

Romijn, H.J., De Jong, B.M., and Ruijter, J.M. (1988) A procedure for culturing rat neocortex explants in a serum-free nutrient medium. *J. Neurosci. Methods*, 23:75-83.

Rose, J.E. (1942a) The ontogenetic development of the rabbit's diencephalon. *J. Comp. Neurol.*, 77:61-129.

Rose, J.E. (1942b) The thalamus of the sheep: Cellular and fibrous structure and comparison with pig, rabbit and cat. *J. Comp. Neurol.*, 77:469-523

Roux, W. (1881) *Der Kampf der Theile im Organismus*, Wilhelm Engelmann Verlag, Leipzig.

Roux, W. (1894) The problems, methods and scope of developmental mechanics. An introduction to the "Archiv für Entwicklungsmechanik der Organismen", trans. W.M. Wheeler. In *Biological Lectures of the Marine Biological Laboratory of Woods Hole, Mass.*, Ginn and Company, Boston, 1895, 149-190.

Rowe, M. (1990) Organization of the cerebral cortex in monotremes and marsupials. In *Cerebral Cortex*, vol. 8B: *Comparative Structure and Evolution of Cerebral Cortex, Part II*, eds. E.G. Jones and A. Peters, Plenum Press, New York-London, 263-334.

Saffrey, M.J., Hassall, C.J.S., Hoyle, C.H.V., Belai, A., Moss, J., Schmidt, H.H.H.W., Forstermann, U., Murad, F., and Burnstock, G. (1992) Colocalisation of nitric oxide synthase and NADPH-diaphorase in cultured myenteric neurons. *Neuroreport*, 3:333-336.

Sagar, S.M. (1986) NADPH-diaphorase histochemistry in the rabbit retina. *Brain Research Reviews*, 13:119-137.

Sandell, J.H. (1985) NADPH-diaphorase cells in the mammalian inner retina. *J. Comp. Neurol.*, 238:466-472.

Sandell, J.H. (1986) NADPH-diaphorase histochemistry in the macaque striate cortex. *J. Comp. Neurol.*, 251:388-397.

Sandell, J.H., and Masland, R.H. (1988) Photoconversion of some fluorescent markers to a diaminobensidine product. *J. Histochem. Cytochem.*, 36:555-559.

Sanderson, K.J., and Weller, W.L. (1990) Gradients of neurogenesis in possum neocortex. *Dev. Brain Res.*, 55:269-274.

Scannell, J.W., Burns, G.A.P.C., O'Neill, M.A., Hilgetag, C.C., and Young, M.P. (1997) The organization of the thalamo-cortical network of the cat. *Soc. Neurosci. Abstr.*, 23: (in press).

Scherer-Singler, U., Vincent, S.R., Kimura, H., and McGeer, E.G. (1983) Demonstration of a unique population of neurons with NADPH-diaphorase histochemistry. *J. Neurosci. Methods*, 9:229-234.

Schlaggar, B.L., and O'Leary, D.D.M (1991) Potential of visual cortex to develop an array of functional units unique to somatosensory cortex. *Science*, 252:1556-1560.

Schlaggar, B.L., and O'Leary, D.D.M (1993) Patterning of the barrel field in somatosensory cortex with implications for the specification of neocortical areas. *Perspectives on Developmental Neurobiology*, 1:81-91.

Schlaggar, B.L., and O'Leary, D.D.M. (1994) Early development of the somatotopic map and barrel patterning in rat somatosensory cortex. *J. Comp. Neurol.*, 346:80-96.

Schlaggar, B.L., Fox, K., and O'Leary, D.D.M. (1993) Postsynaptic control of plasticity in developing somatosensory cortex. *Nature*, 364:623-626.

Schnell, L., and Schwab, M.E. (1990) Axonal regeneration in the rat spinal cord produced by an antibody against myelin associated neurite growth inhibitors. *Nature*, 343:269-272.

Scholes, J. (1979) Nerve fiber topography in the retinal projection of the tectum.

Nature, 278:620-624

Schousboe, A., Frandsen, A., Wahl, P., and Krogsgaard-Larsen, P. (1991) Excitatory amino acid induced cytotoxicity in cultured neurons: Role of intracellular Ca^{2+} homeostasis in glutamate. In *Cell Death and Memory*, eds. P. Ascher, D.W. Choi and Y. Christen, Springer-Verlag, Berlin-Heidelberg-New York, 137-152.

Schreyer, D.J., and Jones, E.G. (1982) Growth and target finding by axons of the corticospinal tract in prenatal and postnatal rats. *Neuroscience*, 7:1837-1853.

Schwab, M.E., and Caroni, P. (1988) Oligodendrocytes and CNS myelin are nonpermissive substrates for neurite growth and fibroblast spreading in vitro. *J. Neurosci.*, 8:2381-2393.

Schwartz, M.L., Dekker, J.J., and Goldman-Rakic, P.S. (1991) Dual mode of corticothalamic synaptic termination in the mediodorsal nucleus of the rhesus monkey. *J. Comp. Neurol.*, 309:289-304.

Seitz, R.J., Huang, Y., Knorr, U., Tellmann, L., Herzog, H., and Freund, H.J. (1995) Large-scale plasticity of the human motor cortex. *Neuroreport*, 27:742-744.

Sendemir, E., Erzurumlu, R.S., and Jhaveri, S. (1996) Differential expression of acetylcholinesterase in the developing barrel cortex of three rodent species. *Cerebral Cortex*, 6:377-387.

Shatz, C.J. (1981) Inside-out pattern of neurogenesis of the cat's lateral geniculate nucleus. *Soc. Neurosci. Abstr.*, 7:140.

Shatz, C.J. (1983) The prenatal development of the cat's retinogeniculate pathway. *J. Neurosci.*, 3:482-499.

Shatz, C.J. (1990) Impulse activity and the patterning of connections during CNS development. *Neuron*, 5:745-756.

Shatz, C.J. (1992a) Dividing up the neocortex. *Science*, 258:237-238.

Shatz, C.J. (1992b) How are specific connections formed between thalamus and cortex? *Current Opinions in Neurobiology*, 2:78-82.

Shatz, C.J., and Kirkwood, P.A. (1984) A prenatal development of functional connections in the cat's primary visual cortex. *J. Neurosci.*, 6:3655-3668.

Shatz, C.J., and Luskin, M.B. (1986) Relationship between the geniculocortical afferents and their cortical target cells during development of the cat's primary visual cortex. *J. Neurosci.*, 6:3655-3668.

Shatz, C.J., and Rakic, P. (1981) The genesis of efferent connections from the visual cortex of the fetal rhesus monkey. *J. Comp. Neurol.*, 196:287-307.

Shatz, C.J., Chun, J.J.M., and Luskin, M.B. (1988) The role of the subplate in the development of the mammalian telencephalon. In *Cerebral Cortex*, vol. 7: *Development and Maturation of Cerebral Cortex*, eds. A. Peters and E. G. Jones, Plenum Press, New York-London, 35-58.

Shatz, C.J., Ghosh, A., McConnell, S.K., Allendoerfer, K.L., Friauf, E., and Antonini, A. (1990) Pioneer neurons and target selection in cerebral cortical development. *Cold Spring Harbor Symposia on Quantitative Biology*, 55:469-480.

Sheng, X-M., Marotte, L.R., and Mark, R.F. (1991) Development of the laminar distribution of thalamocortical axons and corticothalamic cell bodies in the visual cortex of the wallaby. *J. Comp. Neurol.*, 307:17-38.

Sheppard, A.M., Hamilton, S.K., and Pearlman, A.L. (1991) Changes in the distribution of extracellular matrix components accompany early morphogenetic events of mammalian cortical development. *J. Neurosci.*, 11:3928-3942.

Shering, A.F., and Lowenstein, P.R. (1994) Neocortex provides direct synaptic input to interstitial neurons of the intermediate zone of kittens and white matter of cats: A light and electron microscopic study. *J. Comp. Neurol.*, 347:433-443.

Sherman, S.M., and Guillery, R.W. (1996) Functional organization of thalamocortical relays. *J. Neurophysiol.*, 76:1367-1395.

Shibuki, K., and Okada, D. (1991) Endogenous nitric-oxide release required for long-term synaptic depression in the cerebellum. *Nature*, 439:326-328.

Shiells, R., and Falk, G. (1992) Retinal on-bipolar cells contain a nitric oxide-sensitive guanylate cyclase. *Neuroreport*, 3:845-848.

Shoukimas, G.M., and Hinds, J.W. (1978) The development of the cerebral cortex in the embryonic mouse: An electron microscopic serial section analysis. *J. Comp. Neurol.*, 179:795-830.

Sidman, R.L., Miale, L., and Feder, N. (1959). Cell proliferation and migration in the primitive ependymal zone: An autoradiographic study of histogenesis in the nervous system. *Exp. Neurol.*, 1:322-333.

Simmons, P.A., and Pearlman, A.L. (1982) Retinotopic organization of the striate cortex (area 17) in the *reeler* mutant mouse. *Dev. Brain. Res.*, 4:124-126.

Simon, D.K., and O'Leary, D.D.M. (1990) Limited topographic specificity in the targeting and branching of mammalian retinal axons. *Dev. Biol.*, 137:125-134.

Simon, D.K., and O'Leary, D.D.M. (1991) Relationship of retinotopic ordering of axons in the optic pathway to the formation of visual maps in central targets. *J. Comp. Neurol.*, 307:393-404.

Simon, D.K., and O'Leary, D.D.M. (1992) Influence of position along the medial-lateral axis of the superior colliculus on the topographic targeting and survival of retinal axons. *Dev. Brain Res.*, 69:167-172.

Singer, W. (1987) Developmental plasticity—Self-organisation or learning? In *Imprinting and Cortical Plasticity: Comparative Aspects of Sensitive Periods*, ed. J.P. Rauschecker and P. Marler, Wiley, New York-Chichester-Brisbane-Toronto-Singapore. (Wiley Series in Neuroscience, 1), 171-175.

Skaliora, I., Adams, R. and Blakemore, C. (1995) Time-lapse recordings of growing axons in the developing thalamocortical pathway. *Soc. Neurosci. Abstr.*, 21:799.

Skaliora, I., Betz, H., and Puschel, A.W. (1996) Semaphorin expression in the developing mammalian brain. *Soc. Neurosci. Abstr.*, 22:1472.

Snow, D.M., Watanabe, M., Letourneau, P., and Silver, J. (1991) A chondroitin sulfate proteoglycan may influence the direction of retinal ganglion cell outgrowth. *Development*, 113:1473-1485.

Snyder, E.Y., and Kim, S.U. (1979) Hormonal requirements for neuronal survival in culture. *Neurosci. Lett.*, 13:225-230.

Somogyi, P., Hodgoson, A.J., Smith, A.D., Nunzi, M.G., Gorio, A., and Wu, J.-Y. (1984) Different populations of GABAergic neurons in the visual cortex and hippocampus of cat contain somatostatin- or cholecystokinin-immunoreactive material. *J. Neurosci.*, 4:2590-2603.

Soriano, E., Dumesnil, N., Auladell, C., Cohen-Tannoudji, M., and Sotelo, C. (1995) Molecular heterogeneity of progenitors and radial migration in the developing cerebral cortex revealed by transgene expression. *Proc. Natl. Acad. Sci. USA*, 92:11676-11680.

Spatz, W.B. (1989) Loss of ocular dominance columns with maturity in the monkey *Callithrix jacchus*. *Brain Res.*, 488:376-380.

Spemann, H. (1938) *Embryonic Development and Induction*, Yale University Press, New Haven.

Sperry, R.W. (1950) Neuronal specificity. In *Genetic Neurology: Problems of the Development, Growth and Regeneration of the Nervous System and of its Functions*, ed. P. A. Weiss, University of Chicago Press, Chicago, 232-239.

Sperry, R.W. (1963) Chemoaffinity in the orderly growth of nerve fiber patterns and connections. *Proc. Natl. Acac. Sci. USA*, 50:703-709.

Squier, M., Hannan, A., Servotte, S., Katsnelson, A., Blakemore, C. and Molnár, Z. (1997) Characterisation of subcortical neuronal heterotopias in children with

epilepsy. *Soc. for Neurosci. Abstr.*, 23: (in press).
Sretavan, D.W. (1993) Pathfinding at the mammalian optic chiasm. *Current Opinion in Neurobiology*, 3:45-52.
Sretavan, D.W., and Reichardt, L.F. (1993) Time-lapse video analysis of retinal ganglion cell axon pathfinding at the mammalian optic chiasm: Growth cone guidance using intrinsic chiasm cues. *Neuron*, 10:761-777.
Stanfield, B.B., and Cowan, W.M. (1979) The development of the hippocampus and dentate gyrus in normal and *reeler* mice. *J. Comp. Neurol.*, 185:423-460.
Stanfield, B.B., and O'Leary, D.D.M. (1985) Fetal occipital cortical neurons transplanted to the rostral cortex can extend and maintain a pyramidal tract axon. *Nature*, 313:135-137.
Steinberg, M.S. (1963) Reconstruction of tissues by dissociated cells. *Science*, 141:401-408.
Steindler, D.A., and Colwell, S.A. (1976) *Reeler* mutant mouse: Maintenance of appropriate and reciprocal connections in the cerebral cortex and thalamus. *Brain Res.*, 105:386-393.
Steindler, D.A., Settles, D., Erickson, H.P., Laywell, E.D., Yoshiki, A., Faissner, A., and Kusakabe, M. (1995) Tenascin knockout mice: Barrels, boundary molecules, and glial scars. *J. Neurosci.*, 15:1971-1983.
Steriade, M., and Contreras, D. (1995) Relations between cortical and thalamic cellular events during transition from sleep patterns to paroxismal activity. *J. Neurosci.*, 15:623-642.
Sterling, R.V. (1991) Molecules, maps and gradients in the retinotectal projection. *Trends in Neurosci.*, 14:509-512.
Stewart, G.R., and Pearlman, A.L. (1987) Fibronectin-like immunoreactivity in the developing cerebral cortex. *J. Neurosci.*, 7:3325-3333.
Street, V.A., Bosma, M.M., Demas, V.P., Regan, M.R., Lin, D.D., Robinson, L.C., Agnew, W.S., and Tempel, B.L. (1997) The type 1 inositol 1,4,5-triphosphate receptor gene is altered in the *opistotonus* mouse. *J. Neurosci.*, 17:635-645.
Stryker, M.P., and Harris, W. (1986) Binocular impulse blockade prevents the formation of ocular dominance columns in cat visual cortex. *J. Neurosci.*, 6:2117-2133.
Stuermer, C.A.O. (1992) The formation of topographically ordered connections during development and regeneration of the vertebrate visual system. In *Development and Plasticity of the Visual System*, ed. J. Cronly-Dillon, Macmillan, Basingstoke (Vision and Visual Dysfunction, 11), 88-111.
Sur, M., Pallas, S.L., and Roe, A.W. (1990) Cross-modal plasticity in cortical development: Differentiation and specification of sensory neocortex. *Trends in Neurosci.*, 13:227-233.
Swindale, N.V., Vital-Durand, F., and Blakemore, C. (1981) Recovery from monocular deprivation in the monkey. III. Reversal of anatomical effects in the visual cortex. *Proc. R. Soc. Lond. B Biol. Sci.*, 213:435-450.
Szentágothai, J. (1975) The 'module concept' in cerebral cortex architecture. *Brain Res.*, 95:475-496.
Szentágothai, J. (1977) *Functionalis Anatomia. Az ember anatomiája, fejlödéstana, szövettana és tájanatomiája*, III. kötet, harmadik javított kiadás, Medicina Könyvkiadó, Budapest.
Taira, E., Tanaha, N., and Miki, N. (1993) Extracellular matrix proteins with neurite promoting activity and their receptors. *Neurosci. Res.*, 17:1-8.
Tan, S.S., and Breen, S. (1993) Radial mosaicism and tangential cell dispersion both contribute to mouse neocortical development. *Nature*, 362:638-640.
Tan, S.S., Faulkner-Jones, B., Breen, S.J., Walsh, M., Bertram, J.F., and Reese, B.E. (1995) Cell dispersion patterns in different cortical regions studied with an X-inactivated transgenic marker. *Development*, 121:1029-1039.
Tessier-Lavigne, M., and Placzek, M. (1991) Target attraction: Are developing axons

guided by chemotropism? *Trends in Neurosci.*, 14:303-310.

Thomas, E., and Pearse, A.G.E. (1961) The fine localisation of dehydrogenases in the nervous system. *Histochemistry*, 2:226-282.

Thomas, E., and Pearse, A.G.E. (1964) The solitary active cells: Histochemical demonstration of damage-resistant nerve cells with a TPN-diaphorase reaction. *Acta neuropath.*, 3:238-249.

Tigges, M., and Tigges, J. (1991) Parvalbumin immunoreactivity of the lateral geniculate nucleus in adult rhesus monkeys after monocular eye enucleation. *Visual Neurosci.*, 6:375-383.

Tonegawa, S., Li, Y., Erzurumlu, R.S., Jhaveri, S., Chen, C., Goda, Y., Paylor, R., Silva, A.J., Kim, J.J., Wehner, J.M., et al. (1995) The gene knockout technology for the analysis of learning and memory, and neural development. *Prog. Brain Res.*, 105:3-14.

Tóth, K., Freund, T.F., and Miles, R. (1997) Disinhibition of rat hippocampal pyramidal cells by GABAergic afferents from the septum. *J. of Physiol.*, 500:463-474.

Toyama, K., Komatsu, Y., Yamamoto, N., Kurotani, T., and Yamada, K. (1991) *In vitro* approach to visual cortical development and plasticity. *Neurosci. Res.*, 12:57-71.

Trevelyan, A.J., and Thompson, I.D. (1992) Altered topography in the geniculocortical projection of the golden hamster following neonatal monocular enucleation. *Eur. J. Neurosci.*, 4:1104-1111.

Tuttle, R., Schlaggar, B.L., Braisted, J.E., and O'Leary, D.D.M. (1995) Maturation-dependent upregulation of growth-promoting molecules in developing cortical plate controls thalamic and cortical neurite growth. *J. Neurosci.*, 15:3039-3052.

Ulinski, P.S. (1983) *Dorsal Ventricular Ridge*, Wiley, New York.

Ulinski, P.S. (1990) The cerebral cortex of reptiles. In *Cerebral Cortex*, vol. 8A: *Comparative Structure and Evolution of Cerebral Cortex, Part I*, eds. E.G. Jones and A. Peters, Plenum Press, New York-London, 139-215.

Uylings, H.B.M., and van Eden, C.G. (1990) Late generation of GABA-ergic subplate neurons in the rat. *Soc. Neurosci. Abstr.*, 16:647.

Uylings, H.B.M., van Eden, C.G., Parnavelas, J.G., and Kalsbeek, A. (1990) The prenatal and postnatal development of rat cerebral cortex. In *The Cerebral Cortex of the Rat*, eds. B. Kolb, R.C. Tees, MIT Press, Cambridge Mass., 35-76.

Uzman, L.L. (1960). The histogenesis of the mouse cerebellum as studied by its tritiated thymidine uptake. *J. Comp. Neurol.*, 114:137-160.

Vaccaro, T.M., Cobcroft, M.D., Provis, J.M., and Mitrofanis, J. (1991) NADPH-diaphorase reactivity in adult and developing cat retinae. *Cell and Tissue Research*, 265:371-379.

Valverde, F. (1967) Apical dendritic spines of the visual cortex and light deprivation in the mouse. *Exp. Brain Res.*, 3:337-352.

Valverde, F. (1968) Structural changes in the area striata of the mouse after enucleation. *Exp. Brain Res.*, 5:274-292.

Valverde, F., and Facal-Valverde, M.V. (1988). Postnatal development of interstitial (subplate) cells in the white matter of the temporal cortex of kittens: A correlated Golgi and electron microscopic study. *J. Comp. Neurol.*, 269:168-192.

Valverde, F., De Carlos, J.A., and Lopez-Mascaraque, L. (1995a) Time of origin and early fate of preplate cells in the cerebral cortex of the rat. *Cerebral Cortex*, 5:483-493.

Valverde, F., Facal-Valverde, M.V., Santacana, M., and Heredia, M. (1989) Development and differentiation of early generated cells of sublayer VIb in the somatosensory cortex of the rat: A correlated Golgi and autoradiographic study. *J. Comp. Neurol.*, 290:118-140.

Valverde, F., López-Mascaraque, L, Santacana, M., and De Carlos, J.A. (1995b) Persistence of early-generated neurons in the rodent subplate: Assessment of cell death in neocortex during the early postnatal period. *J. Neurosci.*, 15:5014-5024.

Van der Loos, H., and Woolsey, T.A. (1973) Somatosensory cortex: Structural alterations following early injury to sense organs. *Science*, 179:395-376.

Van Essen, D.C. (1997) A tension-based theory of morphogenesis and compact wiring in the central nervous system. *Nature*, 385:313-318.

Vaney, D.I., and Young, H.M. (1988) GABA-like immunoreactivity in NADPH-diaphorase amacrine cells of the rabbit retina. *Brain Res.*, 474:380-385.

Vercelli, A., Assal, F., and Innocenti, G.M. (1992) Emergence of callosally projecting neurons with stellate morphology in the visual cortex of the kitten. *Exp. Brain Res.*, 90:346-358.

Verley, R., and Axelrad, H. (1975) Postnatal ontogenesis of potentials elicited in the cerebral cortex by afferent stimulation. *Neurosci. Lett.*, 1:99-104.

Vincent, S.R. (1986) NADPH-diaphorase histochemistry and neurotransmitter coexistence. In *Neurohistochemistry: Modern Methods and Applications*, eds. P. Panula, H. Pivarinta and S. Soinila, Alan R. Liss, New York, 375-396.

Vincent, S.R., and Hope, B.T. (1992) Neurons that say NO. *Trends in Neurosci.*, 15:108-113.

Vincent, S.R., and Kimura, H. (1992) Histochemical mapping of nitric oxide synthase in the rat brain. *Neuroscience*, 46:755-784.

Vincent, S.R., Johansson, O., Hokfeld, T., Skirboll, L., Elde, R.P., Terenius, L., Kimmel, J., and Goldstein, M. (1983) NADPH-diaphorase: A selective histochemical marker for striatal neurons containing both somatostatin- and avian pancreatic polypeptide (APP)-like immunoreactivities. *J. Comp. Neurol.*, 217:252-263.

Vitalis, T., Cases, O., Seif, I., and Gaspar, P. (1996) Inhibition of monoamine oxidase A: Determination of a critical period to produce an effect on barrelfield development. *Eur. J. Neurosci., Suppl.*, 9:31.

Von Economo, C. (1929) *The Cytoarchitectonics of the Human Cerebral Cortex*, Humphrey Milford-Oxford University Press, London.

Von Economo, C., and Koskinas, G.N. (1925) *Die Cytoarchitectonik der Hirnrinde des Erwachsenen Menschen*, Springer, Berlin.

Wall, P.D., and Egger, M.D. (1972) Formation of new connections in adult rat brains after partial deafferentation. *Nature*, 232:542-545.

Wallace, M.N.J., and Fredens, K. (1992) Activated astrocytes of the mouse hippocampus contain high levels of NADPH-diaphorase. *Neuroreport*, 3:953-956.

Wallis, R.A., Panizzon, K.L., Henry, D., and Wasterlain, C.G. (1992) Neuroprotection against nitric oxide injury with inhibitors of ADP-ribosylation. *Neuroreport*, 5:245-248.

Walsh, C., and Cepko, C.L. (1992) Widespread dispersion of neuronal clones across functional regions of the cerebral cortex. *Science*, 255:434-255.

Walsh, C., and Cepko, C.L. (1993) Clonal dispersion in proliferative layers of developing cerebral cortex. *Nature*, 362:632-635.

Walsh, C., and Guillery, R.W. (1985) Age-related fiber order in the optic tract of the ferret. *J. Neurosci.*, 5:3061-3069.

Wässle, H.M., Chun, M.H., and Muller, F. (1987) Amacrine cells in the ganglionic cell layer of the cat retina. *J. Comp. Neurol.*, 265:391-408.

Watanabe, E., Aono, S., Matsui, F., Yamada, Y., Naruse, I., and Oohira, A. (1995) Distribution of a brain-specific proteoglycan, neurocan, and the corresponding mRNA during the formation of barrels in the rat somatosensory cor-

tex. *Eur J. Neurosci.*, 7:547-554.
Weidman, T.A., and Kuwabara, T. (1968) Postnatal development of the rat retina. An electron microscopic study. *Arch. Ophthalmol.*, 79:470-484.
Weiss, P.A. (1941) Nerve patterns. The mechanics of nerve growth. *Third Growth Symposium*, 5:163-203.
Weiss, P.A. (1947) The problem of specificity in growth and development. *Yale J. Biol. Med.*, 19:235-278.
Welker, E., and van der Loos, H. (1986) Quantitative correlation between barrel-field size and the sensory innervation of the whiskerpad: A comparative study in six strains of mice bred for different patterns of mystical vibrissae. *J. Neurosci.*, 6:3355-3373.
Welker, E., Armstrong-James, M., Bronchti, G., Ourednik, W., Gheorghita-Bächler, F., DuBois, R., Guernsey, D.L., van der Loos, H., and Neumann, P.E. (1996) Altered sensory processing in the somatosensory cortex of the mouse mutant *barrelless. Science*, 271:1864-1867.
Whale, P., and Meyer, G., (1987) Morphology and quantitative changes of transient NPY-ir neuronal populations during early postnatal development of the cat visual cortex. *J. Comp. Neurol.*, 261:165-192.
Wilkemeyer, M.F., and Angelides, K.J. (1995) Adenovirus mediated expression of reporter gene in thalamocortical cocultures. *Brain Res.*, 703:129-138.
Wilkemeyer, M.F., and Angelides, K.J. (1996) Addition of tetrodotoxin alters the morphology of thalamocortical axons in organotypic cocultures. *J. Neurosci. Res.*, 43:707-718.
Williams, C.V., Davenport, R.W., Dou, P., and Kater, S.B. (1993) *In vitro* induction of backbranches: Implications for an alternate mechanism of target location. *Soc. Neurosci. Abstr.*, 19:1084.
Williams, R.W., and Rakic, P. (1985) Dispersion of growing axons within the optic nerve of the embryonic monkey. *Proc. Natl. Acad. Sci. USA*, 82:3906-3910.
Williams, R.W., Bastiani, M.J., Lia, B., and Chalupa, L.M. (1986) Growth cones, axons and development fluctuations in the fiber population of the cat's optic nerve. *J. Comp. Neurol.*, 246:32-69.
Williams, R.W., Borodkin, M., and Rakic, P. (1991) Growth cone distribution patterns in the optic nerve of fetal monkeys: Implications for mechanisms of axon guidance. *J. Comp. Neurol.*, 11:1081-1094.
Wilson, H.V. (1907) On some phenomena of coalescence and regeneration in sponges. *J. Exp. Zool.*, 5:245-258.
Windle (1935) Neurofibrillar development of cat embryos: Extent of development in the telencephalon and diencephalon up to 15mm. *J. Comp. Neurol.*, 63:139-171.
Windrem, M.S., and Finlay, B.L. (1991) Thalamic ablations and neocortical development: Alterations of cortical architecture and cell number. *Cerebral Cortex*, 1:230-240.
Wise, S.P., and Jones, E.G. (1976) The organization and postnatal development of the comissural projection of the somatic sensory cortex of the rat. *J. Comp. Neurol.*, 168:313-344.
Wise, S.P., and Jones, E.G. (1977) Cells of origin and terminal distribution of descending projections of the rat somatic sensory cortex. *J. comp. Neurol.*, 175:129-158.
Wise, S.P., and Jones, E.G. (1978) Developmental studies of thalamocortical and comissural connections in the rat somatic sensory cortex. *J. Comp. Neurol.*, 163:313-344.
Wise, S.P., Fleshman, J.W., Jr., and Jones, E.G. (1979) Maturation of pyramidal cell form in relation to developing afferent and efferent connections of rat

somatic sensory cortex. *Neuroscience*, 4:1275-1297.

Wise, S.P., Hendry, S.H.C, and Jones, E.G. (1977) Prenatal development of sensorimotor cortical projections in cats. *Brain Res.*, 138:538-544.

Wolff, J.R., Böttcher, H., Zetzsche, T., Oertel, W.H., and Chronwall, B.M. (1984) Development of GABAergic neurons in the rat visual cortex as identified by glutamate decarboxylase-like immunoreactivity. *Neurosci. Lett.*, 47:207-212.

Wolfson, B., Gutnick, M.J., and Baldino, F., Jr. (1989) Electrophysiological characteristics of neurons in neocortical explant cultures. *Exp. Brain Res.*, 76:122-130.

Wong, R.O.L. (1993) The role of spatio-temporal firing patterns in neuronal development of sensory systems. *Current Opinion in Neurobiology*, 3:595-601.

Woo, T.U., and Finlay, B.L. (1991) Depletion of layer IV has no effect on survival of subplate neurons. *Soc. Neurosci. Abstr.*, 17:766.

Woo, T.U., and Finlay, B.L., (1996) Cortical target depletion and ingrowth of geniculocortical axons: Implications for cortical specifications. *Cerebral Cortex*, 6:457-469

Woo, T.U., Beale, J.M., and Finlay, B.L. (1990) Dual fate of subplate neurons in the rodent. *Soc. Neurosci. Abstr.*, 16:836.

Wood, J.G., Martin, S., and Price, D.J. (1992) Evidence that the earliest generated cells of the murine cerebral cortex form a transient population in the subplate and marginal zone. *Dev. Brain Res.*, 66:137-140.

Wood, P.L., and Rao, T.S. (1991) A review of in vivo modulation of cerebellar cGMP levels by excitatory amino acid receptors: role of NMDA, quisqualate and kainate subtypes. *Prog. Neuropsychopharmacol. Biol. Psychiatry*, 15:229-235.

Woodward, W.R., and Coull, B.M. (1984) Localisation and organisation of geniculocortical and corticofugal fiber tracts within the subcortical white matter. *Neuroscience*, 12:1089-1099.

Woodward, W.R., Chiaia, N., Teyler, T.J., Leong, L., and Coull, B.M. (1990) Organization of cortical afferent and efferent pathways in the white matter of the rat visual system. *Neuroscience*, 36:393-401.

Woolsey, T.A., and Van der Loos, H. (1970) The structural organization of layer IV in the somatosensory region (S1) of mouse cerebral cortex. *Brain Res.*, 17:205-242.

Woolsey, T.A., and Wann, J.R. (1976) Areal changes in mouse cortical barrels following vibrissal damage at different postnatal ages. *J. Comp. Neurol.*, 170:53-66.

Woolsey, T.A., Anderson, J.R., Wann, J.R., and Stanfield, B.B. (1979) Effects of early vibrissae damage on neurons in the ventrobasal (VB) thalamus of the mouse. *J. Comp. Neurol.*, 184:363-380.

Woolsey, T.A., Welker, C., and Schwartz, R.H (1975) Comparative anatomical studies of the SmI face cortex with special reference to the occurrence of "barrels" in layer IV. *J. Comp. Neurol.*, 164:95-104.

Yamamoto, M., Boyer, A.B., Crandall, J.E., Edvards, M.A., and Tanaka, H. (1986) The distribution of stage-specific neurite-associated proteins in the developing murine nervous system recognized by a monoclonal antibody. *J. Neurosci.*, 6:3576-3594.

Yamamoto, N., and Toyama, K. (1995) Repulsive and attractive mechanisms for the formation of corticofugal projections. *Neuroreport*, 6: 1517-1520.

Yamamoto, N., Higashi, S., and Toyama, K. (1997) Stop and branch behaviors of geniculocortical axons: A time-lapse study in organotypic cocultures. *J.*

Neurosci., 17:3653-3663.

Yamamoto, N., Higashi, S., Sugihara, H., and Toyama, K. (1992) Axonal growth and branch formation of geniculate fibers in visual cortex studied in coculture preparations. *Soc. Neursoci. Abstr.*, 18:224.

Yamamoto, N., Kurotani, T., and Toyama, K. (1989) Neural connections between the lateral geniculate nucleus and visual cortex in vitro. *Science*, 245:192-194.

Yamamoto, N., Yamada, K., Kurotani, T., and Toyama, K. (1991) Laminar specificity of extrinsic cortical connections studied in culture preparations. *Soc. Neurosci. Abstr.*, 17:767.

Yamamoto, N., Yamada, K., Kurotani, T., and Toyama, K. (1992) Laminar specificity of extrinsic cortical connections studied in culture preparations. *Neuron*, 9:217-228.

Yang, T.T., Gallen, C.C., Ramachandran, V.S., Cobb, S., Schwartz, B.J., and Bloom, F.E. (1994) Noninvasive detection of cerebral plasticity in adult human somatosensory cortex. *Neuroreport*, 5:701-704.

Yhip, J.P.A., and Kirby, M.A. (1990) Topographic organization of the retinocollicular projection in the neonatal rat. *Visual Neurosci.*, 4:313-329.

Yntema, C.L. (1968) A series of stages in the embryonic development of *Chelydra Serpentina*. *J. Morph.*, 125:219-252.

Young, M.P. (1992) Objective analysis of the topological organization of the primate cortical visual system. *Nature*, 358:152-154.

Young, M.P., and Scannell, J.W. (1993) Analysis and modelling of the organization of the mammalian cerebral cortex. In *Experimental and Theoretical Advances in Biological Pattern Formation*, ed. P. Maini, J.D. Murray and H.G. Othmer, Plenum, New York-London.

Yuasa, S., Kitoh, J., and Kawamura, K. (1994) Interactions between growing thalamocortical afferent axons and the neocortical primordium in normal and *reeler* mutant mice. *Anat. Embryol.*, 190:137-154.

Yurkewicz, L., Valentino, K.L., Floeter, M.K., Fleshman, J.W., Jr., and Jones, E.G. (1984) Effects of cytotoxic deletions of somatic sensory cortexin fetal rats. *Somatosens. Res.*, 1:303-327.

Yuste, R., Peinado, A., and Katz, L.C. (1992) Neuronal domains in developing neocortex. *Science*, 257:665-669.

Zeki, S. (1990) Parallelism and functional specialization in human visual cortex. *Cold Spring Harbor Symposia on Quantitative Biology*, 55:651-661.

Zilles, K. (1985) *The Cortex of the Rat. A Stereotaxic Atlas*, Springer-Verlag, Berlin-Heidelberg-New York-Tokyo.

Zilles, K., Zilles, B., and Schleicher, A. (1980) A quantitative approach to cytoarchitectonics. VI. The areal pattern of the cortex of the albino rat. *Anat. Embryol.*, 159:335-360.

Zimmer, J., and Gähwiler, B.H. (1984) Cellular and connective organization of slice cultures of the rat hippocampus and fascia dentata. *J. Comp. Neurol.*, 228:432-446.

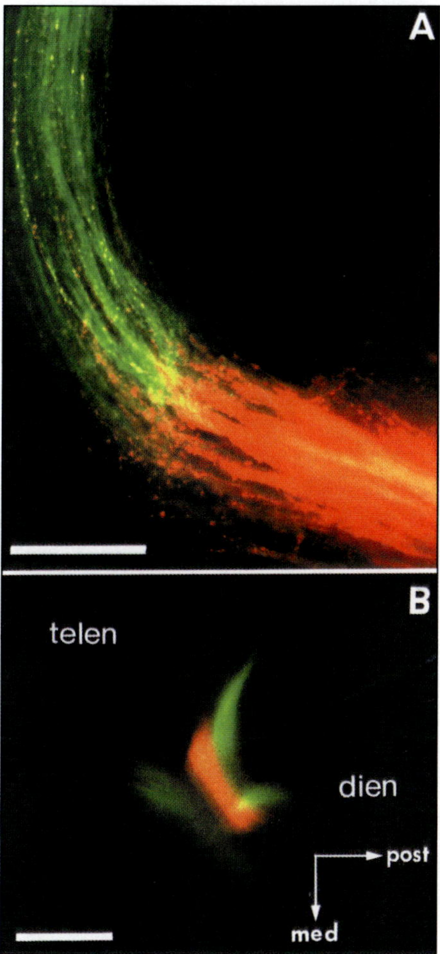

Fig. 6.5. The 'handshake' between thalamic and preplate axons and the precise topography of the early thalamocortical projection. A: In an embryonic day 15.5 (E15.5) fetal rat brain, a crystal of DiI was placed into the dorsal thalamus (putative lateral geniculate nucleus, LGN) and a DiA crystal placed into what was presumed to be the 'matching' region of the occipital cortex (putative area 17), so that both thalamocortical and corticofugal preplate fibers were labelled. At this stage, very few if any thalamic fibers have reached the occipital cortex (no cells backlabelled with DiA were observed in the thalamus), and the DiI-labelled fiber population was exclusively thalamocortical (no cortical cells were backlabelled with DiI). In this coronal section, many DiA-labelled (yellow/green) and DiI-labelled fibers (orange) are associated closely in the same plane of focus, indicating that the preplate and thalamocortical fibers become associated intimately when they meet each other. Scale bar, 100 µm. B: This fluorescence photomicrograph was taken from a 100 µm thick horizontal section of an E20 rat brain after placement of three crystals of different carbocyanine dyes, DiA, DiI and DiAsp, in a parasagittal row along the right hemisphere. Since thalamic fibers have already arrived at the cortex at E16, each placement labelled a mixed bundle of thalamocortical and corticofugal axons. Six weeks' incubation at room temperature was used to enable full anterograde and retrograde diffusion. Three distinct bundles are clearly visible passing through the primitive internal capsule without substantial mixing or crossing. Abbreviations: dien, diencephalon; med, medial; post, posterior; and telen, telencephalon. Scale bar: 500 µm. Reproduced from Molnár and Blakemore (1995a) with kind permission of Elsevier Science Ltd, Kidlington, England.

Development of Thalamocortical Connections,
by Zoltán Molnár. © 1998 Springer-Verlag and R.G. Landes Company.

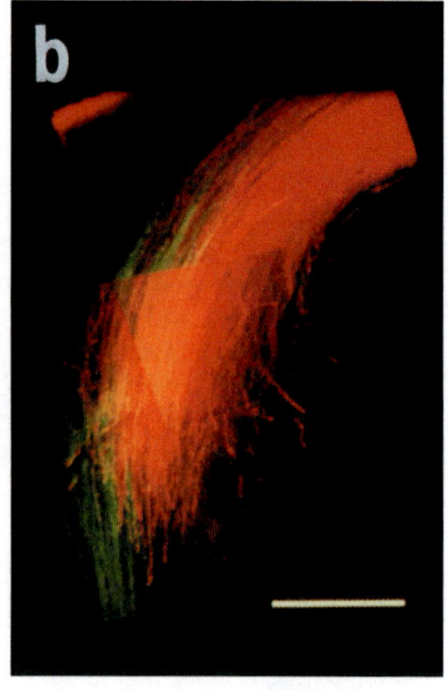

Fig. 7.6. A small crystal of DiI was placed in the convexity of the cortex and a small crystal of DiO in the dorsal thalamus of the same hemisphere of an E13.5 old mouse. The labelled fibers were imaged separately with different photomultipliers using confocal microscopy from the region illustrated with a box on the line drawing. The earliest corticofugal projections (red) fan out from the intermediate zone and loosen up as they turn under the striatum. The fibers end in large growth cones. The thalamic fibers (green) heading towards the cortex have just reached the same region. There is a considerable overlap between the differentially labelled fibers. Scale bar: 200 mm. From Molnár et al (1993a).

Fig. 9.5. (Opposite) A: Development of the columnar activation pattern from a broad activation pattern at postnatal day 2 studied with optical recording. Thalamocortical slices (400 µm thick) were stained with voltage sensitive dye (RH-482 NKK; 20 minutes with 0.05-0.1 mg/ml) and placed in a warmed (32°C) interphase-type recording chamber on the stage of an inverted microscope. Stimulus induced changes in the intensity of transmitted light (700 +/- 30 nm) were measured with a 128 x 128 pixel array of photosensors every 0.6 ms (Fujifilm HR Deltaron 1700) for up to 300 ms at 8 bits resolution (all 8 bits are devoted to measure changes in transmitted light intensity relative to a pre-stimulus reference image). Pseudocolor figures show relative changes in transmitted light intensity $\Delta T/T$ (where T is the intensity of transmitted light) with lower cutoff at twice the standard deviation of the pre-stimulus noise and superimposed on the gray scale reference image of the slice. The response to a single 100 µm stimulus of the thalamus using a bipolar stimulating electrode was averaged 4 to 16 times; intervals between stimuli were at least 30 seconds. During the optical recording extracellular electrodes were used to continuously monitor field potentials. Stimulation of the thalamus at birth or at postnatal day 1 (P1) results in broad activation of the cortex, without clusters or columns of response. In contrast, at P2, the depolarizing response in cortex is arrayed in columns extending from the dense cortical plate (DCP) through layer 6, with individual columns of response separated by sparse responding inter-columnar regions. On the left frame (frame A and B) is the response from a P1 slice, with stimulation from slightly different positions, using opposite polarities of the same bipolar electrode. The black and white inset shows the spatial profile of the response across layer 6 for each stimulus polarity superimposed. There is only a single peak, which shifts slightly when switching polarity of the stimulation, reflecting the topographic mapping of thalamocortical afferents. In contrast, at P2 (frames C and D), there are two distinct columns of response, each of which can be preferentially activated depending upon the polarity used for the thalamic stimulation, indicating topographic order in the thalamocortical connections. The inset shows the superimposed laminar profile for the two responses (2.25 mm in width), revealing the separable peaks for each column. (D is dorsal; P/M is posterior/medial; MZ is marginal zone; DCP is dense

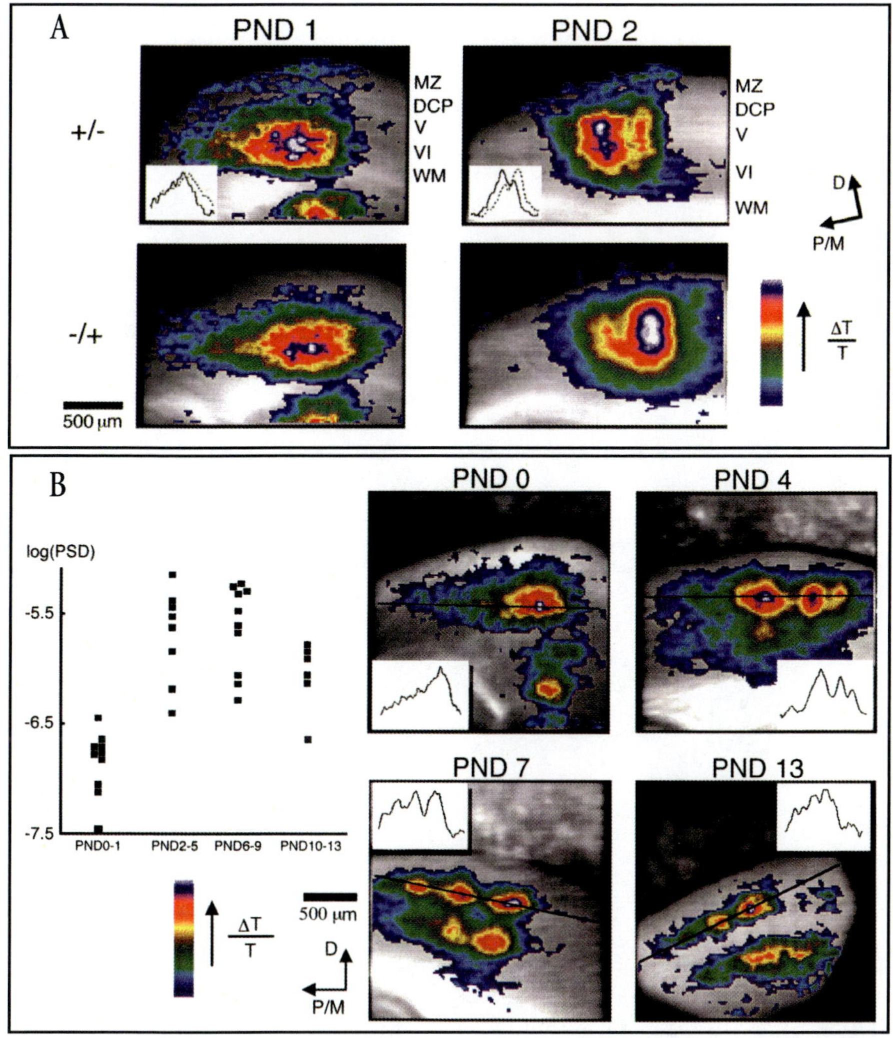

cortical plate; WM is white matter). B: Development of barrel modules. Examples for cortical activation patterns at four ages between P0 and P13. Black and white insets in each color picture depict the response profile along the line through layer 6 of the cortical plate (indicated with black on each panel). The images were taken with the same magnification, all 2.25 mm in width. At birth (PND 0), there is a broad single peak. In contrast, on postnatal day (PND) 4, 7 and 13, there are barrel clusters seen as repetitive peaks and troughs in the laminar profile along layer IV. For each slice, the degree of barrel patchiness was quantified using one-dimensional fourier techniques (results shown in left diagram). The amplitude of the power spectral density is significantly lower at PND 0 and 1 than at all other ages tested. Reproduced from Kurotani et al (1996) with kind permission of Kyoritsu Shuppan Co. Ltd. Publishers, Tokyo.

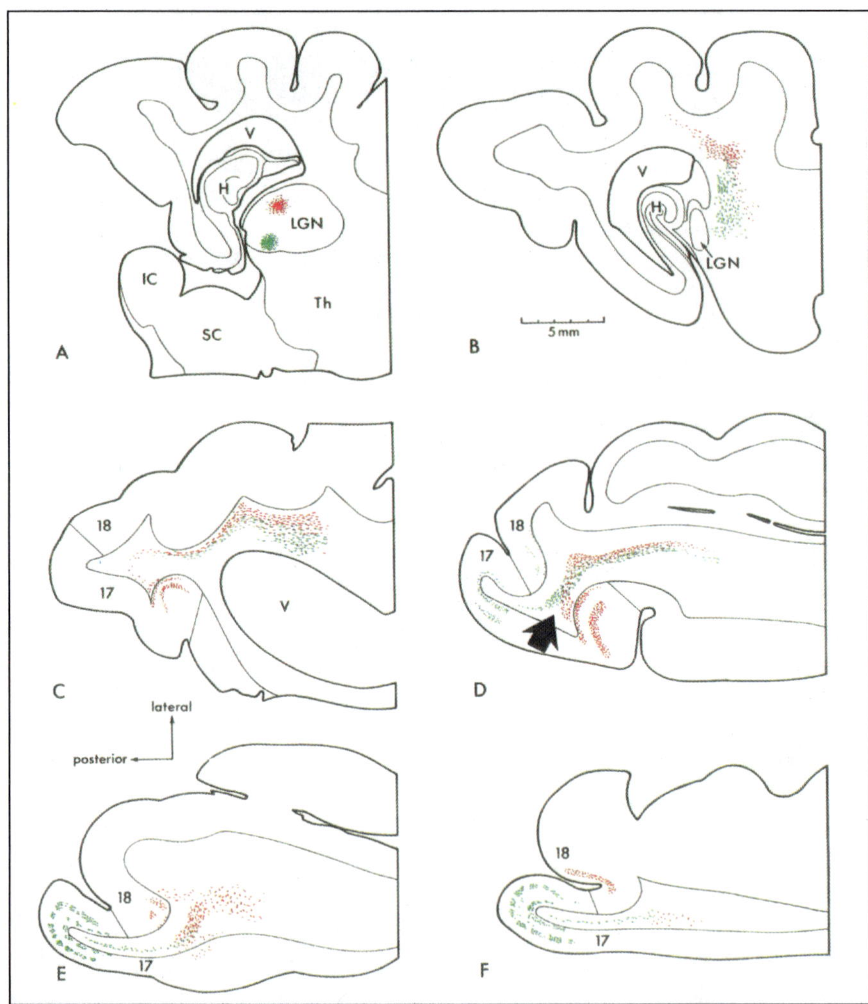

Fig. 8.2. Reconstruction of optic fibers labelled from a mediolateral pair of injections of peroxidase-conjugated wheat-germ agglutinin (WGA)(red) and [^{3}H]proline (green) into the lateral geniculate nucleus. Horizontal sections oriented so that lateral is up, posterior is to the left, and presented ventral to dorsal (A-F). The crossing of the two sets of fibers occurs shortly before reaching area 17, shown in D (arrow). Optic fibers showed no such reversal when labelled from anteroposterior pairs of injections. Figure reproduced from Nelson and LeVay (1985) with kind permission of John Wiley & Sons, Inc, New York.

Index

D

H

I

J

K

L

M

T

V

W

X